Environmental Data Analysis

Carsten Dormann

Environmental Data Analysis

An Introduction with Examples in R

 Springer

Carsten Dormann
Biometry and Environmental System Analysis
University of Freiburg
Freiburg, Germany

ISBN 978-3-030-55022-6   ISBN 978-3-030-55020-2   (eBook)
https://doi.org/10.1007/978-3-030-55020-2

This Springer imprint is published by the registered company Springer Nature Switzerland AG
The registered company address is: Gewerbestrasse 11, 6330 Cham, Switzerland

Preface

This book aims at empowering the reader to carry out statistical analyses. To a non-statistician, many decisions and approaches in statistics look arbitrary. Indeed, non-mathematical statistic books often offer statistical recipes to guide the user through such decisions. Here, instead, I try to offer a look "under the hood" of statistical approaches, in the hope that by explaining statistical analyses from one or two levels lower down, the reader will understand rather than merely follow such recipes.

The vantage point of this book is the Generalised Linear Model (GLM). It is a particularly reasonable vantage point on statistical analyses, as many tests and procedures are special cases of the GLM. The downside of that (and any other) vantage point is that we first have to climb it. There are the morass of unfamiliar terminology, the scree slopes of probability and the cliffs of distributions. The vista, however, is magnificent. From the GLM, t-test and regression, ANOVA and χ^2-test neatly arrange themselves into regular patterns, and we can see the paths leading towards the horizon: to time series analyses, Bayesian statistics, spatial statistics and so forth.

At the same time, there is quite a bit of craftsmanship involved applying GLM-like analyses to real-world data. There is the software implementation (in our case in R), the model checking, the plotting, the interpretation, and, most importantly, the design of experiments and surveys. The book thus purposefully see-saws between the odd-numbered chapters on theory, and the even-numbered chapters on implementing said theory in a software. I tried to avoid excessive mathematical details; however, statistics is part of mathematics, and there is no way around it, particularly not if we want to later expand beyond the GLM. I tried to create an intuition, a feeling for these equations. Statisticians may forgive me when I employ didactic simplification, particularly in the early chapters, in order to get going.

When we asked our Master's students in environmental sciences, a few years ago, what they thought was underdeveloped in their university education, we got a rather unanimous verdict: statistics. Although they did not appreciate this early in their studies, after a Bachelor thesis or half-way through their final Masters thesis many of them realised that environmental sciences lack the luxury of data that look anything like those dealt with in statistics books. Our data are sparse (often less than 100 data points), messy (missing values, unbalanced designs and correlated predictors), and anything but normally distributed (although given the low level of replication, one would be hard pressed to be sure that they were *not* normally distributed). The situation hasn't hugely changed since in some fields, but in others "big data" appear. Working with large data sets is in many ways simpler than with small data sets, as big data are actually solving some of the issues we may have. On the other hand, often such data are structured, for example, thousands of geographical locations from a single animal, but only 12 such animals in the sample. Understanding the statistical challenges that come with such data requires a sound basis. While we'll not exceed thousand-odd data points in this book, or tackle the issues that come with structured data, the reader will be able to understand them from the GLM vantage point. Promise!

Many colleagues grew up harbouring a deep suspicion about statistical analyses. They chucklingly quote the much-cited quip of Winston Churchill: "I only believe in statistics that I doctored myself."[1] It annoys me deeply to think that the same people who discredit statistical analyses apparently merrily invest their money into funds predicted to do well by banks, trust their GPs on diagnoses, and let their GPS-systems guide them through traffic and wilderness. Of course, one can cheat with statistics. Understanding statistical analyses, and my attempt to lay out a structured way to do so, assumes the reader to be interested, focused and honest.

Statistics, in my way of thinking, are a formal approach to prevent being swayed by artefacts, change and prejudice. Too many scientific publications may report irreproducible findings due to the way the well-meaning analyst carried out the analysis (Ioannidis 2005). But without an understanding of the underlying ideas, it is very easy to do unintentionally wrong (and is still easy enough with quite a bit of experience).

The important thing, for this book, is that you *understand* its content. If some phrases or topics are confusing, or poorly explained, I strongly encourage you to seek additional advice. I love books,[2] but you may prefer online video snippets,[3] a Wikipedia page on the topic, a forum discussion,[4] or one of the increasing number of excellent online lectures.[5]

Thanks: This book was translated from its German precursor, which I use and experimented with for several years at the University of Freiburg. Many Bachelor's and Master's students have contributed to improving it, and they and several colleagues have corrected typos and mistakes. The translations were undertaken by Carl Skarbek, who I am indebted to, but I'm sure I managed to sneak a few errors past him. The code was written in R, the text in LaTeX.[6] Both open-source projects are amazing feats of selflessness and competence.

P.S.: For corrections and comments, please feel free to contact me: `carsten.dormann@biom.uni-freiburg.de`.

Freiburg, Germany   Carsten Dormann
2019

References

Bolker, B. M. (2008). *Ecological Models and Data in R*. Princeton, NJ: Princeton University Press.
Harrell, F. E. (2001). *Regression Modeling Strategies - with Applications to Linear Models, Logistic Regression, and Survival Analysis*. New York: Springer.
Hastie, T., Tibshirani, R. J., & Friedman, J. H. (2009). *The Elements of Statistical Learning: Data Mining, Inference, and Prediction*. Berlin: Springer, 2nd edition.
Ioannidis, J. P. A. (2005). Why most published research findings are false. *PLoS Medicine*, 2(8), 0696–0701.

---

[1]Ironically, here is a German writing about an English statesman, whose alleged words were apparently fabricated by the Nazis to discredit him (see `https://en.wikiquote.org/wiki/Winston_Churchill`). The historical roots of this quip appear somewhat murky. Suffices to say that whoever made this statement is being quoted too often.

[2]For example, those of Bolker (2008), Hastie et al. (2009) or Harrell (2001).

[3]Such as those on `http://www.khanacademy.org/#statistics`.

[4]On CrossValidated: `https://stats.stackexchange.com/`.

[5]Such as the introduction to statistical learning by Trevor Hastie and Brad Tibshirani of Stanford University: `https://www.r-bloggers.com/in-depth-introduction-to-machine-learning-in-15-hours-of-expert-videos/`.

[6]LaTeX (`http://ctan.org`) is for word processing what R is for statistics: simply the best. And while we are at it, Wikipedia, Python, Windows, Excel, RStudio, macOS, Google, Genstat, Stata, SAS, Mathematica, Matlab and Libre/OpenOffice are registred trademarks. Their mentioning does not indicate endorsement.

The Technical Side: Selecting a Statistical Software

Life is repetitive—use the command line.

Gita Benadi

At the end of this chapter ...

... R should be installed and working on your computer.

... The pros and cons of both *point-and-click* and code-based Software for statistical analysis should be clear.

... You should have a burning desire to (finally) do some calculations on your own.

Simple tests and calculations can be made with an old-fashioned calculator or a spreadsheet calculation program (i.e., Excel or similar program). For more sophisticated and complex analyses, such as regressions, we need a more specialized statistical software (Excel and friends can sometimes accomplish this, but I wouldn't trust the results due to trivial problems such as calculation errors, row limits, and automatic reformatting). Instead of using such spreadsheet programs, we can use "point-and-click"-statistics software (e.g., SPSS, Statistica, and Minitab), or alternatively, we can enter the world of code-based programs (e.g., Stata, S-plus, Matlab, Mathematica, Genstat, R, and Python). Such programs have two big advantages: their functionality is fundamentally unlimited, since it can always be expanded through further programming; and the code makes the analysis understandable, interpretable and easily repeatable.

This second point cannot be emphasized enough. For example, if we were to find a typo or data entry error in our raw data after a lengthy analysis, we would have to repeat all of our steps one by one if using a point-and-click-software. With code-based programs, we can simply correct the error and rerun the code—and the analysis is updated.

Code-based software is used for all automated processes. The monthly reports from a blood donation database as well as the sophisticated analyses by Google are generated using code-based software. Which program to use is more a matter of taste ... and money.

The learning curve for code-based software is steep at the beginning. It's like learning a new language, with lots of new vocabulary and grammar rules, and as with learning any new language, mistakes will be made. But, this new language also opens up a world of possibilities that are simply not available with point-and-click. As a matter of opinion, I find that there is nothing worse than doing something wrong, just because we are too lazy to learn something new!

Statistics are not for free (in terms of "no effort"). There are, however, two statistical systems that won't cost you a penny: R and Python.[7] At the time of writing, R offers the most complete range of functions specifically made for statistical analysis. Perhaps it will look different in 10 years, but for now, there is nothing more useful.

So R it is.

---

[7] www.python.org; Haslwanter (2016) offers a good introduction to statistics with Python.

Here is a short description of where you can obtain R and how to install it. How to use the software for different statistical applications is described in later chapters.

Downloading and Installing R

R is a statistics and visualisation software (Fig. 1) that is coordinated by the R Foundation (R Core Team 2017). It is platform independent (i.e., works with versions of Linux, Windows, and macOS). The code is free and open source. For technical details and more on the history and backstory of R, check out: www.r-project.org.

When it comes to installation, let's first turn our attention to CRAN, the *Comprehensive R Archive Network*, where the installation files are located. Under the CRAN heading (on the left-hand side of the screen), click on Mirrors, where you can choose a mirror server close to you (for example, `https://www.stats.bris.ac.uk/R/`). The contents on these mirror servers are identical to the main server in Vienna (hence the name "mirrors").

To install R on your computer, you need to have administrator authorisation!

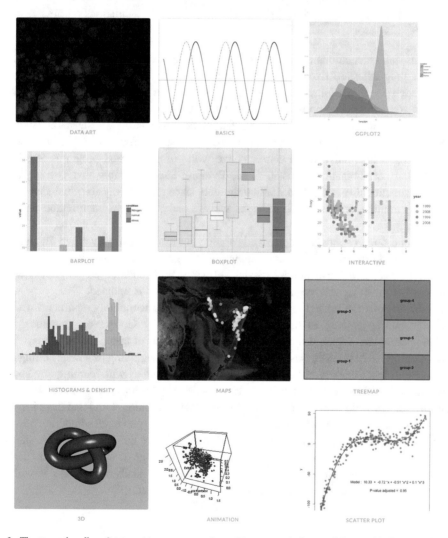

Fig. 1 The R-graph gallery (`http://www.r-graph-gallery.com`) shows off the graphical capabilities of R

Depending on the operating system you are using, click on the correct `Download for Linux/macOS X/Windows` link in the top box, and on the next page, click on base. We are then directed to another page that looks different depending on the selected operating system.

- For Linux, we need to choose between Debian/Ubuntu and Red Hat/Suse (for Linux systems that do not use .deb or .rpm packages, you can click on download source code on the previous page and compile it yourself). A **simpler** way is to use the software management tool within Linux, where you can download and install all `r-base` and `r-cran` packages (including dependencies).
- For macOS, we download `R-X.YY.Z.pkg` (X.YY.Z stands for the current version) and install it by double-clicking the file in the *Downloads* folder.
- For Windows, we download `R-X.YY.Z.pkg` (X.YY.Z stands for the current version) by clicking on `Download R X.YY.Z` for Windows and then install it by double-clicking on the file in the *Downloads* folder.

If everything goes smoothly, R is now on your computer, perhaps in the form of an icon on your desktop, in the start menu, or in the dock.

A Short Test in R

When we start R (by clicking/double-clicking), the following screen should appear (Fig. 2). Exactly how it looks may differ slightly for each operating system. The parameters concerning functionality, such as the R version, the architecture (i.e., 32- or 64-bit) and whether it works with your computer are most important. We can check these three things by typing the following after the prompt (>):

```
> sessionInfo()
```

After pressing the return key, we get the following output:

```
R version 4.0.0 (2020-04-24)—"Arbor Day"
Copyright (C) 2020 The R Foundation for Statistical Computing
Platform: x86_64-apple-darwin17.0 (64-bit)

R is free software and comes with ABSOLUTELY NO WARRANTY.
You are welcome to redistribute it under certain conditions.
Type 'license()' or 'licence()' for distribution details.

Natural language support but running in an English locale

R is a collaborative project with many contributors.
Type 'contributors()' for more information and
'citation()' on how to cite R or R packages in publications.

Type 'demo()' for some demos, 'help()' for on-line
help, or 'help.start()' for an HTML browser interface to help.
Type 'q()' to quit R.

[R.app GUI 1.71 (7827) x86_64-apple-darwin17.0]
>
```

These lines tell us that we are dealing with R version 4.0.0 for macOS in 64-bit mode, is being used with GB-English settings and that seven *packages* are loaded.

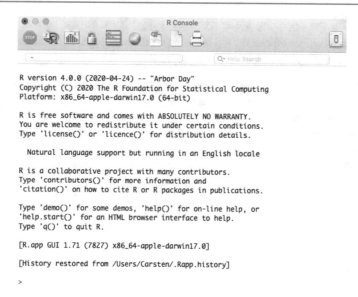

Fig. 2 R, specifically: the R Console after starting the program. Although we should try to keep our R software up to date, the update frequency, consisting of two major releases (4.0.0, 4.1.0, ...) per year and two patches (4.0.1, 4.0.2, ...), is quite high. A two-year-old version isn't necessarily obsolete, but new packages may not be compatible with older versions and may require you to update to the most recent R version

With a couple more quick tests, we can check that everything else is in order.[8] We'll type the following lines separately in the console (using the return key between lines to send the code):

```
> install.packages(c("e1071", "plotrix", "vegan"))
> library(plotrix)
> ?std.error
> help.start()
```

With the first line (install.packages), we install some new packages (**e1071** etc., we will need these later for our analysis). For this to work properly, you must have a functioning internet connection. With the second line (library), we load the **plotrix** package that we just installed. With the ? command on the next line, we open the help page for a certain function (here for the function std.error). The last line (help.start) opens the *local* help page for R. At this point, it may be helpful to do a little tour through R, which you can do by clicking the link An Introduction to R at the top left of the local help page (see the tutorial at the end of the chapter).

Once we have checked that everything is working, we can install all of the other R packages that are used or mentioned in this book. This way, we won't have to worry about this step later. Don't enter the "+"; it's R's way of saying: "Hang on, there should be more to come!":

---

[8]If it is not working properly, then please check the installation instructions for the appropriate operating system on the download page on the CRAN website.

```
> install.packages(c("ADGofTest", "AER", "AICcmodavg", + "Barnard", "car",
"DHARMa", "e1071", "effects", "faraway", + "FAwR", "Hmisc", "lawstats", "lme4", &
"lmerTest", "moments", + "MuMIn", "nlstools", "nortest", "plotrix", "polycor", +
"psych", "raster", "sfsmisc", "vegan", "verification", "VGAM"), + dependencies
=TRUE)
```

Editors and Environments for R

As mentioned in the introduction, a big advantage of code-based software is that you can easily recall the code *in toto*. To do this, however, we must save the code. This means that in addition to the R Console (the window with the >-prompt), we also need to use an editor.

R comes with a built-in editor. With `Ctrl/Cmd-N`, we can open a new editor window.[9] Using a keyboard shortcut (`Ctrl-R` or `Cmd-Return`, depending on the operating system), we can send code directly from the editor to the console.

R code is saved as a simple text file, with the suffix .r or .R.

Other than the default built-in editor, which is sparse and not very user-friendly, there are others with syntax support. Such editors check to make sure that there are the right number of brackets in the code, that text entered in quotations appears in a different colour, and that commands can be auto-completed using the `<tab>` key (see the built-in editor for R on macOS, Fig. 3).

In addition to these built-in solutions, there are two (in my opinion at least) other noteworthy alternatives: RStudio and JGR. Both are Java programs that integrate R into a new interface (thus, often referred to as an IDE, *integrated development environment*). Both are highly recommended.

RStudio (`http://rstudio.org/`) is the most advanced IDE for R (Fig. 4). In addition to the editor window (upper left) and the console (lower left), there is a window that shows existing defined objects (upper right) as well as a window for figures and the help menu (lower right), among other things. These windows can be configured and moved around freely. Another advantage of RStudio is that it looks practically identical on all operating systems. RStudio is free and accessible to all, and is built by competent developers (see Verzani 2011, for details). RStudio can be installed **after installing R**.

JGR (pronounced *jaguar*, `http://www.rforge.net/JGR/`) is an older Java IDE for R (Helbig 2005). It more closely follows the original R design, with independent windows for the editor, console, graphics and help. Together with the *Deducer* extension (`www.deducer.org`[10]), JGR attempts to appeal to *point-and-click* users. The reading of data and statstical tools are available via *pull-down* menus (see Fig. 5). The installation of JGR for Windows and macOS can be completed by installing a file from the rforge website for the software: `http://rforge.net/JGR/files/`.[11] After unzipping the files, JGR installs (with confirmation) multiple R packages that are necessary for its operation. Then the interface is started, which looks like the screen in Fig. 5.[12] In the lower window, we will now install the **Deducer** package:

---

[9]Or using the mouse: File → New script or similar (depending on the operating system); a new window opens where we can write R-code.

[10]The installation instructions on `http://www.deducer.org/pmwiki/index.php?n=Main.DeducerManual?from=Main.HomePage` don't always work. Therefore, the more complicated process is described here.

[11]For installation for Linux systems, see `http://rforge.net/JGR/linux.html`.

[12]Depending on how old your operating system is, there may be some operational problems. If so, the following packages need to be installed manually: **colorspace, e1071, ggplot2, JavaGD, lpSolve, plyr, rJava**: `install.packages(c("colorspace", "e1071", "ggplot2", "JavaGD", "lpSolve", "plyr", "rJava"))`. Next, close R and restart it. JGR should now be functioning.

```
● ● ●                              ⊙ GnA_chap1_sampleStatistics.R
5    ash <- read.csv("ashesGirths.csv")
6    attach(ash)

7
8    pdf(file="../figures/ashes_hist1.pdf")
9    par(mar=c(5,5,1,1))
10   hist(girth[group==1], ylab="number of trees per class" , xlab="girth
     class [cm]", las=1, main="", col="grey", cex.lab=1.5)
11   dev.off()

12
13   par(mfrow=c(3,3))
14   for (i in 2:10) hist(girth[group==i], ylab="number of trees per class" ,
     xlab="girth class [cm]", las=1, main="", col="grey")

15
16   par(mar=c(5.5.1.1))
hist(x, ...)
```

Fig. 3 R code with the built-in R editor for macOS. For Windows, the colourful syntax support is unfortunately not available

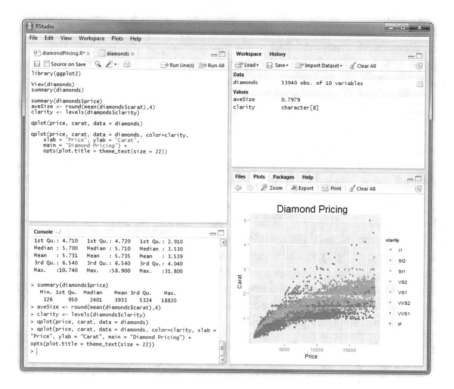

Fig. 4 RStudio under windows

> *install.packages("Deducer")*

Finally, via the menu Packages & Data →Package Manager place a check mark for Deducer in the loaded column. You are now ready to use one of the best modern non-commercial point-and-click interfaces for R!

The main disadvantage of JGR is that many standard R commands have been replaced with commands from the **iplots** package (such as hist being replaced by ihist). When you click to generate the R code (and hopefully also learn in the process), these commands are not always the ones that would be used in R or RStudio.

Furthermore, there are numerous other solutions. The nice RKWard (pronounced: *awkward*) was primarily developed for Linux systems, but now has experimental versions for both

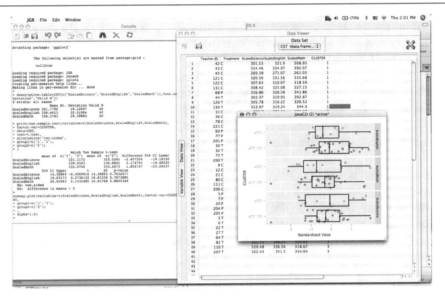

Fig. 5 JGR with deducer

Windows and macOS.[13] Eclipse provides an R-integration (StatET), and so do Emacs and Vim. There are also other R-internal variants, such as the R-Commander (package **Rcmd**). This ugly duckling of an interface is easily expanded through plugins and is also fitted with pull down menus. Without naming all R editors out there, some other noteworthy ones include rattle, SciViews, ade4TkGUI, R2STATS (only for GL(M)Ms), and Orchestra (for jedit).

I've had different issues with many of these options (GTK functions with rattle, slow processing speed with SciViews, speed and stability with Eclipse/StatET, strange shortcut keys with Emacs/ESS). Currently, my favourite is RStudio or the editor delivered with macOS.

Now we can begin.

Tutorial

Most of the questions that you have about R have most likely been asked by others as well. The *Frequently Asked Questions* for R can be found at `http://cran.r-project.org/doc/FAQ/R-FAQ.html`.[14] But, what if we don't have an Internet connection at the moment?

R also comes with some *offline* help options. In this tutorial, you should become familiar with three different methods regarding help in R:

1. Built-in help: This is available for each function and data set for each package installed from CRAN (depending on the quality of the package, the documentation may be more or less user-friendly). To show the help file for the function FUN, we simply type ?FUN in the console. The help for the function `install.packages` that we used earlier can be accessed by entering ?install.packages. For special characters (+ or $) and other special functions (such as the function ? itself), quotations must be used: ?"?". If this is too confusing, you can also simply type help("?"). ? is simply a shortcut for help.

---

[13]`https://rkward.kde.org/`.

[14]For Windows-specific questions, see `http://cran.r-project.org/bin/windows/rw-FAQ.html`, or for macOS, see `http://cran.r-project.org/bin/macosx/RMacOSX-FAQ.html`.

Using the build-in help, check out the help for the functions `mean`, `??` and `library`, and the data set `varespec` from the **vegan** package (which you must first install and then load with the `library` function). Answer the following questions: What is the difference between `?` and `??` (easy), and between `require` and `library` (more difficult)?

2. Package documentation: Using `help.start` you can open a page in your web browser that contains all of the delivered help files. Under the *Packages* menu item, you will find a list of all the pacakges you currently have installed. Click on *Packages*, then scroll down to *vegan* and click on it. The second bullet there is *Overview of user guidesand package vignettes*.

   Check out what is there.

3. Documentation for R: On the main help page (the one opened with `help.start()`) under **Manuals** are six important PDFs, that document R. Open An Introduction *to R*, and scroll down to *Appendix A: A sample session*.

   Type the R code for this example session piece by piece in your R environment. Now you have a first impression of the variety offered by R.

References

Haslwanter, T. (2016). *An Introduction to Statistics with Python with Applications in the Life Sciences*. Berlin: Springer.

Helbig, M., Theus, M., & Urbanek, S. (2005). JGR: Java GUI for R. *Statistical Computing and Graphics Newletter*, 16, 9–12.

R Core Team (2017). *R: A Language and Environment for Statistical Computing*. R Foundation for Statistical Computing http://www.R-project.org, Vienna, Austria.

Verzani, J. (2011). *Getting Started with RStudio*. Sebastopol, CA: O'Reilly Media.

Contents

Samples, Random Variables—Histograms, Density Distribution

My sources are unreliable, but their information is fascinating.
—Ashleigh E. Brilliant

At the end of this chapter...
... you should know what a sample is.
... the terms mean, median, standard deviation, variance, and standard error should be so familiar to you that you can recall their formulas from memory. You should also be comfortable with the concepts of skew, kurtosis and interquartile range.
... you should be familiar with the histogram as a simple and useful graphical representation.
... you should understand what a frequency distribution is.
... you should be able to tell the difference between the frequency and density of a value.
... you should want to get started with doing something practical.

The starting point for any statistical analysis is a (scientific) question. The data you may have at hand to address this question come second. We will return to the necessity of a question in more detail later (Chap. 13) as well as to sound ways the data should be gathered (Chap. 14). For now, we start by seeking to answer a simple question: what is the diameter of an ash tree?

As a first step, we measure the diameter of a tree in the park, at say 37 cm, and this is our first data point. *Must* this tree have *this* specific diameter? Of course not. If it were planted earlier or later, or if it received more light or fertilizer, or if people had not carved their initials into the bark, then the tree would be perhaps thicker or thinner. But at the same time, it could not have *any* diameter. An ash tree would never be 4 m thick. My point is that a data point is just a *single* realisation of many possible ones. From *many* measurements we gain an idea how thick a typical 50 year old tree is. But each individual ash tree is not exactly *that* thick.

In many words, I have described the basic idea of the *random variable*: a random variable is function that produces somewhat unpredictable values for a phenomenon. In our example, a stochastic environment makes tree diameters unpredictable, within limits, and thus natural tree growth is a random variable. Each data set is one possible realisation of such random variables. So, say we send 10 groups of imaginary students into a park. Each group should measure the diameter at breast height (referred to in the following also as "girth") for 10 ash trees. These 10 ash trees are then 10 random observations of the random variable "ash tree diameter". We call such groups of values a *sample*.[1]

The results of these 10 times 10 measurements (= observations) are shown in Table 1.1.

So now we have 100 realisations of the random variable "ash tree diameter". A common and useful form for graphically displaying such data is the histogram (Fig. 1.1).

In the histogram above, the trees are divided into diameter *bins* (categories), here in bins of 10 cm. The bins are all the same size (i.e. each bin has a range of 10 cm). If they were not all the same size, it would be impossible to compare the height of the columns. How many trees there are per bin is shown on the *y*-axis. This type of histogram is called a frequency histogram, since we are counting how many elements there are in each bin. Many statistics programs automatically define a

---

[1] Values, observations, and realisations are used synonymously here. "Realisation" or "random variate" is the most technical, and "observation" perhaps the most intuitive term. In any case, a random variable is a hypothetical construct, from which we can only observe its realisations and measure them randomly.

© Springer Nature Switzerland AG 2020
C. Dormann, *Environmental Data Analysis*,
https://doi.org/10.1007/978-3-030-55020-2_1

Table 1.1 Diameter at breast height for 100 ash trees in cm, measured by 10 different groups.

Group 1	Group 2	Group 3	Group 4	Group 5	Group 6	Group 7	Group 8	Group 9	Group 10
37	54	33	82	57	34	60	65	62	44
80	58	38	6	72	49	50	69	66	62
68	66	51	10	62	49	47	21	40	58
77	48	58	49	22	42	42	49	72	65
47	45	64	61	36	36	57	65	48	68
58	38	57	27	79	90	43	29	61	47
98	49	64	51	35	54	14	79	53	93
60	36	62	31	25	56	41	50	51	38
39	47	74	20	62	57	71	44	57	55
39	74	73	64	82	61	24	39	26	41

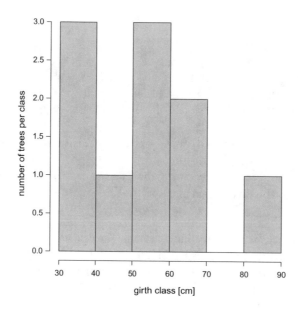

Fig. 1.1 Histogram of ash diameters from group 1

default bin classification, but you can also control this parameter yourself. Too many bins does not make sense, since each class would then only have 0 or 1 tree. Too few bins also does not provide us with useful information, because then each bin would show a very wide range of tree diameters. So in this case, let's trust the software algorithms.[2]

The histograms from the other groups look remarkably different (Fig. 1.2).

Just by looking, we can see that the columns are all arranged a bit differently. We see that the columns have different heights (note that the *y*-axes have different scales) and that the diameter bins have different widths. What we can take from this is that each of these 10 data sets is a bit different and gives a varying impression of ash tree diameter.

What you hopefully also see is that histograms are a very quick and efficient method for summarising data. We can easily see the value range, where the mean lies, etc. For example, in the upper left histogram of Fig. 1.2, we can see that there was apparently one tree with a diameter of <20 cm. In Table 1.1, we see that the value is actually 6 cm. In this way, we can also spot extreme values that may have been the result of a typo during data entry or an error that occurred while measuring in the field. In this case, it was just a tiny tree!

But the park was large and the groups never crossed paths while measuring and of course did not copy each other's data. This means that we actually have measured values for 100 ash trees, not just from 10. The histogram of all data looks like Fig. 1.3.

---

[2]The algorithm used in R is based on the Sturges formula: there are k bins, with $k = \lceil 1 + \log_2 n \rceil$, for n data points ($\lceil \ \rceil$ indicates rounding). R varies this, using the number of unique values instead of n itself. For fewer than 30 data points, the results can be quite ugly. Then there is Scott's suggestion: $k = \frac{3.5s}{n^{1/3}}$. But since we haven't learned about s yet, let's leave it for now.

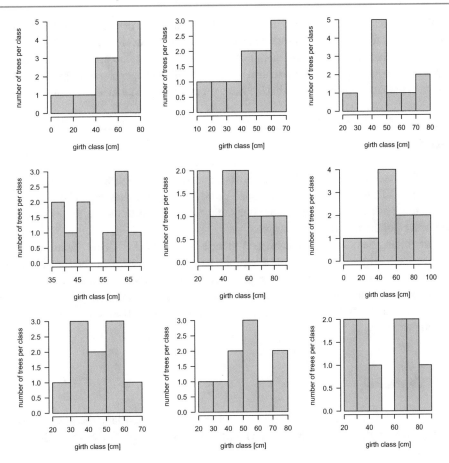

Fig. 1.2 Histogram of ash tree diameter for groups 2 through 10

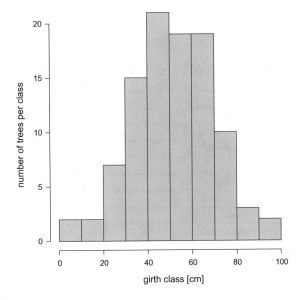

Fig. 1.3 Histogram for all 100 ash tree diameters

This new histogram tells a different story than what we saw from the histograms for the individual groups. Average sized tree diameters appear much more often than the extremes. The columns rise to the middle and then fall smoothly—none of the up-and-down we saw in Fig. 1.2. (An alternative to the histogram, the box plot, will be introduced in a few pages, since it builds on the idea of sample statistics.)

Of course, we can also do calculations to provide us with numbers to describe the data (more on that soon), but as they say, "a picture is worth 1000 statistical data points". Here we can easily see if the data are skewed, which values occur most often, if certain value ranges are missing or under-represented, how extremely the data are distributed and much more. Our brains are wired to look for patterns, so using graphics to visualise data (especially in the exploratory phase) is highly recommended! The visualisation of data is the most powerful tool that we have for conveying data measurements.

1.1   Sample Statistics

We typically do not care for the sample in itself, but we are very interested in the girth of *a typical ash*. If we knew the girth of *all* ash trees, which is statistically called the "population" of ash girths, now that would be a valuable information! Not having the time to measure *all* ash girths, we instead use our sample of ash trees to give us an *estimate* of the typical ash: we estimate some feature of the population from our sample.[3]

Some statistics for describing samples have established themselves as standards (see Table 1.2). Here, we differentiate between statistics for the *location* and the *spread* of a sample. Common measures of the location include the mean and the median. For the spread of the sample values, we can calculate the standard deviation or coefficients of variation. These measurements are essential, and anyone who does any kind of statistics needs to know them. Other measures for location and spread discussed here are meant for specific applications, and are therefore only used for certain disciplines. Here, they are listed for the sake of completeness.

1.1.1   Measures of Centrality

In order to describe the diameter of a typical ash tree, one value is particularly important: the **mean**. This is the most common and the most important sample statistic. This value can be identified for a group of n measured values (x_i) as follows:

$$\bar{x} = \frac{\sum_{i=1}^{n} x_i}{n} \tag{1.1}$$

For the first 10 ash tree diameters (sample diameter $n = 10$), the mean would be $\bar{x} = (37 + 54 + 33 + 82 + 57 + 34 + 60 + 65 + 62 + 44)/10 = 52.8$.

Table 1.2 A compilation of possible sample statistics (from Quinn and Keough 2002, with additions)

Name	Abbreviation
Mean	$\bar{x}$
Median	
Mode	
Huber's mean estimate	
Variance	s^2
Standard deviation	s
Standard error of the mean	
Mean absolute deviation	
Coefficient of variation	CV
95% confidence interval	95% CI
95% quantiles	
Interquartile range	IQR
Range	
Skewness	
Kurtosis	

---

[3]Confusingly, a value that describes our data is called a "statistic". Thus, statistics (as in the field of mathematics) deal with the computation of statistics (as in numbers that describe features of our data).

While the mean represents the average value, the **median** is the middlemost value, in the sense that half of the values in the sample are larger and half are smaller. Depending on whether there is an even or odd number of data points, it can be calculated as follows:

$$\text{Median} = \begin{cases} x'_{(n+1)/2}, & \text{for an odd } n; \\ (x'_{n/2} + x'_{(n/2)+1})/2, & \text{for an even } n. \end{cases} \tag{1.2}$$

where by x'_n is the nth value of the size-sorted (= ordered) numerical values of sample x. $x'_{(n+1)/2}$ is therefore the middlemost value.[4]

The median is therefore the middlemost measured value (for an odd n) or the average of the two middlemost values (for an even n).

The median is a special case of the *quantile*. A *p*-quantile divides a sample into two parts, one where the values are smaller than $100 \cdot (1 - p)$ percent of the observed values, and one where the values are greater than or equal to this value. The 10%-quantile is therefore the value at which 10% of the sample values are smaller.[5] Accordingly, the median is the 0.5-quantile.

Finally, there is the measurement of the *mode*. This is simply the most often occurring value (or values) in a sample.

Robust estimators (Huber 1981) are a more seldom used measurement of centrality. They calculate the mean by giving a lower weight to outliers. Robust statistics are considered to be more conservative, i.e. they are not as easily influenced by strong divergences. Despite their usefulness, they are hardly ever mentioned in the applied statistics literature.

1.1.2 Measures of Spread

In addition to measures of centrality, we also need statistics that describe the spread of our data. The most important such measure is the **standard deviation**:

$$s = \sqrt{\frac{\sum_{i=1}^{n} (x_i - \bar{x})^2}{n - 1}} \tag{1.3}$$

If the data correspond to a normally distributed random variable, or more concisely, if "the data are normally distributed", then around 68% of the data points are ± 1 standard deviation from the mean and around 95% of the data points are ± 2 standard deviations from the mean. With skewed histograms (those that are *not* normally distributed), the standard deviation has to be larger in one direction that the other. For this reason, the percentages here only apply to normally distributed data.

Its counterpart is the **variance**, s^2, which is simply the square of the standard deviation.

$$s^2 = \frac{\sum_{i=1}^{n} (x_i - \bar{x})^2}{n - 1} \tag{1.4}$$

The variance is not often used to describe a sample, because its dimensions are not exactly intuitive. The standard deviation has the same dimension as the mean (like cm for our tree diameters), while the variance would be the quadratic term (cm^2).

A real comparability of spread across different data set is achieved if we standardise the standard deviation s with the mean $\bar{x}$:

$$\mathbf{CV} = \frac{s}{\bar{x}} \tag{1.5}$$

This measure, the **coefficient of variation** (CV), is directly comparable between data sets, because it is independent of the absolute values of the sample. For small sample sizes, the CV is *biased*, i.e. incorrect. The corrected version is:

$$CV^* = \left(1 + \frac{1}{4n}\right)\frac{s}{\bar{x}} \tag{1.6}$$

---

[4]Example: Our measured values are $x = (4, 2, 7, 1, 9)$. In order, they would be $x' = (1, 2, 4, 7, 9)$. x'_1 has the value 1, $x'_4 = 7$. The median is then $x'_{(5+1)/2} = x'_3 = 4$.

[5]The calculation of this often requires interpolation, and statistics programs may differ (sometimes strongly) in the way that they calculate quantiles; see the next chapter for examples.

The following robust but seldom used variant, the **quartile-dispersion coefficient**, is based on the relationship between the interquartile range (see below) and the median:

$$\text{Quartile-dispersion coefficient} = \frac{Q_3 - Q_1}{Q_3 + Q_1}, \tag{1.7}$$

where Q_1 and Q_3 are the first and third quartile (= 25 and 75% quantile).

Another often reported sample statistic is the **standard error of the mean**, sem.[6] In this case, we are **not** dealing with a measure of the spread of the sample! The standard error of the mean is rather a description of the accuracy of how well we can estimate the population mean through our sample mean calculation. If we have a large number of data points, we can be relatively certain that they also describe well the mean of our population, but this is not the case if we have a small sample size. The standard error of the mean quantifies this:

$$\text{sem} = \frac{s}{\sqrt{n}} \tag{1.8}$$

This measure is interpreted similarly to the standard deviation, not in how it relates to the sample, but how it relates to the mean calculated from the sample. The true (but unknown) population mean lies with 95% probability within ± 2 standard errors around $\bar{x}$. This is *not* the same as saying that 95% of the *data points* lie ± 2 standard deviations around the mean! In this respect, the standard error does not fit into the measure of spread, but is rather an accuracy measurement of the mean.

Another way of describing the 95%-confidence interval is that the population mean lies within it with a probability of 95%. If we collected 100 data sets, and computed a (different) SEM for each of them, then in 95% of cases the true population mean would lie within the 95% confidence interval.[7] Since this point is easy to misunderstand, let's look at a short example. Our 10 groups each measured 10 ash trees. Each group can measure a mean, standard deviation and the standard error of the mean for their collected data points. Since I invented the data, I know the population mean is 52.2. For group 1, $\bar{x}_1 = 60.3$, $s_1 = 20.49$ and $se_1 = 20.49/\sqrt{10} = 6.48$, yielding a 95%-confidence interval = $[60.3 - 2 \cdot 6.48, 60.3 + 2 \cdot 6.48] = [47.33, 73.26]$. We can compute the same for all other groups, yielding 10 different 95%-confidence intervals ($[43.99, 59.01]$, $[48.86, 65.94]$, ...). Now we expect that in 95% of the groups, the population mean lies within these confidence intervals! Or, in other words, for 100 groups going out to measure ash girths, we would have 5 groups were the population mean falls *outside* their 95%-confidence interval. In our few groups, the mean of all 100 trees of 52.2 happens to always be within the 95%-CI.

So to reiterate: the mean ± 2 standard error of the mean describes the range of values in which we expect the true population mean to lie with a 95% probability.[8]

The **interquartile range**, IQR, is to the median what the standard deviation is to the mean. It is the range of values in which 50% of the data points lie (namely those between the 0.25- and the 0.75-quantile, a.k.a. 25% and 75% percentile).

The *median absolute deviation* (MAD) is a robust variant of the interquartile range. The specifics of its derivation are not important here, but it is more robust than the IQR because it collapses the values on both sides of the median and then recalculates the median—i.e. the middle of the collapsed values. This is then multiplied by a constant, so that the MAD coincides with the standard deviation for normally distributed data[9]:

$$\text{MAD} = \text{Median}(|x - \text{Median}_x|) \cdot 1.4826 \tag{1.9}$$

The bars mean "absolute value", or the removal of the minus sign. x represents a vector with the measured values: $x = (x_1, x_2, \ldots, x_n)$.

Finally, you can also simply provide the smallest and largest values for a sample (the interval of the data), or the difference between these two values, which is known as the *range*.

If a sample is not as beautifully symmetrical as our ash tree diameter example, it may be useful to measure other properties of the sample, i.e. other statistics. For example, if the histogram is drawn out to the right, with more larger values than

---

[6]It is no longer common to speak simply of the *standard error*, but rather of the standard error of the mean.

[7]The standard error of the mean is a constant source of confusion. It is easily misunderstood, and miscommunicated, as it builds on the idea that our observed data set is only one of many possible. Across all of these repeated samples, 95% would have their confidence interval comprise the true population mean. The idea of making statements about the mean's accuracy "under repeated sampling" is due to Jerzy Neyman.

[8]As a completely analogue measure, you can also calculate the standard error for other sample statistics, for example, the standard error of the standard deviation: $se_s = s^2 \sqrt{\frac{2}{n-1}}$.

[9]The factor is equivalent to the probability density of the standard normal distribution at 0.75 (1/qnorm(3/4)). The background is described in Tukey [1977].

smaller values, we would call it right-skewed. Or, if the histogram is *very* pointy, we would could compute a measure called *kurtosis* that describes the peak's curvature. In our ash tree example, these skew and kurtosis values are −0.073 and −0.054, respectively. Values close to zero (such as these), tell us that the skew and the kurtosis are not very different than those of a normally distributed sample.

Skew and kurtosis can be calculated in multiple ways. The basis for the **skew** is g, as it is calculated in older textbooks:

$$g = \sqrt{n} \frac{\sum_{i=1}^{n} (x_i - \bar{x})^3}{\left(\sum_{i=1}^{n} (x_i - \bar{x})^2\right)^{3/2}} \tag{1.10}$$

with n = number of data points and $\bar{x}$ = mean of $\boldsymbol{x}$.

Today, these two values are usually provided with one of two possible corrections for small samples (Joanes and Gill 1998).[10] All three formulas are correct and can be used ("unbiased under normality").

$$g_2 = g \cdot \frac{\sqrt{n(n-1)}}{(n-2)} \tag{1.11}$$

$$g_3 = g \cdot ((n-1)/n)^{3/2} = g(1 - 1/n)^{3/2} \tag{1.12}$$

Similarly, there are also three formulas for **kurtosis**. Since only one of these formulas is unbiased under normality, it is the only one shown here (see Joanes and Gill 1998):

$$k = \frac{\left((n+1)\left(\frac{n\sum_{i=1}^{n} x_i^4}{(\sum_{i=1}^{n} x_i^2)^2} - 3\right) + 6\right)(n-1)}{(n-2)(n-3)} \tag{1.13}$$

Sample statistics can be made for *any* sample of numerical values.[11] Their job is to describe the sample, e.g. its centrality and spread. For our ash trees, the mean $\bar{x} = 52$ cm and the standard deviation $s = 17.9$ cm. It is common to provide one more decimal place for the standard deviation than for the mean. Common sense should also be used when deciding how many decimal places to report for the mean. In our case, since a measuring tape hardly ever has an accuracy greater than 0.5 cm, I think that we should avoid using a decimal here. Exactness should not be pretended.

Another way to *graphically* display a sample is by using a box-and-whisker-plot, or box plot for short (see Fig. 1.4). As opposed to the histogram, where the raw data are simply displayed, the box plot is based on the sample statistics that we just learned. Unfortunately, statistics programs differ in what all is contained within a box plot. As a rule, a box plot shows a measure of centrality (usually the median) as a thick horizontal line. Around this line is a box, which (again, usually) contains the range between the 1st and the 3rd quartiles, or the central 50% of the data points. The error bars (the *whiskers*) show the extreme values, but only if they are within 1.5 times the length of the box (= 1.5 IQR). Otherwise, these extreme values are either left out (common) or shown as points (less common).

There is another variant in which the box around the median contains notches. These notches are the non-parametric equivalent of the standard error of the mean, and are calculated as $\pm 1.58 \cdot \text{IQR}\sqrt{n}$. According to Chambers et al. (1983, S. 62), overlapping notches show that two data sets are not significantly different from one another.

The box plot summarizes the information more concisely than the histogram, but may be less informative. On the other hand, by using box plots, multiple data sets can be directly compared with one another. Figure 1.5 shows the same data from Fig. 1.2 as a box plot, but with a more efficient use of space.

Here we see that the notches can sometimes be larger than the box, leading to humanoid looking graphics. Regardless of whether we think this plot looks ridiculous, we can see that all of the samples are relatively similar to each other.

When trying to gain information from a box plot, we should focus on two things in particular: (1) Does the median lie near the middle of the box? If this is the case (as for groups 5 and 8), then the sample data are rather symmetrically distributed. If this, in turn, happens to be the case, then the mean will be similar to the median, and there is a good chance that the data are normally distributed. (2) Are the error bars, the whiskers, similarly long? The interpretation is then the same as for (1).

---

[10]The first correction (g_2) is common when using SAS or SPSS and is the default for R in **e1071**. Minitab uses option 2, or g_3.

[11]By "numerical values", I mean metric variables, i.e. those for which the numerical value has a quantitative meaning. Some people think that the calculation of these values is based on the assumption of a normal distribution. This is not the case. $\bar{x}$ and s are *sample statistics*, not distribution parameters. This does not mean that the mean and standard deviation always provide *sensible* information regarding a sample.

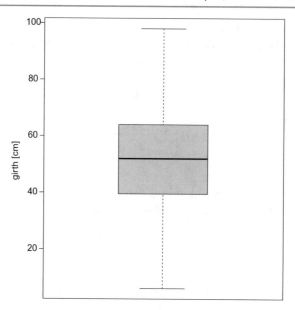

Fig. 1.4 Box plot of 100 ash tree diameters. While the horizontal line in the box shows the median, the cross shows the mean. With symmetrically distributed data (such as the data here), these two measures are nearly identical

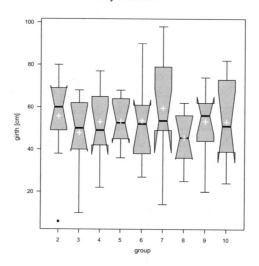

Fig. 1.5 Box plot of the ash tree diameters from groups 2–9. The extreme value in group 2 lies outside 1.5-times the IQR and is shown as a point. For the other groups, extreme values lie outside of the box, but within the 1.5-times IQR. The notches, which are a visual measure of difference between the samples, sometimes extend like arms and legs from the box. The white crosses show the mean of the group

In the example case, the data from group 8 are symmetrical while the data from group 3 are clearly skewed. If we directly compare this to Fig. 1.2, the box plots seem less insightful and more difficult to interpret. Then again, the sample size of 10 data points per group is also very small.

For the sake of completeness, it should be noted that there are a number of hybrid forms that are part histogram/part box plot that show the distribution of the data, e.g. in the form of a violin-shape (used e.g. in Dormann et al. 2010). A nice overview of such visualizations is provided by Ibrekk and Morgan (1987), Cleveland (1993), Robbins (2005).

1.1.3 Sample Statistics: An Example

Our example data set here contains observations of red mason bees while stocking their brood cells. The mason bees fly away, collect pollen and then store the pollen in a brood cell. When this cell is full, an egg is laid inside and the cell is sealed with clay. The data set here shows the duration of the pollen collection flights in a fruit orchard (in minutes). In total, 101 observations were made.

```
1.79 2.79 1.65 2.15 2.98 1.88 1.93 1.86 2.00 2.18 2.71 1.80 1.71
2.05 1.71 2.06 1.88 1.37 2.22 2.01 2.53 2.56 2.84 2.44 2.49 2.00
2.81 2.86 1.86 1.79 2.26 2.64 3.30 2.70 2.85 2.56 1.73 1.42 1.49
2.06 2.89 1.80 2.09 3.09 1.93 2.37 1.77 1.93 1.83 2.09 2.84 2.20
2.60 1.88 2.07 1.76 2.46 2.07 2.09 2.22 1.69 2.51 1.89 2.34 1.82
1.98 1.39 1.99 1.64 2.00 2.03 2.02 2.27 2.13 2.30 2.05 2.57 2.17
2.20 1.89 2.34 1.49 2.57 2.11 2.42 1.84 3.41 1.93 2.09 1.91 2.55
1.71 2.37 2.53 2.58 2.29 1.98 1.90 2.04 2.09 1.42
```

First and foremost, we will have a look at the histogram of the data (Fig. 1.6). Here, we combine the histogram with a horizontal box plot for direct comparison. We see that most of the flights last for around 2 min, and rarely last less than 1.5 min or more than 3 min. The box plot shows a clear deviation of symmetry in both the box and the whiskers.

Now we can calculate the sample statistics from Table 1.2. Some of the values are redundant, unusual, or not intuitive. Let's have a look at the whole flurry of centrality and spread measures, just so we have an overview. Here they are listed according to importance. First, the centrality measures of mean, median, mode and Huber's robust mean:

Mean = $\bar{x} = 2.16$
Median = 2.07
Mode = 2.09
Huber's Mean = 2.14
Standard error of the mean $sem = 0.042$
95% confidence interval of the mean = [2.079, 2.245]

And now for the spread measures:

Standard deviation $s = 0.422$
Huber's standard deviation = 0.400

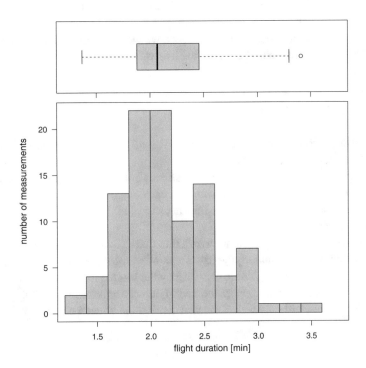

Fig. 1.6 Duration of the pollen collection flights of the red mason bee (*Osmia bicornis*) in an apple orchard. The box plot is shown horizontally above the histogram to ease comparison of the two types of visualisation

Variance $s^2 = 0.177$
Coefficient of variation CV $= 0.20^{12}$
95% quantiles $= [1.420, 3.035]$
Interquartile range IQR $= 0.58$
Median absolute deviation MAD $= 0.400$
Interval $= [1.37, 3.41]$ with a range of 2.04
Skew $= 0.569$
Kurtosis $= 0.026$.

So many numbers, but what do they tell us?

First, let's have a look at the difference between mean and median (2.16 versus 2.07). These values are relatively close to one another, the mean is slightly higher, since there are more longer flights than shorter ones (the histogram is slightly right skewed). Based on so few data, the mode is not able to tell us much. Huber's mean is close to the arithmetic mean, telling us that our mean is not distorted by extreme values. Finally, the standard error of the mean tells us that we expect the population mean not to be far from $\bar{x}$ (since *sem* has a small value). It defines the 95%-confidence interval, within which the true (population) mean lies in 95% of repeated measurements (although we do not know whether this is one of the 5% or not): $[\bar{x} - 2se, \bar{x} + 2se] = [2.078, 2.246]$.[13]

The measurements of spread quantify how strongly the values differ from each other. Only some of these measures can be directly interpreted without more information. The coefficient of variance of 0.2 (or 20%) tells us that the values only vary moderately relative to the mean. In other words, the standard deviation of $s = 0.422$ is not particularly large (or small) for a mean of $\bar{x} = 2.16$. CV values below 0.05 show very high precision, while those over 0.2 show low precision. In ecological applications, such measures can occasionally be over 1, since we usually have a low number of data points and are probing a highly variable system.

The difference between the 95% confidence interval and the 95% quantiles can be easily seen in these statistics. The confidence interval describes the accuracy of the mean estimator $\bar{x}$, based on its standard error. The quantiles, on the other hand, describe the range of values within the data: 95% lie between 1.42 and 3.04.

At 0.58, the interquartile range is larger than the standard deviation (0.4). This tells us that the data are not symmetric, because the IQR gets larger in one direction while that standard deviation changes only slightly. And, in fact, we can see this not only in the histogram, but also in the skew, which at 0.57 is rather different than 0 (symmetric). Positive values indicate right skewness (= left peak), while negative values indicate left skewness (= right peak).

In summary, we can say that mason bees need an average of 2.16 ± 0.042 (mean $\pm$ standard error) minutes for a pollen collection flight. If we were to now take additional measurements, we would expect that each bee would take 2.16 min, and that it would be very likely that each flight would be within the interval $[1.42, 3.04]$ min.

1.2   Frequency, Density and Distribution

In Chap. 3 we will explore distributions more intensely. Here, we will now try to make the transition from our sample (displayed, for example, as histogram) to a distribution.

Up to now, our histograms have shown the number of observations per bin on the y-axis (see Fig. 1.3). In other words, the histogram is based on the *frequency* of the data. Alternatively, we can compute the *density*. This is the frequency divided by the number of measurements and the bin size. The sum of the density of all bins is equal to 1. The only thing that changes here is the scale of the y-axis (see Fig. 1.7).

The division into bins is arbitrary. There are some algorithms that provide particularly nice looking divisions, but a categorisation into bins always happens! The underlying data are, however, continuous; an ash tree can have any girth within the measured interval. Therefore, it would be logical to also display the sample in a continuous manner. Right?

No, actually not! The sample is only a realisation of the underlying distribution. Accordingly, you can't simply assign a distribution to any sample. Our sample will only reflect the true distribution when we have a *large* number of data points.

---

[12]This number is often multiplied by 100 and reported as a %, so here it would be 20%.

[13]The slight difference to the reported 95%-CI is due to the actual value to multiply the standard error of the mean with is 1.96, not 2. If this makes a difference for your data, you should consider collecting more data!

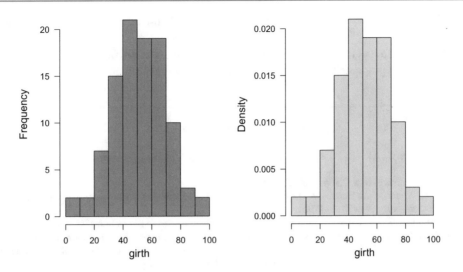

Fig. 1.7 Measurements of all 100 ash tree diameters as frequency- and density-histograms

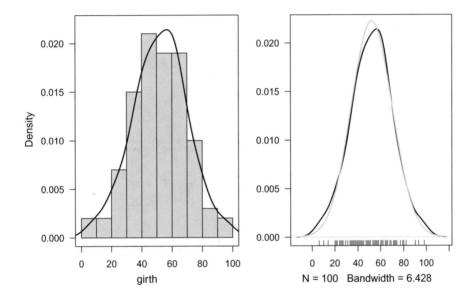

Fig. 1.8 Measurements of all 100 ash tree girths as an empirical density curve over the density histogram (left) and with the underlying normal distribution as a grey line (right) The ticks on the right (the total of which are known as a *rug*) show where the data points lie

With this in mind, let's look at how the density histogram would look as a density distribution (Fig. 1.8 on the left). This curve shows how a continuous density of data would look. The density curve is calculated using a sliding kernel.[14] The information under the *x*-axis gives the number of data points (N = 100) as well as the width of the kernel (here, 6.4 cm). At this point, we will assume that the method knows what it is doing.

As a result, we no longer have a representation of our sample data, but rather a continuous distribution based on our sample. One thing to notice is that the density curve extends beyond the range of our measurements (if only slightly). We hope that this distribution is very similar to the true distribution of all ash tree griths, i.e. the population of ash girths.

To illustrate this point, we can also add a normal distribution to the picture (Fig. 1.8, right). This has two parameters (see Sect. 3.4.1), namely the mean μ and the standard deviation σ. For reasons explained in that chapter, we calculate the mean of our sample $\bar{x}$ (= 52.2) and the standard deviation s (= 17.94) and assume that these are good estimates for μ and σ. As

---

[14]The math behind this calculation is surprisingly complicated. The idea is a sliding calculation of density, whereby only the points currently in the window contribute to the calculation.

we see in the figure, the two curves lie very close to each other. We can confidently say that our ash tree girths are normally distributed.

But why do we care so much about the normal distribution? What about other distributions? Can we somehow statistically test if a certain distribution is appropriate? These are questions that we will explore in Chap. 3!

References

1. Chambers, J. M., Cleveland, W. S., Kleiner, B., & Tukey, P. A. (1983). *Graphical methods for data analysis*. New Plymouth, NZ: Wadsworth & Brooks/Cole.
2. Cleveland, W. S. (1993). *Visualizing data*. Corona, CA: Hobart Press.
3. Dormann, C. F., Gruber, B., Winter, M., & Herrmann, D. (2010). Evolution of climate niches in European mammals? *Biology Letters, 6,* 229–232.
4. Huber, P. J. (1981). *Robust statistics*. New York: Wiley.
5. Ibrekk, H., & Morgan, M. G. (1987). Graphical communication of uncertain quantities to nontechnical people. *Risk Analysis, 7,* 519–529.
6. Joanes, D. N., & Gill, C. A. (1998). Comparing measures of sample skewness and kurtosis. *The Statistician, 47,* 183–189.
7. Quinn, G. P., & Keough, M. J. (2002). *Experimental design and data analysis for biologists*. Cambridge, UK: Cambridge University Press.
8. Robbins, N. B. (2005). *Creating more effective graphs*. Hoboken, N.J: John Wiley & Sons.
9. Tukey, J. W. (1977). *Exploratory data analysis*. Reading: Addison-Wesley.

I hear, and I forget,
I see, and I remember
I do, and I understand.
—Unknown

At the end of this chapter...
... it should be clear to you that data collection should be planned in advance and you should know what an ideal data sheet looks like.
... you will be able to read data into R.
... you will have the skills to transform raw data in R into a histogram.
... you will be able to calculate the sample statistics we learned in Chap. 1 in R.
... you will have finally done something practical!

2.1 Data Collection

Data are usually collected in the field, in the laboratory or from a survey, and data points are often recorded on a standardised data form. The purpose of such a form is to ensure that all relevant information is recorded. If we take our ash tree girths example, each tree would have its own row for recording corresponding data. Additionally, we might want to record the GPS coordinates, the location and name of the park, the name of the person doing the data collection, the date etc. (see Fig. 2.1). Such forms can be easily prepared in a spreadsheet program (see below).

In a laboratory setting, data are often recorded on printouts from analysis machines (e.g., a CN-analyser), in large special files (e.g., near-infrared spectroscopy) or by hand in a lab notebook. Data in special formats can (in most cases) be directly imported into R (see below for further details).

These data are the raw data for our analysis.

2.2 Data Entry

Theoretically, you could enter raw data directly into R (see Sect. 2.3.1 on p. 18). A more practical solution is to use a spreadsheet program. The most well known spreadsheet program is Microsoft Excel, but there are a number of open access (and free) software solutions as well. Here, for example, we use LibreOffice Calc,[1] which are very similar to Excel.

The core element of such programs is the spreadsheet (Fig. 2.2).

---

[1] Available for all operating systems and in many languages from www.libreoffice.org.

© Springer Nature Switzerland AG 2020
C. Dormann, *Environmental Data Analysis*,
https://doi.org/10.1007/978-3-030-55020-2_2

Ash girth measurements   sheet no:

number	Girth (at breast height)
1	
2	
3	
4	
5	
6	
7	
8	
9	
10	
11	
12	
13	
14	
15	
16	
17	
18	
19	
20	

Date/time:
Location:

GPS:

Surveyer:

Comments:

Fig. 2.1 An example data form for the measurement of ash tree girth. It is important that there is a space for comments and that the spaces are large enough for the data to be collected. In this example form, it is no good that there is no space available to make a comment for each tree. Such data forms are usually only one sided so that the back side can be used for additional notes. To be sure that no data are missed during data entry, **P.T.O.** (*please turn over*) should be written on the lower right corner if the back is used!

Fig. 2.2 A spreadsheet program with an empty sheet (LibreOffice Calc)

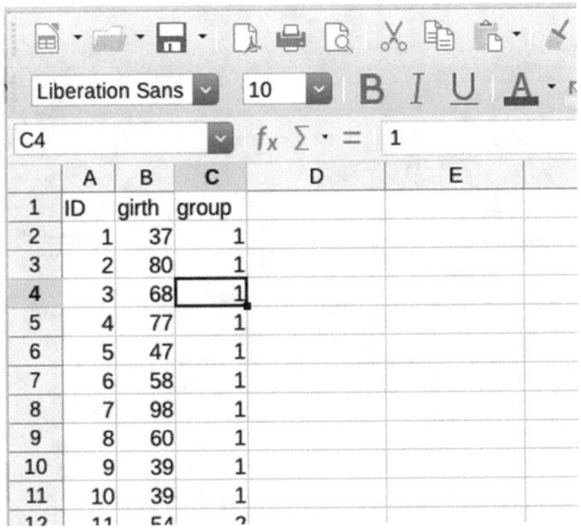

Fig. 2.3 The ash tree girths in a simple spreadsheet. All 100 values are in a single column, with the group number noted in another column. This is the preferred long format, whereas Table 1.1 shows the wide format

This is where we enter our data. During this process, typos regularly happen—here are a few tips that may help minimise such errors:

1. Keep the data sheet simple. Start in the upper left corner and do not leave any blank rows or columns. Write comments in a separate column.
2. Use the number pad on your keyboard! If you do not have one (the case for many laptop computers), borrow one.
3. Try to enter the numbers *blind*—i.e. only move your eyes from the data form to the screen and let your fingers work without looking at them. After a short time, your fingers will know where to go and the transfer of numbers from your paper form to the electronic sheet will go quickly and with surprisingly few errors.
4. Enter all of the data for a row/column/block/set, take a quick break, and then go through the data again to check for errors.
5. Make a histogram of the data, column by column. This will allow you to see obvious outliers and identify potential typos.[2]
6. Make data entry a team activity: One person dictates, one types. The reader should read the numbers one after the other ("four-two-point-one-seven"). The person typing repeats the numbers while typing.

At then end, the spreadsheet might look something like Fig. 2.3.

For some data, there is more than one way to enter it. Vegetation surveys are often entered in the so-called "wide format": each column is a species, whereas each row corresponds to a location (Fig. 2.4).

The alternative is the "long format", where each piece of information has its own row. This makes the table much longer, and potentially more confusing. However, data in this format are easier to read into R and other databases, and be easily transformed, and are more accessible for statistical analysis. Unless we have good reasons otherwise, we try to always use the long format.

In ecology, the wide format is the most common for applications such as vegetation surveys (or other samplings such as pitfall traps) or for environmental data that correspond to a particular location (such as pH, elevation, soil moistpidure, etc.).

Transforming the long to the wide format is very easy in R[3] and you should therefore **not** cut and copy data in your spreadsheet just to get it into a wide format. This can cause unnecessary errors and your time will be better spent collecting more data.

Once we have entered and double-checked the data, we *never* change this spreadsheet again! All future reformatting, correcting etc. is done in R, thus recording all changes we make as R-code for maximal reproducibility (see, e.g., British Ecological Society 2015, Cooper and Hsing 2017).

---

[2]There are special R packages meant to help with quality control, such as **qcc**.

[3]Using the `cast` command and back with `melt`, in the **reshape** package.

row.names	Cal.vul	Emp.nig	Led.pal	Vac.myr	Vac.vit	Pin.syl	Des.fle	Bet.pub	Vac.uli	Dip.mon	Dic.sp	Dic.fus	Dic.pol	Hyl.spl	Ple.sch	Pol.pil	Pol.jun	Pol.com	Poh.nut	Pti.cil	Bar.lyc
18	0.55	11.13	0	0	17.8	0.07	0	0	1.6	2.07	0	1.62	0	0	4.67	0.02	0.13	0	0.13	0.12	0
15	0.67	0.17	0	0.35	12....	0.12	0	0	0	0	0.33	10.92	0.02	0	37.75	0.02	0.23	0	0.03	0.02	0
24	0.1	1.55	0	0	13....	0.25	0	0	0	0	23...	0	1.68	0	32.92	0	0.23	0	0.32	0.03	0
27	0	15.13	2.42	5.92	15....	0	3.7	0	1.12	0	0	3.63	0	6.7	58.07	0	0	0.13	0.02	0.08	0.08
23	0	12.68	0	0	23....	0.03	0	0	0	0	0	3.42	0.02	0	19.42	0.02	2.12	0	0.17	1.8	0.02
19	0	8.92	0	2.42	10....	0.12	0.02	0	0	0	0	0.32	0.02	0	21.03	0.02	1.58	0.18	0.07	0.27	0.02
22	4.73	5.12	1.55	6.05	12.4	0.1	0.78	0.02	2	0	0.03	37.07	0	0	26.38	0	0	0	0.1	0.03	0
16	4.47	7.33	0	2.15	4.33	0.1	0	0	0	0	1.02	25.8	0.23	0	18.98	0	0.02	0	0.13	0.1	0
28	0	1.63	0.35	18.27	7.13	0.05	0.4	0	0.2	0	0.3	0.52	0.2	9.97	70.03	0	0.08	0	0.07	0.03	0
13	24.13	1.9	0.07	0.22	5.3	0.12	0	0	0	0.07	0.02	2.5	0	0	5.52	0	0.02	0	0.03	0.25	0.07
14	3.75	5.65	0	0.08	5.3	0.1	0	0	0	0	0	11.32	0	0	7.75	0	0.3	0.02	0.07	0	0
20	0.02	6.45	0	0	14....	0.07	0	0	0.47	0	0.85	1.87	0.08	1.35	13.73	0.07	0.05	0	0.12	0	0
25	0	6.93	0	0	10.6	0.02	0.1	0.02	0.05	0.07	14...	10.82	0	0.02	28.77	0	6.98	0.13	0	0.22	0
7	0	5.3	0	0	8.2	0	0.05	0	8.1	0.28	0	0.45	0.03	0	0.1	0	0.25	0	0.03	0	0
5	0	0.13	0	0	2.75	0.03	0	0	0	0	0	0.25	0.03	0	0.03	0.18	0.65	0	0	0	0
6	0.3	5.75	0	0	10.5	0.1	0	0	0	0	0	0.85	0	0	0.05	0.03	0.08	0	0	0.08	0
3	0.03	3.65	0	0	4.43	0	0	0	1.65	0.5	0	0.55	0	0	0.05	0	0	0	0.03	0.03	0
4	3.4	0.63	0	0	1.98	0.05	0.05	0	0.03	0	0	0.2	0	0	1.53	0	0.1	0	0.05	0	0
2	0.05	9.3	0	0	8.5	0.03	0	0	0	0	0	0.03	0	0	0.75	0	0.03	0	0	0.03	0
9	0	3.47	0	0.25	20.5	0.25	0	0	0	0.25	0	0.38	0.25	0	4.07	0	0.25	0	0.25	0.25	0
12	0.25	11.5	0	0	15.8	1.2	0	0	0	0	0.25	0.25	0	0	6	0	0	0	0.25	0	0
10	0.25	11	0	0	11.9	0.25	0	0	0	0	0	0.25	0.25	0	0.67	0	0.25	0	0.25	0	0
11	2.37	0.67	0	0	12.9	0.8	0	0	0	0	0	0.25	0.25	0	17.7	0.25	0.25	0	0.25	0.67	0
21	0	16	4	15	25	0.25	0.5	0.25	0	0	0.25	0.25	3	0	2	0	0.25	0.25	0.25	10	3

Fig. 2.4 Wide format of a vegetation survey (data set `varespec` from the R-package **vegan**, shown in the R Data Editor using `edit`). The plant species are in the columns (such as *Calluna vulgaris*, *Empetrum nigrum*, etc.), the survey number is shown in the first column `row.names`. The number values here are the plant coverage in ridiculously pseudo-accurate percent

2.3 Importing Data to R

2.3.1 Entering Small Data Sets Manually

If we are just dealing with a few values, we can simply directly enter these in R. We don't necessarily need the help of a spreadsheet program. In the simplest case, we are dealing with a data *vector*:

```
> Tree <- c(25, 23, 19, 33)
> Tree

[1] 25 23 19 33
```

We have to save the data as an object ("`Tree`"). The "`c`" (for concatenate) defines a vector in R.

If we want to enter a mini table with multiple columns, we need a `data.frame`:

```
> diversity <- data.frame("site"=c(1,2,3,4), "birds"=c(14,25,11,5),
+     "plants"=c(44, 32, 119,22), "tree"=Tree)
> diversity

  site birds plants tree
1    1    14     44   25
2    2    25     32   23
3    3    11    119   19
4    4     5     22   33
```

Each column is entered into the `data.frame` as a vector. Vectors that have already been defined can be easily included (see the last column of the object `diversity`).

Within R, various commands can be used to access this table and interactively change values in a cell (e.g. `fix`, `edit`). Do *not* do that! When re-running the analysis, you have to do such manual changes again, which greatly impedes reproducibility and speed. If you discover a typo, use R-commands to correct them. In this example, imagine site 4 should have been site 14. In the R-script, we would then write:

```
> diversity$site[4] <- 14 # typo: wrongly entered as "4"
```

This means, replace the 4th value in column "site" of data set "diversity" with the new value of 14. Do add a comment as reminder for later.

Finally, in some cases a vector data set is available in a "Ready for R" format, for example in a tutorial or in the R help. If this is the case, then you can simply copy the numbers and paste them in the command line (or the Editor) of R:

```
> tryit <- c(#paste the numbers here#)
```

Do not forget that numbers should **not** be placed in quotation marks, and that individual values should be separated by a comma.[4]

2.3.2   Read in Larger Data Sets from a File

R can read many different file formats. And yet, it is still surprising that most data sets are so unstructured and disorganized in a table that they are not directly usable in *any* software—not even in R. Therefore, it is both important and incredibly time saving to pay attention to the data entry suggestions given in the last section.

The most common and simplest way to import data into R is by using an old but highly standardised table format. R can read in data from Excel and LibreOffice Calc using supplementary packages, but this requires more lines of R code. The old standard table formats are the ASCII-Text (recognisable from the .txt file extension) and comma separated values (.csv). The correct file extension is neither necessary nor sufficient for the format! As an example, we will export the ash tree diameter data from LibreOffice Calc as a .csv file ("ashes.csv"; Fig. 2.5).

The most important R-function for reading in data is `read.table`. It reads a file that must contain plain text. Using arguments, the format can be specified (see the R help for this function by entering `?read.table`). In addition to `read.table`, there are specific `read.csv` and `read.delim` functions that already have the necessary options set for .csv and .txt files, respectively. For data sets where commas are used instead of decimal points and semicolons are used to separate data points, `read.csv2` and `read.delim2` may be used.

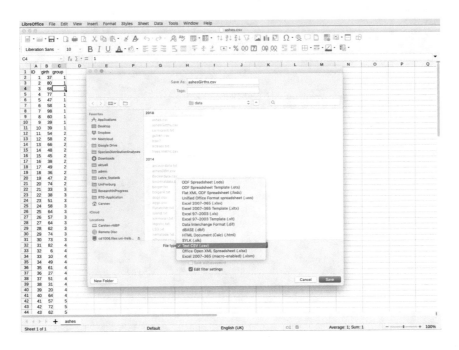

Fig. 2.5 Saving a spreadsheet as a .csv file. Only the current visible spreadsheet is saved! It should contain no images or descriptions – it should be as bare as possible

---

[4]In the **psych** package, the function `read.clipboard` is available, which can read data from the copy-paste clipboard.

Fig. 2.6 A quick check of the file format using a primitive editor

Just to be certain, you should look at the exported data with a simple Editor program[5] to ensure that the data format is what it should be (Fig. 2.6).

By default, numbers are written as such, but character strings are placed in quotation marks. This setting can be changed when exporting from the spreadsheet program, but I would not recommend deviating from the default settings, which are usually compatible with R.

Now we can finally read the data into R and view them there. To do this, we use the following lines of code:

```
> ash <- read.csv("ashesGirths.csv")
> ash
```

	ID	girth	group
1	1	37	1
2	2	80	1
3	3	68	1
4	4	77	1
5	5	47	1
6	6	58	1
7	7	98	1
8	8	60	1
9	9	39	1
10	10	39	1
11	11	54	2
12	12	58	2
13	13	66	2
14	14	48	2
15	15	45	2

...

---

[5]In Windows, a program called Editor, in macOS, TextEdit. A faster solution for all operating systems is the open source program muCommander (www.mucommander.com).

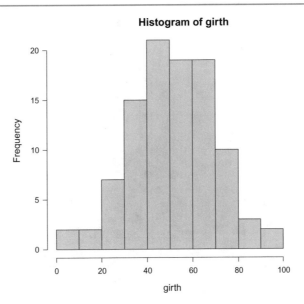

Fig. 2.7 The histogram for all values using the `hist` command. With the `col` argument, we can choose a color, `las` determines the orientation of the text description (see `?par` and `?hist`)

With the arrow (`<-`) we assign the imported data to an object (called "ash").[6] We can now view the object (= print it in the R console) by simply calling the object (i.e. typing in the name)

If the data are in a different format (such as a dbase, SAS, JSON, ncdf, ArcGIS-shape, raster or virtually anything else), there is usually a `read` command for reading in the data. These specific commands can be found by searching on the R help website and downloading the required package. There are hundreds of special file formats that can be read by R, including satellite binaries, sequencing data, climate data and many more. Also web data can be directly scraped in different ways and formats (see https://cran.r-project.org/web/views/WebTechnologies.html).

If a certain format proves to be particularly tricky, it is often possible to use `scan` or `readLines` to read the lines of data in one by one. There are help files related to this on the R websites and from R mailing lists.

2.4 Simple, Descriptive Statistics and Graphic Representation

2.4.1 Graphic Representation of Sample Statistics with R

The distinctively important histogram can be created using the `hist` command:

```
> attach(ash) # makes the variable available as the respective column name
> hist(girth, col="grey", las=1)
```

In my experience, the bin sizes suggested by R tend to be excellent, especially for data exploration purposes. For a figure in a publication, however, it makes sense to customize the bin sizes to an ideal value. Let us first calculate the range of the data (i.e. the minimum and maximum) using `range` and define a sequence of bin widths that fits our taste (for example, every 5 cm) using the `breaks` argument (Fig. 2.8):

```
> range(girth)

[1]  6 98
```

---

[6]If you don't like the arrow (officially: assign arrow), you can also use an equals sign to assign objects (`=`). In addition to using an arrow in one direction `<-`, we can also turn it around and use it the other way (`->`). So we could have written `read.csv("ashes.csv") -> ash`. This does make sense logically ("read in the data and put it in the object ash"), but is rather unconventional. I have never seen such a line of code. See also the R help for `?"<-"`.

Histogram of girth

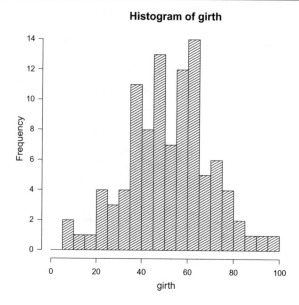

Fig. 2.8 Histogram of ash tree girths. Similar to Fig. 2.7, but with self defined bins and hatching (defined with the arguments density and angle)

```
> hist(girth, las=1, breaks=seq(0, 100, by=5), density=20, angle=30)
```

With this summary, we have lost sight of the fact that each group only collected 10 data points. So, if we wanted to make a histogram for a single group, then we need to make use of a helpful programming tool: the square brackets ([]). Using these, we can do what is called *indexing*. This concept is best explained using an example.

```
> girth[group==3]
```

```
[1] 33 38 51 58 64 57 64 62 74 73
```

The double equals sign (==) is a logical query (just like <, > and !=; the last one meaning "not the same"). We therefore use the logical query, "is the value for group equal to 3?" to index those values for which the query applies.[7] The histogram for this group can then be generated by entering hist(girth[group==3]).

boxplot is a similarly simple R command. Here as well, the visualization is determined by optional arguments, such as horizontal=TRUE. Using the command windows(.) (or quartz() in macOS and X11() in Linux) we can open a new window and can also determine the width and height (in inches, so 2.54 cm). In doing this, we can make the plot slimmer, which is advisable for an individual box plot. With par(.), we can define other graphical parameters, in this case with mar=c(.) we define the margin size (beginning from the bottom, then going clockwise). The argument whisklty makes the dashed default error bar into a solid one. Fig. 2.9 shows the result.

```
> X11(width=5, height=2) # or windows (for windows); quartz (for macOS)
> par(mar=c(4,1,1,1))
> boxplot(girth, horizontal=T, las=1, col="grey", xlab="girth [cm]", whisklty=1)
> points(mean(girth), 1, pch="+", col="white", cex=2)
```

Histograms and box plots can be combined with the help of the hist.bxp function from the **sfsmisc** package. Figure 1.6 was created with a more complex code, based on the layout function, which allows the plotting region to be split into differently sized sub-regions. For all enthusiastic R users, here are the details:

```
> pollen <- read.csv("../data/pollenflight.csv")
> durations <- pollen$x
```

---

[7]More exactly, within the square brackets there is a vector made up entirely of TRUE and FALSE. The square brackets selects only those values from the vector named before the brackets for the vector in the brackets that are TRUE. You can easily see this if you enter group==3 in the R-console.

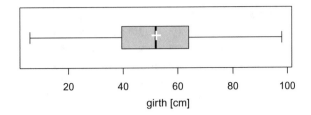

Fig. 2.9 Box plot of the ash tree diameters, customized using multiple arguments

```
> def.par <- par(no.readonly = TRUE) # save default, for resetting...
> xlims <- range(durations) * c(0.9, 1.1)
> layout(matrix(c(1,2), 2,1), c(1,1), c(1,3))
> par(mar=c(1,4,0.5,0.5))
> boxplot(durations, col="grey", las=1, horizontal=T, ylim=xlims, axes=F)
> axis(side=1, labels=F)
> box()
> par(mar=c(4,4,0,0.5))
> hist(durations, las=1, col="grey", main="", xlab="flight duration [min]",
+    xlim=xlims, ylab="number of measurements")
> box()
> par(def.par)#- reset to default
```

2.4.2 Descriptive Statistics with R

The simple descriptive statistics described in the previous chapter (mean, median, standard deviation, etc.) are implemented in R as functions. You can recognize functions, because they are always followed by parentheses "()". Table 2.1 shows the R-functions for the respective sample statistics.

We can calculate the mean of our ash tree diameters with:

```
> mean(ash$girth)
```

```
[1] 52.19
```

Using the $, we can access the column names when a data.frame is not attached. Since the data set ash has the diameter values in the column girth, these values can be called with ash$girth:

```
> girth
```

```
 [1] 37 80 68 77 47 58 98 60 39 39 54 58 66 48 45 38 49 36 47 74
[21] 33 38 51 58 64 57 64 62 74 73 82  6 10 49 61 27 51 31 20 64
[41] 57 72 62 22 36 79 35 25 62 82 34 49 49 42 36 90 54 56 57 61
[61] 60 50 47 42 57 43 14 41 71 24 65 69 21 49 65 29 79 50 44 39
[81] 62 66 40 72 48 61 53 51 57 26 44 62 58 65 68 47 93 38 55 41
```

Alternatively, we can make these column names directly retrievable by adding them to the search path. This can be done using the attach command:

```
> attach(ash)
> girth
```

```
[1]   37 80 68 77 47 58 98 60 39 39 54 58 66 48 45 38 49 36 47 74
[21]  33 38 51 58 64 57 64 62 74 73 82  6 10 49 61 27 51 31 20 64
[41]  57 72 62 22 36 79 35 25 62 82 34 49 49 42 36 90 54 56 57 61
[61]  60 50 47 42 57 43 14 41 71 24 65 69 21 49 65 29 79 50 44 39
[81]  62 66 40 72 48 61 53 51 57 26 44 62 58 65 68 47 93 38 55 41
```

With `detach(ash)` we can remove the name from the search path. Analog to the mean, we can also calculate the median and the standard deviation:

```
> median(girth)

[1] 52

> sd(girth)

[1] 17.94137
```

We can calculate many different statistics in this way, but the most common ones can be found in Table 2.1.

In addition to the median, there are two other quartiles. The 1st quartile is the value at which 25% of the measured values are smaller. The idea is the same as with the median (50% of the measured values are smaller), and also holds true for the 3rd quartile (75% of all values are smaller). You can retrieve any quantile with the `quantile` function. However, there are many different variants for how the quantiles are calculated. Therefore, you should be sure to read the help file for this function before using it! Here is an example with the 95% sample range in addition to the quartiles:

```
> quantile(girth, c(0.025, 0.25, 0.5, 0.75, 0.975))

 2.5%  25%    50%   75% 97.5%
16.85 39.75  52.00 64.00  86.20
```

The value that SPSS would calculate can be found using "type 6":

Table 2.1 A collection of possible sample statistics and the corresponding function in R

Name	Abbr.	R Function
Mean	$\bar{x}$	mean
Median		median
Mode[a]		
Huber's M		huber in **MASS**
Variance	s^2	var
Standard deviation	s	sd
Standard error of the mean		std.error in **plotrix**
Median absolute deviation		mad
Coefficient of variance	CV	cv in **raster**[b]
95% Confidence interval	CI	t.test(x)$conf.int
95% Quantiles		quantile(., c(0.025, 0.975))
Interquartile range	IQR	IQR
Range		range
Skewness		skewness in **e1071** or **fUtilities**
Kurtosis		kurtosis in **e1071** or **fUtilities**

[a]As far as I know, there is no R-function for calculating the mode, probably because the mode is seldom used. You could define it for a data set thusly: `Mode <- function(x) {b <- sort(table(x), decreasing=TRUE);as.numeric(names(which.} {max(b)))}`
[b]It seems like overkill to load such a large package such as **raster** just for calculating the coefficient of variance, when you can easily program this yourself: `CV <- function(x) sd(x, na.rm=TRUE)/mean(x, na.rm=TRUE)`. The argument `na.rm=TRUE` deletes missing values (NAs) before the mean and standard deviation are calculated

```
> quantile(girth, c(0.025, 0.25, 0.5, 0.75, 0.975), type=6)

   2.5%    25%    50%    75%  97.5%
 12.100 39.250 52.000 64.000  91.425
```

Here there is no right or wrong, simply different definitions. The details can be found on the `?quantile` help page.

If we want to calculate values for which the functions in Table 2.1 need to be loaded from other packages, this can be done as follows:

```
> library(plotrix)
> std.error(girth)

[1] 1.794137
```

The confidence intervals can also be retrieved as the by-product of a t-test. We will learn exactly what a t-test is later (Sect. 11.1). The function `t.test` gives us an R-object (of the class `htest`) with multiple entries (a "list"). We can view these with the help of the `str` function. Using the `$`, we can access individual entries, in this case, the 95% confidence interval limits.

```
> t.test(girth)

    One Sample t-test

data:  girth
t = 29.0892, df = 99, p-value < 2.2e-16
alternative hypothesis: true mean is not equal to 0
95 percent confidence interval:
 48.63004 55.74996
sample estimates:
mean of x
    52.19
```

```
> str(t.test(girth))

List of 9
 $ statistic  : Named num 29.1
  ..- attr(*, "names")= chr "t"
 $ parameter  : Named num 99
  ..- attr(*, "names")= chr "df"
 $ p.value    : num 2.65e-50
 $ conf.int   : atomic [1:2] 48.6 55.7
  ..- attr(*, "conf.level")= num 0.95
 $ estimate   : Named num 52.2
  ..- attr(*, "names")= chr "mean of x"
 $ null.value : Named num 0
  ..- attr(*, "names")= chr "mean"
 $ alternative: chr "two.sided"
 $ method     : chr "One Sample t-test"
 $ data.name  : chr "Umfang"
 - attr(*, "class")= chr "htest"
```

```
> t.test(girth)$conf.int
```

```
[1] 48.63004 55.74996
attr(,"conf.level")
[1] 0.95
```

In addition to the confidence interval itself, the confidence level (conf.level) is also shown. We might also be interested in the 50 or 99% confidence interval. We can acquire these values by adding an additional argument to the t.test function (for an overview of all possible arguments and their meaning, see ?t.test):

```
> t.test(girth, conf.level=0.5)$conf.int
```

```
[1] 50.97541 53.40459
attr(,"conf.level")
[1] 0.5
```

For the 50% confidence interval, the values are of course much tighter around the mean than those for the 95% interval.

Again, it should be pointed out that the confidence interval is a measure of the accuracy of the mean value, not a measure for the spread of the data points. Now, we want to calculate the spread measures from Table 2.1.

```
> var(girth)
```

```
[1] 321.8928
```

```
> sd(girth)
```

```
[1] 17.94137
```

```
> mad(girth)
```

```
[1] 17.7912
```

```
> library(raster)
> cv(girth)
```

```
[1] 0.3437703
```

```
> quantile(girth, c(0.025, 0.975))
```

```
 2.5% 97.5%
16.85 86.20
```

```
> IQR(girth)
```

```
[1] 24.25
```

```
> range(Umfang)
```

```
[1]   6 98
```

```
> library(e1071)
> skewness(girth)
```

```
[1] -0.07268684
```

```
> kurtosis(girth)
```

```
[1] -0.05421861
```

Of course we would not want to calculate *all* of these measures for each data set. A simple summary of the most important statistics can be made with the function `summary`:

```
> summary(girth)
```

```
 Min. 1st Qu.  Median    Mean 3rd Qu.    Max.
 6.00   39.75   52.00   52.19   64.00   98.00
```

It shows the mean and the median as well as the range (`Min.` and `Max.`) and the 1st and 3rd quartiles. By comparing the mean and median we can get a rough idea about the skewness, which we can validate by looking at the distances between the median and the quartiles: for a symmetric distribution, the 1st quartile will be as far from the median as the 3rd.[8]

Since we have already plotted the histogram, we roughly know what these values should look like.

2.4.3   Density Histogram and Empirical Density

For the sake of completeness, here is the code for plotting a density histogram and the empirical density distribution. We use a boatload of arguments for each function. The meaning of each of these arguments can be found in the help file for the respective functions.

If two functions are nested, i.e. one is called within another (such as `plot(density(girth))`), then it can get quite confusing with the different arguments. Therefore, it is important to pay attention to the parentheses!

```
> par(mfrow=c(1,2), mar=c(4,4,1,1))
> hist(girth, col="grey50", main="", las=1)
> hist(girth, col="grey80", freq=F, main="", las=1)
```

This produces exactly what we see in Fig. 1.7 (shown again here as Fig. 2.10). With the `par` function, we set the graphical parameters. In this case, we define that two figures should be placed next to each other (`mfrow` = multiple frames, row-wise), more exactly, that we have one row with two columns of graphics. `mar` defines the margins around the graphics, `oma` defines the outer margins (for both graphics together), starting at the bottom and going clockwise. We already know the arguments from `hist`, only `freq=F` is new here. With this argument, we tell `hist` that we want to display densities and not frequencies.

For Fig. 2.11 we need these functions plus a couple of others. In order to illustrate the density of the data points, we can draw these in with the `rug` function. Then, `mtext` (margin text) draws in a legend in the margin next to the figures.

```
> par(mar=c(4, 3,1,1), mfrow=c(1,2), oma=c(0,3,0,0))
> hist(girth, col="grey80", freq=F, main="", las=1, ylim=c(0,0.022))
```

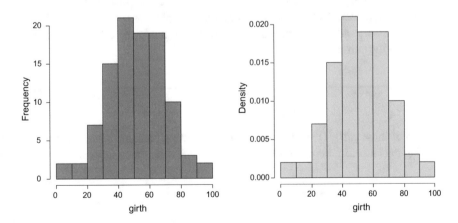

Fig. 2.10   Ash tree girths as frequency and density histograms

[8] An even more condensed output is provided by the `fivenum` function, which shows the minimum, maximum, and the three quartiles in between. Just enter: `fivenum(girth)`.

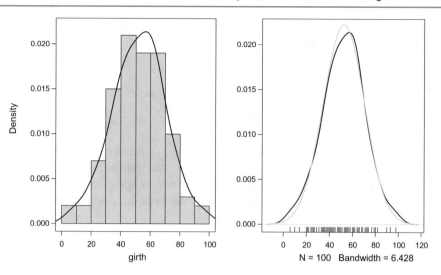

Fig. 2.11 Ash tree girths as an empirical density curve (left) and with an underlying normal distribution curve (grey, right)

```
> lines(density(girth), main="", lwd=2)
> box()
> # now for the figure on the right:
> plot(density(girth), main="", ylim=c(0,0.022), lwd=2, ylab="", las=1)
> curve(dnorm(x, mean=52.2, sd=17.94), col="grey", lwd=2, add=T)
> rug(Umfang)
> mtext("density", side=2, line=1, outer=T, cex=1.5)
```

The core element of these functions is the density function. As you find out when reading the help page (?density), there is quite a bit of mathematics behind this function. The field of *kernel density estimation*[9] (see Härdtle et al. 2004) is important for many applications in statistics. In my experience, however, it is such a well developed field, that we can use the functions without giving too much thought to how they exactly work. Whoever wants to learn more about this can look at the literature cited in the help file as a starting point!

We use the argument ylim, to define the value range of the y-axis. Since the normal distribution in the figure on the right reaches a bit higher, we have to increase the value range to 0.022. After the first plot command, there are three lines for the graphic on the right. First we make an empty plot (type="n" plots nothing in the figure). Then we use curve to insert a mathematical function. In this case, we insert the density of a normal distribution (dnorm) with the arguments mean and sd, whose values we just calculated earlier. We have to include add=T so that curve does not make a new figure, but rather is added to the existing one we have defined. lwd (= line width) defines the thickness of the line, which we set to 2 to increase the visibility. Finally, we use lines to add the curve from the plot on the left. lines does not make a new figure, as plot would, but simply adds a line to the current plot.

2.5   Exercises

These exercises are made up of three parts: (1) Collecting data, (2) entering data and saving it in an appropriate format and (3) Visualising and analysing data in R. In this exercise, it should become clear that using care in parts (1) and (2) can save a lot of time in part (3).

---

[9]http://en.wikipedia.org/wiki/Kernel_density_estimation.

1. Collect your own data. If nothing comes to you, feel free to use one of the following ideas as inspiration:
   (a) Count cars: For at least 30 intervals of 2 min, count all (a) red, (b) black, (c) passenger vehicles, (d) by male/female driver.
   (b) Measure the diameter at breast height for at least 30 trees.
   (c) Measure the height, shoe size and sex for 30 of your classmates.
   (d) Count all bikes parked in front of 30 different university buildings.
   (e) Count the interval of pedestrian walk signals for at least 20 different crossings.
   (f) Look at the cost per square meter (or square foot depending on where you live) and total living space for at least 50 rental offers on different streets.
   (g) Look on the Internet for interesting data, like the number of bird species in each country, ecological damage from chemical accidents in the last 10 years, …
2. Enter the data in a spreadsheet program of your choice (OpenOffice/LibreOffice Calc, MS Excel, etc.) and save it in an R-readable format of your choice (this would usually be .csv or .txt, but if you want to get R to read .dbf, .ods, or .xls files, go for it. This is, however, not intended as part of the exercise).
3. Visualise and describe the data in R.
   (a) Read data into R (using `read.table` or another `read.`-function).
   (b) Make histograms of the data. Make box plots of the data.
   (c) Calculate the mean, median, SD and IQR of the data.

References

1. British Ecological Society. (2015). *A guide to data management in ecology and evolution*. London: British Ecological Society. Retrieved from https://www.britishecologicalsociety.org/publications/guides-to/.
2. Cooper, N., & Hsing, P.-Y. (2017). *A guide to reproducible code in ecology and evolution*. London: British Ecological Society. Retrieved from https://www.britishecologicalsociety.org/publications/guides-to/.
3. Härdtle, W. K., Müller, M., Sperlich, S., & Werwatz, A. (2004). *Nonparametric and semiparametric models*. Berlin: Springer.

La théorie des probabilités n'est, au fond, que le bon sens réduit au calcul.
(The theory of probability is basically common sense reduced to calculation.)

—Pierre-Simon Laplace

At the end of this chapter...
... you will know what a distribution is and what properties it has.
... you will understand the concept of likelihood.
... you will understand how to use "maximum likelihood" to achieve the best possible fit of a distribution for empirical data.
... you will know a few important distributions, such as the normal distribution, the Poisson, the Bernoulli, and the binomial distribution.
... you will know where to look when you want to read more about distributions and their parameters.

In Chap. 1, we dealt with the description of samples. In this chapter, we will go a step further and see if we can find typical patterns in a sample, and make statistical use of this pattern.

If you recall, we refer to a sample as one of (infinitely) many possible realisations of a random variable. The thought behind our ash tree example was that it is a complete coincidence that a tree has a certain thickness. If it had rained just a bit more in the last ten years, it could possibly be thicker. If the tree had not been the host for an insect parasite in its third year of life, it would be even thicker still! So, we can say that a specific ash tree girth is a random realisation of a distribution of possible ash tree girth. Or, in more abstract terms, *a measured data point is a realisation of an underlying random variable*.

Most applications of statistics are concerned with analysis in which we can assume an underlying *distribution* for a random variable. Distributions have parameters that describe the distribution. For this reason, this type of statistics is called "parametric statistics": we estimate the parameters of the underlying distribution. Before we have a look at different possible distributions, I want to explicitly point out that distribution-based parametric methods are based on a distribution *assumption*: how the data values actually originate is often unclear, we just assume that a distribution is appropriate. From a technical point of view, there are many advantages when we can make this assumption. However, we have to justify this assumption for each case.

3.1 Distribution

Let us go back to the density distribution of our ash tree girths as compared to a normal distribution (Fig. 3.1). Because of the high overlap of the measured ash tree girths and a normal distribution, it seems plausible here to assume that a normal distribution is indeed the underlying distribution, from which our ash tree girth measurements are random realisations.

The Concept of Distributions is Absolutely Essential For Understanding Statistics.

Everything else builds off of distributions. For this reason, it is important that we clarify some terms and definitions.

© Springer Nature Switzerland AG 2020
C. Dormann, *Environmental Data Analysis*,
https://doi.org/10.1007/978-3-030-55020-2_3

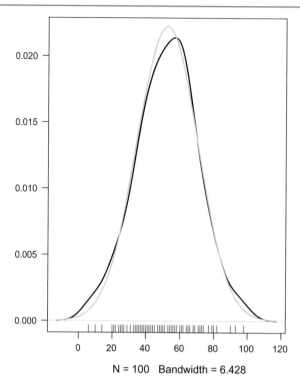

Fig. 3.1 Ash tree girths as an empirical density curve with an overlaid normal distribution curve (grey)

Frequency distributions arise from empirical data. Our histograms and the empirical density distribution are examples of frequency distributions.

Probability distributions are the theoretical counterpart to frequency distributions. As Kass [2011] puts it, data and frequency distributions belong in the *real world*, whereas probability distributions and statistical models belong in the *theoretical world*. Distributions can be seen as human constructs or inventions that we hope will help us in our inspection and analysis of data.

Probability distributions (from here on simply referred to as distributions) have some characteristic properties:

- A distribution is described by a probability density function, also called the density function or simply the density. This function shows the density of the probability at a specific point. Its definite integral (from a to b) shows the probability that an event will occur within this range.
- Distributions describe *independent* events; a value drawn from a distribution has no influence on the next value that is drawn from this distribution.
- There are both continuous and discrete distributions. This means that the values that a variable can take are either continuous, or can be a special value such as a whole number, or integer (with no in-between values).
- The area under the curve of a density function is 1. The sum of the probabilities of a discrete distribution is also 1.
- Distributions can be described as density functions (pdf = probability density function for continuous distributions and pmf = probability mass function for discrete distributions), or as cumulative distribution functions (cdf) (see Fig. 3.2).

There is an inexhaustible number of different distributions (Johnson and Kotz 1970).[1] Each of these has different properties, different applications, and a different history (for an explanation of the 40 most important, see Evans et al. 2000). Some were derived from typical frequency distributions (such as the log normal-distribution), others were simply the result of a theoretical derivation (such as the t-distribution). Some of them help us to describe data (for example, the binomial distribution), whereas others form the basis for statistical tests (such as the F-distribution), which are called test distributions.

---

[1] See https://en.wikipedia.org/wiki/List_of_probability_distributions for a (somewhat) complete list.

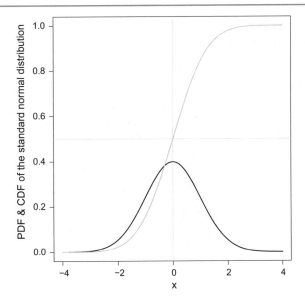

Fig. 3.2 Density function (black) and distribution function ($\widehat{=}$ cumulative density, grey) of the standard normal distribution

3.1.1 Central Limit Theorem

Here, we will have a closer look at distributions in general, and discuss some points in detail by first considering the normal distribution. But why is the normal distribution so important, and what makes it "normal"? Well, the normal distribution, also known as the *Gaussian* distribution or bell curve, is the distribution that results from the summation of all other distributions, when there are enough (a large number of) realisations.

This means that even skewed and discrete distributions will look like a normal distribution if "enough" of them are added together. "Enough" is a vague term, but while some distributions can quite quickly begin to resemble normal distributions, some others require thousands of independent realisations. The normal distribution is therefore the "normal" destiny of any and every distribution with often repeated summation.[2]

Gelman and Hill (2007) describes the normal distribution in a more intuitive way, in which they state that data are ultimately normally distributed when many independent factors contribute to an event. Our ash tree girths are the product of many weather events, pest infestations, soil structures, nutrient availabilities, cuttings, etc. In sum, they all contribute to the girth of the tree, but no single factor is dominant (in our example at least).

The main point is this: the normal distribution has a special meaning as the final distribution of all other distributions under certain conditions. Since these conditions are often *not* met, we also need to be familiar with other distributions, which we will use for statistical analysis.

3.2 Parameters of a Distribution

Distributions can be described mathematically using a distribution function. Here is the density function of the normal distribution:

$$P(x|\mu, \sigma) = \frac{1}{\sigma\sqrt{2\pi}}e^{\frac{-(x-\mu)^2}{2\sigma^2}} \tag{3.1}$$

We can read this as: "The probability of observing a value x is a function of the square of the distance to x and the distribution parameter μ, as well as (in a bit more complicated fashion) the distribution parameter σ." μ and σ determine this function; they are the parameters of the distribution function, or simply the parameters of this distribution.

---

[2]Jaynes [2003] suggested to refer to the normal distribution as the "central" distribution, but his suggestion never caught on.

The left side of Eq. (3.1) can be read as "P of x, given μ and σ", or in real words "The conditional probability of x, given μ and σ."[3] This formula is essential enough to learn it by heart!

Side Note 1: Conditional Probability

We call A and B events. This could be heads/tails of a coin flip, or the height of a randomly selected person. An event is just a single number from a realm of possible numbers: the statistical population S. We observe two events A and B from S. A appears with a certain probability $P(A)$ (the proportion of A within S), and the same is the case for B with $P(B)$.

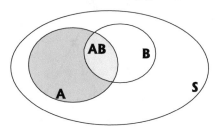

The probability that an event will occur, either A or B, is:

$$P(A \cup B) = P(A) + P(B) - P(A \cap B)$$

From your school years you may remember that $\cup$ symbolizes the union and $\cap$ the intersection. The formula above therefore tells us the following: The probability that A or B occurs is equal to the occurrence of each individual event, minus the intersection of both events. If A and B are independent of each other, the occurrence of both A *and* B is equal to the intersection: $P(A \cap B) = P(A)P(B)$. If there is independence, probabilities are multiplied.

Independence is difficult to show graphically: A and B can still overlap. A is independent from B if $P(A) = P(A|B)$, i.e. if the knowledge about an occurrence of B does not provide us with any information about the occurrence of A. This is the case if the proportion of the area of A from S is exactly the same as the proportion of the intersection $A \cap B$ of B. If we look at the diagram above, we can calculate the probability of observing event A if we have already observed event B. This is the *conditional* probability of A, given B:

$$P(A|B) = P(A \cap B)/P(B)$$

In words: The probability that A occurs when B has already occurred, is the proportion of the intersection $A \cap B$ of B. If the circle B lies completely within circle A (B is a subset of A), then A *must* occur when B occurred: $P(A|B) = 1$. In the context of likelihood, B are assumed values of the distribution, and A is a data set. So under distribution parameters B, a data set A has a certain conditional probability.

(If this way of thinking about probability theory appeals to you, please see the excellent work by Casella and Berger 2002).

Other distributions have different density functions that may be either short and simple or long and ugly. As an example, below you will see the functions for a Poisson distribution (for count data) and a Weibull distribution (often used for dropout or death rates):

$$P(x|\lambda) = \frac{\lambda^x e^{-\lambda}}{x!}; \ x = 0, 1, 2, \ldots \tag{3.2}$$

$$P(x|\lambda, k) = \begin{cases} \frac{k}{\lambda}\left(\frac{x}{\lambda}\right)^{k-1} e^{-(x/\lambda)^k} & \text{if } x \geq 0; \\ 0 & \text{if } x < 0. \end{cases} \tag{3.3}$$

The normal distribution and the Weibull distribution have two parameters (μ, σ and k, λ, respectively), whereas the Poisson distribution only has one (λ).

These distribution functions describe the probability density of x for certain parameter values. Perhaps this will become more clear with an example. Let's plot the normal distribution function for different values of μ or σ (Fig. 3.3). We can see that the distribution parameters determine the height and the width of the distribution.

For discrete distributions, such as the Poisson distribution, the figure looks qualitatively different for different parameter values (Fig. 3.4). Since we only have whole numbers (= integers) in this case, we cannot draw a smooth curve. The term probability mass function, the correct term for such discrete distribution functions, takes this aspect into account.

---

[3] You will have noticed that the x is separated from μ and σ by a bar, rather than a comma. This is because we are really interested in x, and μ and σ are just parameters that have some value when we compute the probability density of x. Such a *conditional* probability is indicated by a vertical line, "|", and implies that the parameters to its right are known or fixed. The notation differs between statistics and other areas of mathematics, where a semicolon is used instead of the bar.

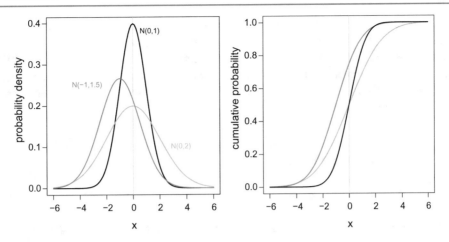

Fig. 3.3 Probability density and cumulative probability of three normal distributions. The standard normal distribution (black) has parameter values of $\mu = 0$ and $\sigma = 1$. This is notated N(0,1). If σ is larger, the distribution becomes more flat. If μ changes, the distribution becomes longer (to the right for $\mu > 0$). The cumulative distribution function (right panel) gives the proportion of the probability density function which is *smaller or equal* than the current x-value, i.e. the partial integral of the density function from $-\infty$ to x

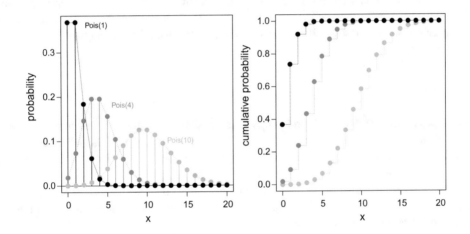

Fig. 3.4 Probability (left) and cumulative distribution function (right) of three Poisson distributions. The black distribution has a parameter value of $\lambda = 1$. This is notated Pois(1). As λ moves along the x-axis, the distribution becomes flatter. For Pois(10), the distribution looks very similar to a normal distribution (an illustration of the Central Limit Theorem). The dotted lines only help to better visualize the differences in the three distribution functions depicted here. In the areas between the actual data points, the function is not defined

Let us assume that we measured a value, let's say 2, and we want to know the probability that this value occurs in a standard normal distribution (with $N(\mu = 0, \sigma = 1)$). The probability that *exactly this value* occurs is 0 (for continuous distributions), since the function integrates over an infinite number of values. If each point had a value >0, the area under the curve would be infinite.

This fact usually leads to confusion. The distribution function is a mathematical function that is defined within this framework. But the probability that a specific value will occur in a continuous distribution still has to be 0.

What we can read from this is the probability *density* for a (infinitesimally) small range around 2, but not the probability itself!

In the left panel of Fig. 3.3, we can now determine the probability density for $X = 2$ where $N(0, 1)$: $P(X = 2|\mu = 0, \sigma = 1) = 0.054$. In the cumulative probability curve, however, we *can* read a probability, namely the probability that a measured value is less than or equal to x. In our case, $P(X \leq 2|\mu = 0, \sigma = 1) = 0.977$. This means that if we measure a value of $X = 2$ in a standard normal distribution, then 97.7% of all values will be smaller or equal to 2. Our value of 2 is therefore an unusual value. If we turn it around, $100 - 97.7 = 2.3\%$ of all values of the normal distribution are >2.

Let's do the same exercise again for the Poisson distribution (Fig. 3.4). Here as well, we observe a value of $X = 2$, but this time with a λ value of 10. $P(X = 2|\lambda = 10) = 0.0022$. This is the probability that our value of exactly 2 will occur in a Poisson distribution when $\lambda = 10$. In fact, 0.27% of all values are less than 2: $P(x \leq 2; \lambda = 10) = 0.00269$.

Probability functions are intended to calculate the probability for certain observed values given specific distribution parameter values. The crux of the matter is the word "specific". How do we know what values the distribution parameters take? How can we then specify these distribution parameters?

3.3   Estimators (for Distribution Parameters)

An estimator is a rule for calculating a value of interest based on data. An estimate is the actual data-derived value, here for a distribution parameter.

For our ash tree data set, we measured 100 tree girths. These data look as if they are normally distributed (Fig. 2.11). Of course, we do not want to plot just any normal distribution, but rather the one that fits best. Since the normal distribution has two parameters (μ and σ), we need to find the values for μ and σ that maximise the fit.

For this, we need two important ingredients: first, a measure of "fit", and second, a method for determining the values of the parameters to achieve the maximal fit. This process is called "fitting a distribution to the data". The results of this process are two estimated values for the parameters – i.e. the parameter estimates. Since we only have one sample, we do not know the true values of μ and σ of the population. What we can actually calculate is only a good *estimate* of these true values. An estimator is a good estimator if it is "consistent", i.e. with an increase in the number of data points, the estimate comes closer to the true value. If the estimator is systematically too high or too low, this would be called bias.[4]

3.3.1   The Likelihood

Likelihood, denoted as $\mathscr{L}$, is sort of a "reverse" probability density. While the density function tells us the probability of a certain combination of data and distribution parameters ($P(x|\theta)$, where x is a data point, θ the parameter(s) of the distribution), the likelihood quantifies the conditional opposite: How plausible are my parameters given the data points $\mathscr{L}(\theta \,|\, \text{Data})$? In fact, the probability density defines the likelihood: $\mathscr{L}(\theta \,|\, \text{Data}) = P(\text{Data} \,|\, \theta)$.

The likelihood is usually aggregated from many data points. It answers the question: How plausible are the distribution parameters for exactly this data set? In this way, it is possible for us to find the distribution whose likelihood is the maximum for our data set, by testing different parameter combinations.

An example: If we toss a coin 100 times, then the density function tells us the probability that "heads" will come up n times. The corresponding distribution in this case is the binomial distribution with two parameters (Fig. 3.5): p = the probability of "heads", and n, the number of coin tosses: $P(X = n|p = p(\text{heads}), n = 100)$. With a coin toss, we assume $p(\text{heads}) = 0.5$; the probability of observing 60 heads among 100 tosses is therefore $P(X = 60|p = 0.5, n = 100) = 0.011$.

We can now vary the value for p, calculate the respective likelihood, and then determine which value of p maximises our likelihood. The answer is intuitively obvious: 0.6. And in fact, if $p(\text{heads}) = 0.6$, then $\mathscr{L}(p = 0.6, n = 100|X = 60) = P(X = 60|p = 0.6, n = 100) = 0.081$, a much higher value than for $p = 0.5$.

Let's go back one more time to our ash tree data set. Here we need two parameter estimates, one for μ and one for σ. At this point, we should really give these parameter proper names to avoid any confusion (even for the Greek scholars among you): μ is the mean and σ is the standard deviation of the normal distribution.

We learned both of these terms in Chap. 1 when learning about how to describe a sample. Now they appear again, this time as distribution parameters. *There is a difference!* In order to make this difference clear, we use different symbols in this case: $\bar{x}$ is the mean of a sample, μ is the mean of a distribution; s symbolizes the standard deviation of a sample, σ the standard deviation of a distribution. While we can (and are allowed to) calculate $\bar{x}$ for a sample with any distribution, the symbol μ implies a normal distribution! This equivalent is true for s and σ.

It is possible to show (see further below) that the mean of a sample ($\bar{x}$) is a good estimator for the expected value of the underlying normal distribution (the mean, μ), and similarly, that the empirical standard deviation is a good estimator for the

---

[4]Notice that the thing we use to get a good value for our distribution parameter is called the "estimator", while the actual value is called the "estimate". As we will soon see, the mean of our sample, $\bar{x}$, is an estimator for the mean of the underlying population mean, μ.

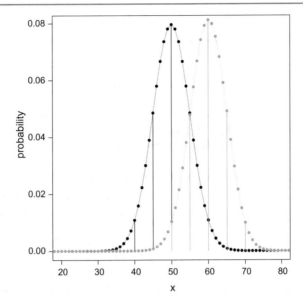

Fig. 3.5 Density of the binomial distribution with the parameters $n = 100$ and $p = 0.5$ (black) and 0.6 (grey). The above figure shows the probability of obtaining x number of heads when tossing a coin 100 times. The observed value of our example is $x = 60$. Here, the probability for $p = 0.6$ is much higher than for $p = 0.5$. For the sake of clarity, the vertical lines are drawn for every 5th value, the value range is restricted to the interval [20, 80], and the points are connected with a line. The binomial distribution is a discrete distribution

standard deviation of the normal distribution, σ. However, this does not mean that the underlying distribution actually *is* a normal distribution. But if it *were*, the $\bar{x}$ would be a good estimator for μ, and s a good estimator for σ.
How do we know this? Is there a simple sample-based estimator for the parameters of all distributions?

3.3.2 Maximizing the Likelihood

Consider the following problem: we collected some data and want to determine the parameters of the distribution from which the data most likely originate, i.e. we want to fit a distribution. In other words, we want to estimate the distribution parameters.

We have just learned that the likelihood quantifies the probability density of a data point, if we are willing to assume a distribution and provide values for its parameter(s). The next step is to move from one data point to multiple data points.[5] Then comes the step of calculating the optimal values for the parameters.

We assumed that all data points, all ash tree girths, are independent from each other. No ash tree is thicker or thinner just because another ash tree has a certain thickness.[6] Therefore, the following rule applies for calculations with probabilities:

If two events, A and B, are independent from one another, the probability of the event $A \cap B$ (A and B) is the same as the product of the probability of A and B: $P(A \cap B) = P(A)P(B)$ (see also Box 3.2). The probability density of *all* ash tree girths is the product of the probability of each individual ash tree girth:

$$P(\text{all girths}) = \prod_{i=1}^{100} P(\text{girth}_i).$$

---

[5]It is logically intuitive that the more data points we have, the better the estimate for the distribution parameters will be. This is the "law of large numbers": if we continuously repeat a measurement, then the relative frequencies converge to the expected probability of these values.

[6]This may not be true *in realitas*! If data points are not independent, we need to perform some statistical contortions to be able to use maximum likelihood in calculations. This is an advanced topic that falls outside of the purview of this book. For *mixed effect models* or modeling of variance-covariance matrices, see Pinheiro and Bates (2000), Gelman (2007), Zuur et al. (2009).

where the large pi symbol ($\prod$) stands for the product, the counter i goes through all 100 values of our ash tree girths. The number P(all girths) is the data's likelihood $\mathscr{L}$. In general, the likelihood is defined as:

$$\mathscr{L} = P\left(\mathbf{x}|\boldsymbol{\theta}\right) = \prod_{i=1}^{N} P\left(x_i|\boldsymbol{\theta}\right) \tag{3.4}$$

The bold symbol $\mathbf{x}$ shows that we are dealing with a vector or a matrix (so not with a single value (scalar), but rather a row (vector) or table (matrix) full of values). $\boldsymbol{\theta}$ stands for the parameter(s) of the distribution, in our case $\boldsymbol{\theta} = (\mu, \sigma)$, i.e. the mean and standard deviation of the distribution.

The likelihood of an individual events is usually small. The product of many numbers becomes so small that calculating it can lead to errors simply due to rounding. Therefore, we use the old trick of calculating the logarithm of the likelihood, the so-called log-likelihood, ℓ. According to the rules of logarithms, $\log(AB) = \log(A) + \log(B)$.[7] Accordingly, from Eq. (3.4) we can derive:

$$\ln(\mathscr{L}) = \ell = \ln\left(P\left(\mathbf{x}|\boldsymbol{\theta}\right)\right) = \ln\left(\prod_{i=1}^{N} P\left(x_i|\boldsymbol{\theta}\right)\right) = \sum_{i=1}^{N} \ln\left(P\left(x_i|\boldsymbol{\theta}\right)\right) \tag{3.5}$$

As an example, let's do the calculation for our ash tree girths. Remember the normal distribution Eq. (3.1). Let's set $\mu = 50$ and $\sigma = 20$: our first data point has a value of 37. Now we can plug this value into the density function of the normal distribution. We get:

$$P(x = 37 \mid \mu = 50, \sigma = 20) = \frac{1}{20\sqrt{2\pi}}e^{\frac{-(37-50)^2}{2\cdot20^2}} = \frac{1}{50.13}e^{\frac{-169}{800}} = 0.0199 \cdot 0.810 = 0.0161.$$

Therefore, $\mathscr{L}_1(\mu = 50, \sigma = 20|x = 37) = P(37 \mid N(50, 20)) = 0.0161$ and $\ell_1 = \ln(0.0161) = -4.13$. We then do this calculation for *all* data points and sum the results. We get $\ell = -431.90$. This is the log-likelihood of our ash tree girth data set under the assumption of a normal distribution with the parameters mean $= \mu = 50$ and standard deviation $= \sigma = 20$.

If we choose other values for the distribution parameters, we get a different value for ℓ, for example $\ell(\mu = 40, \sigma = 30 \mid$ all girths$) = -457.97$. Since this value is *lower* and we want to find the *maximum* ℓ, we know that our first set of parameters was better.

3.3.3 Maximum Likelihood—Analytical

From school you (hopefully) remember problems involving extreme value calculation. In such problems, the minimum and/or maximum had to be determined for a function. This is done by setting the first derivative of the function equal to zero (this is called *root finding*). We'll use such methods in the following section.

3.3.3.1 Analytical Maximum Likelihood-Estimators for the Parameters of the Normal Distribution

We want to find the optimal value for μ and need to set the density function of the normal distribution (Eq. 3.1) for P in Eq. (3.4), differentiate μ (partially) and set it equal to 0. This creates long drawn out equations, but ones that can be solved quite easily!

You may remember some power rules from school: e.g. $\prod e^x = e^{\sum x}$ (perhaps more familiar written as $e^a \cdot e^b = e^{a+b}$). The following applies for constant a: $\prod_{i=1}^{n} a = a^n$.

We begin with the likelihood function of the normal distribution:

$$\mathscr{L}(\mu, \sigma|x) = \prod_{i=1}^{n} \frac{1}{\sigma\sqrt{2\pi}}e^{\frac{-(x_i-\mu)^2}{2\sigma^2}}$$

$$= \frac{1}{(\sigma\sqrt{2\pi})^n}e^{-\frac{1}{2\sigma^2}\sum(x_i-\mu)^2} \tag{3.6}$$

---

[7]This of course applies to logarithms of any base. To calculate the log-likelihood, it is conventional to use the natural logarithm (base e, ln).

Taking the logarithm leads to:

$$\ell(\mu, \sigma | x) = \ln\left[\frac{1}{(\sigma\sqrt{2\pi})^n} e^{-\frac{1}{2\sigma^2}\sum(x_i - \mu)^2}\right] \tag{3.7}$$

$$= \ln\left[\sigma^{-n}(2\pi)^{-n/2}\right] - \frac{1}{2\sigma^2}\sum(x_i - \mu)^2 \tag{3.8}$$

$$= -n\ln\sigma - \frac{n}{2}\ln(2\pi) - \frac{1}{2\sigma^2}\sum(x_i - \mu)^2 \tag{3.9}$$

To find the maximum, we set the derivative of this term equal to zero with respect to μ:

$$\frac{d\ell}{d\mu} = 0 = \frac{1}{\sigma^2}\sum_{i=1}^{n}(x_i - \mu)$$

Since $\sigma > 0$ (otherwise the normal distribution would just be a line to infinity[8]), the rest has to be zero. Therefore,

$$0 = \sum_{i=1}^{n} x_i - \sum_{i=1}^{n} \mu = \sum_{i=1}^{n} x_i - n\mu.$$

Now we have $\sum_{i=1}^{n} \mu = n\mu$, since μ is constant and is added n times. After transforming the equation, we see that

$$\mu = \frac{\sum_{i=1}^{n} x_i}{n}.$$

This equation should look familiar to us: this is the arithmetic mean, i.e. the mean of the sample, $\bar{x}$!

This calculation shows (1) an analytical solution for the optimal estimator for μ, and (2) that this is the arithmetic mean. In other words: If we want to fit the best possible normal distribution to a data set, the mean value of the data is our maximum likelihood estimator for the distribution parameter μ of the normal distribution.

For the variance, we differentiate Eq. (3.9) with respect to σ, where the constant term drops out, and we set the equation equal to zero (remember that the derivative of $\log x = 1/x$ and the derivative of $-1/x^2 = 2/x^3$):

$$\frac{d\ell}{d\sigma} = 0 = \frac{-n}{\sigma} + \frac{\sum_{i=1}^{n}(x_i - \mu)^2}{\sigma^3}$$

When we multiply both sides of the equation by σ^3 we get

$$0 = -n\sigma^2 + \sum_{i=1}^{n}(x_i - \mu)^2$$

$$\sigma^2 = \frac{\sum_{i=1}^{n}(x_i - \mu)^2}{n}$$

Voilà! We also know this formula from Sect. 1.1.2: it is the formula for the variance, with the slight difference that n is in the denominator, while for the sample variance this is $n - 1$.

3.3.3.2 Analytical Maximum Likelihood-Estimator for the Parameters of the Poisson Distribution

We can go through the same steps for Poisson distributed data. Since

$$\mathscr{L}(\lambda | x) = \frac{\lambda^x}{x! e^\lambda}$$

---

[8]This also exists, namely the (Dirac) δ-distribution.

(see Eq. 3.2), we get the following for the likelihood function:

$$\mathcal{L}(\lambda|x_i) = \frac{\lambda^{x_1}}{x_1!e^\lambda} \cdot \frac{\lambda^{x_2}}{x_2!e^\lambda} \cdots \frac{\lambda^{x_i}}{x_n!e^\lambda} = \frac{\lambda^{\sum x_n}}{x_1! \cdots x_n!e^{n\lambda}}$$

We take the log and get

$$\ell(\lambda|x) = -n\lambda + (\ln \lambda)\sum x_i - \ln\left(\prod x_i!\right)$$

differentiate with respect to λ and set the equation equal to zero:

$$\frac{d}{d\lambda}\ell = -n + \frac{\sum x_i}{\lambda} = 0$$

$$\lambda = \frac{\sum_{i=1}^n x_i}{n}$$

For the parameter λ of the Poisson distribution, the arithmetic mean is also the maximum likelihood estimator! Accordingly, λ is often referred to as the "mean" of the Poisson distribution.

3.3.3.3 Analytical Maximum Likelihood Estimators for Other Distributions

It won't be of much use to you to present further analytical derivations here. The point of the last two sections was to show that there are certain analytical estimators for some distributions. Since we will work more with binomial and negative binomial distributions in later sections, here are the parameter estimators for those distributions, simply for the sake of completeness.

For the binomial distribution:

$$P(x|n, p) = \binom{n}{x}p^x(1-p)^{n-x}$$

is the maximum likelihood estimator (MLE) of $p = \bar{x}/n$ and n must be a known value.

For the negative binomial distribution:

$$P(x|r, p) = \binom{x+r-1}{x}p^x(1-p)^r \quad \text{(only if } x \geq 0, \text{ else } P(x; r, p) = 0)$$

is the MLE of $p = rn/(rn + \sum x_i)$. Since p is still dependent on r, the two can not be estimated independently.

3.3.4 Maximum Likelihood—Numerical

As we have just seen with the example of the negative binomial distribution, depending on the circumstances, it may not be possible to find an analytical solution for the MLE of every parameter. In such cases, we need to calculate this numerically, i.e. through trial and error. Since the maximum likelihood usually (but not always!) has a mathematically friendly form (continuous, differentiable), very fast optimisation algorithms can be applied to find the optimal MLE with few iterations.

Optimisation is an advanced topic which won't be covered here in further detail. (see Bolker (2008), for a good introduction to this topic).

3.3.5 Maximum Likelihood—High Praise

The reasons that maximum likelihood is the most widely used method for fitting a distribution to data boil down to the following (Mosteller & Hoaglin 1995):

1. MLEs are **asymptotically unbiased**: with an increase in the sample size ($n \to \infty$) the MLE converges on the true value of θ.

2. MLEs are **consistent**: with an increase in the sample size ($n \to \infty$) MLE continually generates a more precise estimate of the parameter. (This is simply a repeat of point 1 with a bit more detail. Instead of simply saying that MLEs *are* unbiased, here we state that the MLE becomes more exact the bigger n becomes.)
3. MLEs are **efficient**. Among all the unbiased estimators, MLEs have the lowest variance. This means that no other estimator has less asymptotic variance than the MLE. Or, in other words, MLE uses the data more efficiently than any other method.
4. The MLE is **communicative** with every other function of θ. In this way, $\mathrm{MLE}(g(\theta))$ is the same as the function of the MLE for θ, $g(\mathrm{MLE}(\theta))$. For example, the MLE for the standard deviation σ is simply the root of the MLE for the variance. Because $\sigma = \sqrt{\sigma^2}$, therefore $\mathrm{MLE}(\sigma) = \sqrt{\mathrm{MLE}(\sigma^2)}$.

An important alternative[9] to the maximum likelihood for estimating distribution parameters are the *methods of moments*; see the corresponding Side Note 2). Additionally, there are some variations on the maximum likelihood estimation, e.g. *restricted* maximum likelihood or the *quasi*-maximum likelihood (= *pseudo*-maximum likelihood). Some of these methods are much more efficient (in terms of calculation) for certain distributions or are used if a likelihood can not be specified. These cases are used in more advanced applications and are outside of the purview of an introduction. We will learn more about the *quasi*-likelihood when we discuss overdispersion later in this book.

3.4   Some Important Distributions[10]

Traditionally, continuous probability distributions are notated as $P(x)$ and discrete ones as $P(k)$. The notation $P(X \in [x, x + dx])$ represents the probability density that realisation of random variable X has a value of x plus an infinitesimally small amount dx.[11] For discrete data, $P(X = k)$ represents the probability that observed value of random variate X is equal to k. As we here compute probabilities at a finite set of discrete values, each $P(x)$ is actually a probability, not "only" a probability density, as for the continuous case.

3.4.1   Normal (= Gaussian) Distribution

The normal distribution is a continuous distribution with two parameters. It describes the results of very many small contributions, such as Brownian molecular motion. Because of the Central Limit Theorem, the normal distribution is ubiquitous and incredibly useful. Due to its dominance in the literature, however, the appropriateness of other distributions for specific applications is sometimes overlooked. The normal distribution has properties that make a mathematician's heart flutter: definition over a complete range and differentiable. Swoon.

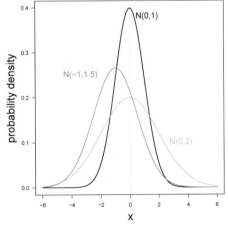

---

[9]For even more alternatives, see http://en.wikipedia.org/wiki/Maximum_likelihood.

[10]In addition to appearing in classic statistics textbooks, the following sections were also written with the help of the excellent wikipedia pages on the subject.

[11]This is always confusing! $P(X = x)$ is *always* 0. If it were larger, then the area under the curve would be infinitely large, because there would be infinite values for X. So, mathematically, we carefully write that we are interested in the density at the location "from x to x plus a little bit", $P(X \in [x, x + dx])$. As dx approaches 0, this term gives the density of $P(X = x)$.

Something that may also be useful is a note about exponents. Here, the deviation square of the value x is the mean value of the distribution μ. As the distance increases, this expression becomes larger, and the "to the-e" part becomes smaller and smaller, and $P(x)$ decreases exponentially with the square of the distance to the mean value. Many optimisations use the "least-square" method, which is motivated by exactly this squared term in the normal distribution.

Distribution Function

$$N(x) = P(X = x | \mu, \sigma) = \frac{1}{\sqrt{2\pi\sigma^2}} e^{-\frac{(x-\mu)^2}{2\sigma^2}}$$

Domain: $x \in \mathbb{R} = (-\infty, \infty)$[12]
Mean: μ
Variance: σ^2.

3.4.2 Bernoulli Distribution

The Bernoulli distribution is a discrete distribution with a single parameter, p. It describes the probability of a binary event, such as tossing a coin, survival/non-survival, or sex at birth. The Bernoulli distribution is very simple and only has two outcomes. In the literature, the binomial distribution (see below) is often used instead, as the Bernoulli distribution is a special case (with parameter $n = 1$).

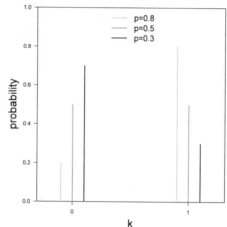

Distribution Function

$$\text{Bern}(k) = P(X = k | p) = \begin{cases} p & \text{if } k = 1, \\ 1 - p & \text{if } k = 0, \\ 0 & \text{else.} \end{cases}$$

Domain: $k \in \{0, 1\}$[13]
Mean: p
Variance: $p(1 - p) = pq$.

3.4.3 Binomial Distribution

The binomial distribution has two parameters, n and p. They describe the probability $P(X = k)$ that with n repetitions, a Bernoulli-event (i.e. 0/1) will occur k times if its probability is p. A typical example involves toxicological research, where, for example, 30 water fleas (*Daphnia*) are placed in water with a certain salt concentration. If the death rate is $p = 0.8$,

---

[12]The round parentheses of the interval mean that we are dealing with an *open* interval, i.e. the endpoints are not included. A closed interval is symbolized with square brackets (e.g. [0, 1]). Mixing round parentheses and square brackets is called a half open or half closed interval.

[13]Curly brackets indicate a *set*, i.e. a list of all possible events: {0, 1} thus means "either 0 or 1".

then the binomial distribution gives us the probability that k water fleas will die. This occurrence is called a *success*, the non-occurrence is called a *failure* (even if the "success" seems to celebrate the death of the research subject).

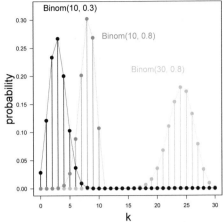

For $p = 0.5$, the binomial distribution is symmetrical.
For $n = 1$ it is the same as the Bernoulli distribution.

Distribution Function

$$\text{Binom}(k) = P(X = k|n, p) = \binom{n}{k} p^k (1 - p)^{n-k}$$

Domain: $k \in \mathbb{N}_0 = \{0, 1, 2, 3, \ldots\}$
Mean: np
Variance: $np(1 - p)$.

3.4.4 Poisson Distribution

The Poisson distribution is a discrete, single-parameter distribution. It describes count data, such as the number of eggs that a hen lays in a week. An important feature is that with the Poisson distribution, the mean is the same as the variance, i.e. random variables with a higher mean have *per definition* a higher variance! With the normal distribution, mean and variance are independent of one another. If a random variable has a higher (or lower) variance than its mean, the distribution is called *overdispersed* (or *underdispersed*). A possibly better distribution for such a case would be the negative binomial distribution (see below).

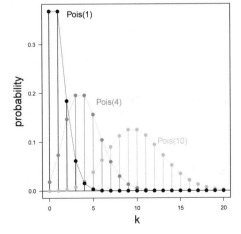

Distribution Function

$$\text{Poisson}(k) = P(X = k|\lambda) = \frac{\lambda^k e^{-\lambda}}{k!}$$

Domain: $k \in \mathbb{N}_0 = \{0, 1, 2, 3, \ldots\}$
Mean: λ
Variance: λ.

3.4.5 Negative Binomial Distribution

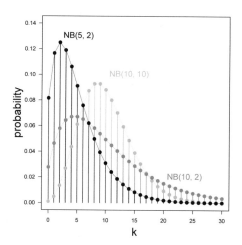

The negative binomial distribution is a discrete distribution with two parameters. Like the Poisson distribution, it describes count data. It describes the probability of r successes in k repetitions of a Bernoulli process.

Confusingly, there are two alternative parameterizations, one (1) more common in environmental science, the other (2) more common in mathematics and statistics.

Possibility 1: Through the parameters μ and r. μ, the mean, is equivalent to λ of the Poisson distribution, whereas r is called the dispersion parameter. Practically, the negative binomial distribution can be seen as an extension of the Poisson distribution. The figure here uses this parameterization.

Possibility 2: Through the parameters p and r. p describes the probability of a success when tossing a coin, r is the number of successes to be reached (for example, $r = 5$ times heads in k coin tosses with $p = 0.5$).

The alternatives can be converted into one another, since $p = r/(r + \mu)$ or $\mu = rp/(1 - p)$.

With $r = 1$, the negative binomial distribution becomes the geometric distribution:

$$\text{Geom}(1 - p) = \text{NB}(1, \ p)$$

For $r \to \infty$ the negative binomial distribution becomes the Poisson distribution:

$$\text{Poisson}(\lambda) = \lim_{r \to \infty} \text{NB}\left(r, \ \frac{\lambda}{\lambda + r}\right)$$

Distribution function

$$P(X = k|r, p) = \binom{k + r - 1}{k}(1 - p)^r p^k$$

Domain: $k \in \mathbb{N}_0 = \{0, 1, 2, 3, \ldots\}$
Mean: μ or $rp/(1 - p)$
Variance: $\mu + \mu^2/r$ or $rp/(1 - p)^2$.

3.4.6 Log-Normal Distribution

The log-normal distribution is the logarithmic variant of the normal distribution, so it is continuous and has two parameters. If you were to take the log of a log-normally distributed value, then the distribution would be normal again. In other words: if you set the normal distribution as en exponent (such as $e^{N(x)}$), this would be the log-normal distribution. One of the resulting characteristics is that the log-normal can only have positive values. This distribution is rarely mentioned in entry level statistics textbooks, since you can simply take the log of the values and then use them like you would in a normal distribution. We'll discuss it here because it often comes up in ecological applications and it is important to know its form. Additionally, people often make the mistake of calculating the mean of the log-normal distribution as $e^{\bar{x}}$, which is **false**!

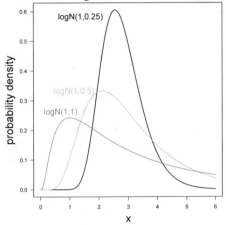

Like the normal distribution, the log-normal distribution has two parameters, the mean μ and the standard deviation σ. In fact, the log-normal distribution is defined in terms of the normal distribution.

Distribution Function

$$\ln N(x) = P(X = x | \mu, \sigma) = \frac{1}{x\sigma\sqrt{2\pi}} e^{-\frac{(\ln x - \mu)^2}{2\sigma^2}}, \quad x > 0$$

Note that x now also appears before the e in the normalization term!

Domain: $x \in \mathbb{R}_0^+ = (0, \infty)$

Mean: $e^{\mu + \sigma^2/2}$; the variance co-determines the mean of the log-normal distribution!

Variance: $(e^{\sigma^2} - 1)e^{2\mu + \sigma^2}$; the mean co-determine the variance!

3.4.7 Uniform Distribution

The uniform distribution (also called the continuous uniform distribution or rectangular distribution) is a continuous distribution with two parameters, a and b, that specify the lower and upper limit. It is a rather trivial distribution and probably does not exist in nature. Nevertheless, it is important for didactic reasons, because it represents the simplest possible distribution.

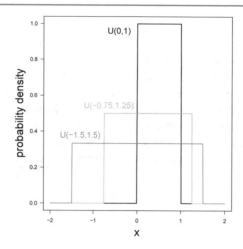

In addition, the uniform distribution is important in Bayesian statistics, because it often uses uninformative distributions and the uniform distribution is quite uninformative within its defined range.

Finally, any empirical frequency distribution can be transformed to a uniform distribution, and this property is interesting for a certain test we can do for a normal distribution (for the Anderson-Darling-Test, see Sect. 4.4).

If you sum two uniform distributions (and then divide by 2), you get a triangular distribution.

Distribution Function

$$U(a, b) = P(X = x | a, b) = \begin{cases} \frac{1}{b-a} & a \leq x \leq b \\ 0 & \text{sonst.} \end{cases}$$

a and b can be any real value, but $b > a$. If $a = b$, then we have the *very* special case of a δ-distribution, which has the value ∞ for a, but is 0 otherwise. For $a = 0$ and $b = 1$ it is called the *standard* uniform distribution.

Domain: $x \in \mathbb{R} = (-\infty, \infty)$

Mean: $(a + b)/2$

Variance: $\frac{1}{12}(b - a)^2$.

3.4.8 β-distribution

The β-distribution (beta distribution) is a continuous distribution that is restricted to the interval $(0, 1)$. Its form varies widely depending on the selection of the two parameters (a = shape, b = scale). This flexibility makes it very interesting for certain applications. With only two parameters, you can generate increasing, decreasing, concave, convex, linear, and unimodal distributions. The β-distribution is often used for modelling in Bayesian statistics.[14] In statistical analysis, it can be used when working with proportions, if the values are between 0 and 1.

---

[14]The reason is that it is the *conjugate prior* of the Bernoulli, binomial and negative binomial distribution. Explaining what this means would require going much too deep into the interesting subject of Bayesian statistics.

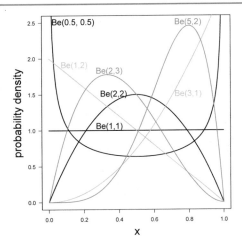

For $a = b = 1$, we get the standard uniform distribution.

Distribution Function

$$\text{Be}(x) = P(X = x | a, b) = \frac{x^{a-1}(1 - x)^{b-1}}{\int_0^1 u^{a-1}(1 - u)^{b-1} \, du} \tag{3.10}$$

$$= \frac{\Gamma(a + b)}{\Gamma(a)\Gamma(b)} x^{a-1}(1 - x)^{b-1} \tag{3.11}$$

a and b can take any value >0.

The second line formulates the β-distribution as a function of the Γ(Gamma)-function. This may be confusing, but mathematically simpler than the integration step in the first line.

Domain: $x \in (0, 1)$

Mean: $\frac{a}{a+b}$

Variance: $\frac{ab}{(a+b)^2(a+b+1)}$.

Additional note: The β distribution has a counterpart in the β *function* $(B(a, b))$. It is possible to express the distribution by means of the function $(\text{Be}(x) = \frac{1}{B(a,b)} x^{a-1}(1 - x)^{b-1})$. This means that the β-function normalizes the β-distribution. Distribution and function are, however, two very different things!

Second note: The β-distribution can be generalized to any interval $[a, b]$, whereby it becomes a 4-parameter distribution.[15]

3.4.9 γ-distribution

The γ-(= gamma-)distribution is a continuous distribution limited to the interval $[0, \infty)$. It has two parameters, k = shape and θ = scale (these are occasionally formulated differently as $p = k$ and $q = \frac{1}{\theta}$; whereby q is then called the *rate*).

---

[15] See http://en.wikipedia.org/wiki/Beta_distribution.

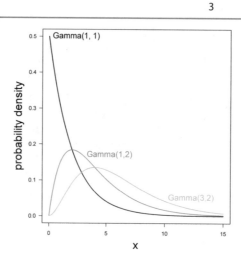

The γ-function comes up when analysing waiting times (such as for the effect of a medication), or leaf area lost to invertebrate herbivores. Due to its extreme level of skew, it is difficult to interpret and hard to fit exactly. On the other hand, the γ-distribution is also one of the "undervalued" distributions, which should really be considered more often.

Distribution Function

$$\mathrm{Gamma}(x) = P(X = x | k, \theta) = x^{k-1} \frac{e^{-x/\theta}}{\theta^k \, \Gamma(k)},$$

with $x \geq 0$ and $k, \theta > 0$. The Γ-function[16] occurs in the density function of the γ-distribution. For the alternative form with p and q we get: $P(x | p, q) = \frac{q^p}{\Gamma(p)} x^{p-1} e^{-qx}$, for $x \geq 0$.

Domain: $x \in \mathbb{R}_0^+ = [0, \infty)$

Mean: $k\theta = p/q$

Variance: $k\theta^2 = p/q^2$.

For integer values of k the distribution is also called the Erlang distribution.

3.4.10 Truncated Distributions

Often, certain values of a distribution may not occur, even though the data seem to come from this distribution. For example, if we count how many eggs a blackbird lays in its nest, the number 0 does not come up at all, but the rest of the data may come from a Poisson distribution. A blackbird does not make a nest and then not lay eggs in it. Accordingly, the expected distribution is a *truncated* Poisson distribution (called zero-truncated or positive Poisson), where $P(k = 0) = 0$.

In the same way, the normal distribution can also be restricted to values $x \geq 0$, a so-called truncated normal distribution.

The statistical trick that we need to perform for truncated distributions is to increase all the values by the fraction that has been truncated so that the overall probability (the area under the curve) is 1 again:

$$\mathrm{trPois}(k > 0, \lambda) = \mathrm{Pois}(k > 0, \lambda) \cdot (1 + \mathrm{Pois}(k = 0, \lambda))$$

If $\mathrm{Pois}(k = 0) = 0.1$, then for the *truncated* Poisson distribution, all trPois values need to be multiplied by $(1 + 0.1)$. By doing this, the total probability is $\sum_{i=1}^{\infty} \mathrm{trPois}(k = i; \lambda) = 1$.

Side Note 2: Moments of a Distribution or Sample

"Moments" describe the form of a distribution or sample. If we take a random sample from a distribution, then we can describe the sample using sample statistics such as mean and standard deviation. Moments can be seen as analytically derived statistics for random variables of a distribution. Moments generalize sample statistics such as mean or variance for any distribution and sample. The thought behind these moments is that together, they can more and more accurately describe the actual distribution.

---

[16]The letter Γ is the Greek capital letter Gamma; for the Γ-function, see http://en.wikipedia.org/wiki/Gamma_function.

Moments are statistically quite abstract. In the theory of probability, however, moments are very important and therefore are briefly presented here. Moments can be calculated for distribution functions as well as for samples. There is a formula that calculates the moments $m_1 \ldots m_n$ for a distribution function $f(X)$: $m_n(c) = \int_{-\infty}^{\infty} f(x)(x-c)^n dx$. For the first moment, $c = 0$, and for all others, $c = m_1$. With $c = 0$, the first moment m_1 is the mean and all other moments are centred around the mean ("central moments").

For a sample, the integral over the whole distribution gives us a sum over all N data points: $m_n(c) = \frac{1}{N} \sum_{i=1}^{N} (x_i - c)^n$. The first moment of a sample, $m_1 = \frac{1}{N} \sum_{i=1}^{N} (x_i - 0)^1 = \frac{1}{N} \sum_{i=1}^{N} x_i$, we immediately recognise as the sample mean, $\bar{x}$.

The second moment of the sample should also be familiar to us, since this is the variance: $m_2 = \frac{1}{N} \sum_{i=1}^{N} (x_i - \bar{x})^2$. The 3rd and 4th moments are called skewness and kurtosis, and we know them from the previous chapters. After these moments, there are no established names.

If we now estimate the moments from a sample, then we have an alternative approach for fitting a distribution to the data—the *method of moments*. Since we cannot estimate all $n = N$ moments with reasonable effort, but rather only the first few, the method of moments typically is inferior to maximum likelihood. In addition, the estimators are often only asymptotically unbiased. If these moments are easy to calculate, however, this approach is much more efficient.

A short example may help to make this idea more tangible. In this case, we are interested in the parameters of the negative binomial distribution, p and r. We can calculate the mean of our sample ($\bar{x}$, as described in Chap. 1), which is calculated as $\bar{x} = \frac{rp}{1-p}$ for the negative binomial distribution; the variance is then $s^2 = \frac{rp}{1-p}^2$ (for both equations, see Sect. 3.4.5). Since we can calculate $\bar{x}$ and s^2 from the sample, we now have two equations with two unknowns that we can solve for r and p. (Then we have $p = 1 - \bar{x}/s^2$ and $r = \bar{x}(1-p)/p = \ldots = \bar{x}^2/(s^2 - \bar{x})$.). Thus we can use the first two moments of the sample to parametrize the negative binomial distribution, that is, to fit the distribution to these data.

(Again the best reference to an "introduction" to the foundations of statistical probability is Casella and Berger 2002.)

3.5   Selecting a Distribution

The typical job of statistics is to fit observed values to a distribution. Which distribution you select for your data depends greatly on the processes that generated the data. Often we know that these specific data originate from, for example, a Poisson process or a belong to a binomial distribution because of the underlying processes and context. In these cases, of course, we take this distribution for our further analysis. If there is no clear idea of which distribution the data "should" fall into, then we can also try a number of different ones and select the most appropriate.

A typical example for when we have an unclear distribution are data that were generated through aggregation or conversion of values. Vegetation cover usually has values between 0 and 100% and could therefore belong to a binomial distribution, since a plant could cover, say 14 of 100% of the ground in a plot. Or we could argue that the β-distribution is more appropriate, since the data are limited to a range of $[0, 1]$. Often, such vegetation coverage data are gathered using the Braun-Blanquet scale, which has many classes for small values, and only few classes for larger values. If these data are then converted to a percentage, we may be at a loss for what distribution we should fit to the data.

Another example deals with ratio values. If we calculate the quotient of two variable values, for example the portion of viviparous reptilian species for each country in the world, then both the number of all reptile species as well as the number of viviparous species varies. Which distribution the quotient should follow is difficult to predict (see Appendix D of Clark 2007, for a probability theory approach to dealing with such mixed distributions).

In order to select something, we first need to have some alternatives. This is why a number of distributions were introduced in the previous section. The unimaginative use of the normal distribution, which may be inappropriate in a number of cases, should be expanded by having a look at the list of all probability distributions.[17] Furthermore, every field of science seems to have a typical favourite distribution, which should not be forgotten.

As we saw with the example of the log-normal distribution, in some circumstances, a transformation of data can lead to the selection of a completely different distribution. In general, this is problematic, because if we transform the data, the variance also changes. This can be either a positive or negative thing. If we take the square-root of data from a Poisson distribution, the data often look to be normally distributed. Since the variance is the same as the mean in a Poisson distribution, the square-root effect on small values is less than on large values. The resulting distribution therefore often has similar variance over the whole range of values, which corresponds to the assumptions of a normal distribution (where the variance σ^2 is constant).

Transforming data *can* be a way make nice looking data from otherwise "wild" looking data. The result must nonetheless correspond to the assumptions of the new distribution![18]

---

[17] For example: https://en.wikipedia.org/wiki/List_of_probability_distributions.

[18] Taking the logarithm is definitely the most common transformation type. In vegetation ecology, the often used arc-sin-square root transformation for vegetation surveys is, on the other hand, unfounded and usually unfavourable (Warton and Hui 2011).

If we fit two different distributions using maximum likelihood to the **exact same data**, then we can compare the *goodness of fit*. To do this, we have two options: The Kolmogorov-Smirnov test for comparing two distributions, as well as comparing the fit via the likelihood.

3.5.1 Comparing Two Distributions: The Kolmogorov-Smirnov Test

The Kolmogorov-Smirnov-test, called the KS-test for short, compares two continuous cumulative probability functions. These could be frequency densities, empirical cumulative distribution functions (ECDF), or parametric density functions. The KS-test simply quantifies the maximum distance between two distributions, D. The bigger D is, the worse the consistency between the two cumulative distributions is.

As an example let's use the bee flight data. We can compare the cumulative values with a fitted normal distribution (Fig. 3.6). In this example, the comparison of the data with the normal distribution does not look perfect. Especially in the middle of the graph around the mean of 2.16 and for very short flight times there are clear discrepancies. However, these differences are not *so* big that we would throw out the normal distribution as a valid candidate. The KS-test returns a p-value, which tells us whether the compared distributions are significantly different. In this case, they are not ($D = 0.123$, $p = 0.096$).

Now let's take another distribution, such as a log-normal distribution (Fig. 3.6, light grey line). This distribution is closer to the actual values in the "problem zones" of the normal distribution. Accordingly, the D value is marginally better ($D = 0.087$, $p = 0.423$). In this direct comparison, the log-normal distribution comes out on top.

Side Note 3: Testing for Normality

Some traditional analytical tools (regressions, ANOVA) build upon the normal distribution. To test for a normal distribution, the KS-test is often used. There are five points to say about this:

1. The assumption that the data are "normally distributed" is semantically confusing. What is important is that the *residuals* are normally distributed. So, if you plug your raw data into a KS-test and find some deviation, you still have learned nothing about whether you meet the assumption of normality.
2. The KS-test has no exact solution if there are *ties* within the data, i.e. if a value is measured multiple times.
3. The KS-test is relatively insensitive for small sample sizes. With an increasing sample size, the p-value decreases linearly. This means that small sample sizes (20–30 data points) are usually not statistically different from a normal distribution, even if they come from a different distribution!
4. The Lilliefors test is a corrected variant of the KS-test that takes into consideration that the parameters of the reference distribution are only estimated. In addition to the Lilliefors/KS-test, there is also the Shapiro-Wilk-test, the Shapiro-Francia-test, the Pearson-χ^2 test, the Anderson-Darling test and the Crámer-van-Mises test, which all have similar issues with sample size.
5. A graphical evaluation, as shown in Fig. 3.6, requires some experience, but is much more revealing than a KS-test. Another useful form of graphical evaluation is the quantile-quantile plot (QQ-plot). When we explore the topic of model diagnostics (Sect. 9.1ff.), we will take a closer look at the assumptions of traditional approaches.

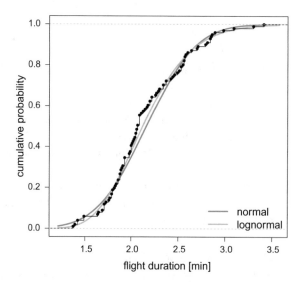

Fig. 3.6 Cumulative distribution of bee flight data (black points) compared to a cumulative normal distribution (dark grey) and a cumulative log-normal distribution (light grey) fitted to the data

3.5.2   Comparing Fits: Likelihood and the Akaike Information Criterion

By fitting a distribution to the data using maximum likelihood, we already have a built-in means of comparison: the likelihood. If one distribution has a log-likelihood (ℓ) of -200 and the other has one of -300, then the first distribution is more suitable, because it has a larger ℓ. We can therefore compare the fit of two distributions by directly comparing the value for ℓ. In the bee flight example above, $\ell_{\text{normal}} = -55.6$ and $\ell_{\text{lognormal}} = -52.3$. The log-normal distribution provides a better fit.

As we saw above, distributions can have different numbers of parameters. When directly comparing a distribution with four parameters with one that only has one parameter, we would expect that the higher flexibility of the 4-parameter distribution would lead to a better fit. For this reason, Akaike [1973] suggested that we should correct for the number of fitted parameters. He derived the AIC ("An Information Criterion", but which is usually referred to as "Akaike's Information Criterion" in his honour). For a more complete introduction, see Burnham & Anderson [2002], for a concise one, see Link and Barker (2006). Here we will simply focus on the essentials.

The AIC is calculated as

$$\text{AIC} = -2\ell + 2p, \tag{3.12}$$

where p is the number of fitted parameters. A lower AIC value suggests a better fit.

A more suitable form is the AICc, which provides a correction for small sample sizes:

$$\text{AICc} = -2\ell + 2p + \frac{2p(p+1)}{n-p-1}, \tag{3.13}$$

where $n =$ number of data points.

For each *additional* parameter that we fit, the AIC value increases by 2 (the AICc increases a bit more: $2 + 4/n$). This multiplier for the number of parameters is called the "penalisation factor". The value of 2 was theoretically derived by Akaike. Some prefer to use other values,[19] since the AIC is *asymptotically biased*, and therefore prefers distributions (and models) with many parameters (see, e.g., Link and Barker (2006). In general, we should really use the AICc instead of the AIC, since it prevents using large models for small data sets, and for larger data sets it is simply the AIC.

The most important alternative penalisation factor of $\log(n)$[20] (instead of 2 in the AIC) gives us the BIC, the Bayesian Information Criterion, also known as Schwarz' Information Criterion. Since the BIC automatically corrects for sample size, it is not necessary to have a separate "BICc".

$$\text{BIC} = -2\ell + p\log(n) \tag{3.14}$$

The AIC value for the fit of the normal distribution on the bee flight data ($\ell = -55.56$) is AIC$= -2 \cdot -55.56 + 2 \cdot 2 = 115.1$. The AICc value is 120.3. The small sample correction for 70 data points is still quite substantial, leading to the difference between AIC and AICc values. The BIC value in this case is only as interesting as the log-likelihood, since $p = 2$ for both distributions and AIC, AICc and BIC follow the information provided by the log-likelihood.

Now let's look back to our ash tree girth example. Here, all of our data are integers, so we can compare the fit of a normal distribution to that of a Poisson distribution (Table 3.1). The latter, as we know, only has a single parameter.

The normal distribution comes out as winner on all fronts here.[21] So, to repeat: while we want to have the highest value possible for ℓ (or the least negative), we want to have the lowest AIC and BIC values.

---

[19]Personally, I do as well.

[20]With n we really symbolize the "limiting sample size" (Harrell 2001, S. 61). For continuously distributed data, this is the same as the number of data points. For binary data (0/1/; Bernoulli distributed), the limiting sample size is calculated differently: $n = \min$ (number of 0s, number of 1s). If we have 100 data points and 90 of them have the value 1, then $n = min(10, 90) = 10$. This tells us something as well: if we only had one 0, then we could only learn something from this one value.

[21]If you like, you can test this difference statistically. The quotient of the likelihoods, or the difference between log-likelihoods and the AICc values, is asymptotically χ^2 distributed. But this goes against the philosophy of the approach.

Table 3.1 Comparison of normal and Poisson distributions as the possible underlying distribution for ash tree girth data

Distribution	ℓ	AICc	BIC
Normal	−430.1	890.9	869.4
Poisson	−618.5	1264.3	1241.6

AIC and similar measures can be used for a number of applications other than calculating the fit of data to a distribution. The "information theoretical approach" (Burnham and Anderson 2002), which we applied here, has dominated ecological statistics literature for the last 15 years, and will most likely be with us for many years to come.

References

1. Akaike, H. (1973). Information theory as an extension of the maximum likelihood principle. In B. Petrov & F. Csaki (Eds.), *Second International Symposium on Information Theory* (pp. 267–281). Budapest: Akademiai Kiado.
2. Bolker, B. M. (2008). *Ecological models and data in R*. Princeton, NJ: Princeton University Press.
3. Burnham, K. P. & Anderson, D. R. (2002). *Model selection and multi-model inference: A practical information-theoretical approach* (2nd ed.). Berlin: Springer.
4. Casella, G., & Berger, R. L. (2002). *Statistical inference*. Pacific Grove, CA: Duxbury Press/Thomson Learning.
5. Clark, J. S. (2007). *Models for ecological data: An introduction*. Princeton, NJ: Princeton University Press.
6. Evans, M., Hastings, N., & Peacock, B. (2000). *Statistical distributions* (3rd ed.). Hoboken, NJ: Wiley.
7. Gelman, A., & Hill, J. (2007). *Data analysis using regression and multilevel/hierarchical models*. Cambridge, UK: Cambridge University Press.
8. Harrell, F. E. (2001). *Regression modeling strategies—With applications to linear models, logistic regression, and survival analysis*. New York: Springer.
9. Jaynes, E. T. (2003). *Probability theory: The logic of science*. Cambridge University Press.
10. Johnson, N. L., & Kotz, S. (1970). *Distributions in statistics* (4 Vols.). New York: Wiley.
11. Kass, R. E. (2011). Statistical inference: The big picture. *Statistical Science, 26*, 1–9.
12. Link, W. A., & Barker, R. J. (2006). Model weights and the foundations of multimodel inference. *Ecology, 87*, 2626–2635.
13. Mosteller, C. F. & Hoaglin, D. C. (1995). Statistics. In R. McHenry & Y. C. Hori (Eds.), *Encyclopaedia Britannica—Macropedia* (15th ed., pp. 28:217–226). Chicago: Encyclopaedia Britannica.
14. Pinheiro, J. C., & Bates, D. M. (2000). *Mixed-effects models in S and S-Plus*. Berlin: Springer. ISBN 0-387-98957-0.
15. Warton, D. I., & Hui, F. K. C. (2011). The arcsine is asinine: The analysis of proportions in ecology. *Ecology, 92*, 3–10.
16. Zuur, A. F., Ieno, E. N., Walker, N. J., Saveliev, A. A., & Smith, G. M. (2009). *Mixed effects models and extensions in ecology with R*. Berlin: Springer.

Actually, I see it as part of my job to inflict R on people who are perfectly happy to have never heard of it. Happiness doesn't equal proficiency and efficiency. In some cases the proficiency of a person serves a greater good than their momentary happiness.
—Patrick Burns

At the end of this chapter...

... you will have been introduced to around 20 continuous and discrete distribution functions and will be able to graphically display each of these in R.

... you will know what the acronym ECDF stands for and can even plot it in R.

... you will be able to use the Kolmogorov-Smirnov test to formally compare samples and multiple continuous distributions.

... you will be able to calculate the log-likelihood and the AIC of a dataset for any distribution.

... you will be able to apply different tests for normal distributions, but you won't want to.

For beginners, the command line interface of R can be somewhat intimidating. In this chapter, however, the superiority over every *point-and-click* software will become clear: R is unbeatable when it comes to distributions. No other software offers so many different distributions (Table 4.1). This list may be long (and even then, still incomplete), but it is nevertheless important, because these distributions are actually used in modern environmental science! You can find out what situations each of these distributions is useful for on your own (e.g. with the help of Wikipedia).

Important skills to learn are:

- how to plot a function, to get a visual picture of it;
- how to calculate the (log-)likelihood of a data set;
- how to fit a distribution to the data.

The first important bit of information is that the likelihood function is the counterpart of the density function. This means that the basis of all calculations for the likelihood of data, given a distribution, is the d... function of this distribution. To illustrate with an example: if we observe a value of 7 and think that this value comes from a Poisson distribution with a mean $\lambda = 5$, in R using the code `dpois(7, lambda = 5)` ($= 0.104$), we can calculate the corresponding probability. For a normal distribution with a mean $\mu = 5$ and standard deviation $\sigma = 2$, the probability density is `dnorm(7, mean = 5, sd = 2) =` 0.121.[1] This can be done for each of the distributions introduced here, if the parameters are given.

---

[1] Warning! This is *not* the probability of the value 7 in the normal distribution, but rather its probability *density*. This can also have values >1 and is only used for comparison, not for the calculation of the probability itself. For discrete distributions (such as the Poisson) the probability density is also the probability, since in this case, all discrete values sum to 1. This is *not* the case for continuous distributions, since these can have an infinite number of x values. Here the values are integrated into 1, although the probability of obtaining a particular value is 0. We noted that earlier, but it is such a common misunderstanding that we like to repeat it.

© Springer Nature Switzerland AG 2020
C. Dormann, *Environmental Data Analysis*,
https://doi.org/10.1007/978-3-030-55020-2_4

Table 4.1 Alphabetical list of important distributions implemented in R. In addition to those listed here, there are some dozens in the **VGAM** package, as well as countless others found sprinkled across other packages (e.g. **fBasics**, **SuppDists**). The `d...` stands for *density* and refers to the density function. For all of these functions, a `p...` (cumulative probability distribution function), `q...` (quantile function) and `r...` (random values from this distribution) are also available. See also `?Distributions`

	Name	R-Function
Continuous	β (beta)	dbeta
	Cauchy	dcauchy
	χ^2 (Chi-square)	dchisq
	Exponential	dexp
	F	df
	γ (gamma)	dgamma
	Log-normal	dlnorm
	Log	dlogis
	Normal	dnorm
	Student or t	dt
	Uniform	dunif
	Weibull	dweibull
	Multivariate normal	dmvnorm mvtnorm
	Wishart	dwish in **MCMCpack**
	Mixed van Mises	dmixedvm in **CircStats**
Discrete	Binomial	dbinom
	Geometric	dgeom
	Hypergeometric	dhyper
	Multinomial	dmultinom
	Negative binomial	dnbinom
	Poisson	dpois

4.1 Displaying Distributions

R provides a variety of graphical functions. For distributions, it is important for us to know if they are continuous, in which case we can draw a line, or if they are discrete, where all values should be represented by points or columns.

4.1.1 Continuous Distributions

Who would know the Cauchy distribution?[2] What does it look like? What happens when we change one of the two parameters? We can take these questions as a motivation for code with which we can graphically represent most continuous distributions.

First, it is important to know the range of validity for the distribution you want to display. To figure this out, we can look at the distribution function or just look it up in a statistics textbook or on the internet. In the case of the Cauchy distribution, the domain is $-\infty$ to ∞. Many other distributions have a restricted domain, e.g only positive values (such as the log-normal distribution: see Sect. 3.4.6) or an upper and lower limit (such as the β-distribution: see Sect. 3.4.8).

Then we need to figure out what the R function for this distribution is called. To do this, we simply enter the following code into R:

```
> ??cauchy
```

---

[2]$Cauchy(x) = P(x|l, s) = 1/(\pi s(1 + ((x - l)/s)^2))$.

Cauchy {stats}   R Documentation

The Cauchy Distribution

Description

Density, distribution function, quantile function and random generation for the Cauchy distribution
with location parameter location and scale parameter scale.

Usage

```
dcauchy(x, location = 0, scale = 1, log = FALSE)
pcauchy(q, location = 0, scale = 1, lower.tail = TRUE, log.p = FALSE)
qcauchy(p, location = 0, scale = 1, lower.tail = TRUE, log.p = FALSE)
rcauchy(n, location = 0, scale = 1)
```

Arguments

x, q	vector of quantiles.
p	vector of probabilities.
n	number of observations. If length(n) > 1, the length is taken to be the number required.
location, scale	location and scale parameters.
log, log.p	logical; if TRUE, probabilities p are given as log(p).
lower.tail	logical; if TRUE (default), probabilities are $P[X \leq x]$, otherwise, $P[X > x]$.

Details

If location or scale are not specified, they assume the default values of 0 and 1 respectively.

The Cauchy distribution with location l and scale s has density

$$f(x) = 1 / (\pi s (1 + ((x-l)/s)^2))$$

for all x.

Fig. 4.1 The top of an R help page for probability distributions, here for the Cauchy distribution (dcauchy)

R now looks through all of the installed packages for help pages that contain the word "cauchy".[3] Capital or lowercase
letters don't make a difference when searching. In this case, the Cauchy distribution can be found in packages such as **circstats**,
stats and **VGAM**, depending on what is installed (**stats** is always loaded).[4]

Now we want to check out the function dcauchy in **stats**. By typing in ?cauchy, we can bring up the help page (Fig. 4.1).
There, we read that the two parameters are called 'location' (l) and 'scale' (s) and by default are set to 0 and 1, respectively.
This may not be the case for all distributions: some distribution arguments have no default values (see, e.g., ?dpois).

Now we can simply display the Cauchy distribution with these default values (Fig. 4.2):

```
> curve(dcauchy(x), from=-2, to=2)
> abline(v=0, col="grey")
```

The curve function draws a mathematical function of x for values between from and to. With abline, we can draw a
reference line (in this case a vertical line at 0 in grey).

If we now want to change the parameters, we simply provide dcauchy with the desired values. Accordingly, we then need
to adjust from and to. To add text along the axis, we can specify ylab and choose an appropriate text size with cex.lab:

```
> curve(dcauchy(x, location=-3, scale=2), from=-15, to=5, lwd=2,
+   ylab="probability density", cex.lab=2)
```

---

[3]While ?? (the shortcut for help.search) searches through the names or titles of all *installed* packages, ? (the shortcut for help) only looks at the
help pages of functions with this name in *loaded* packages.

[4]If you want to be sure that you don't overlook something, you should install the package **sos** and use ??? when searching. To use this function,
you have to be connected to the Internet, because ??? searches using RSiteSearch (http://search.r-project.org). The search result is then displayed
as a webpage.

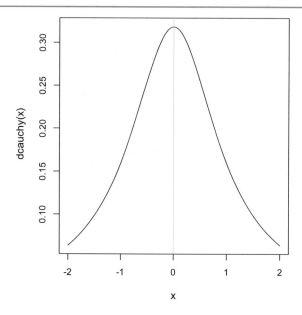

Fig. 4.2 The Cauchy distribution with $l = 0$ and $s = 1$ between -2 and 2

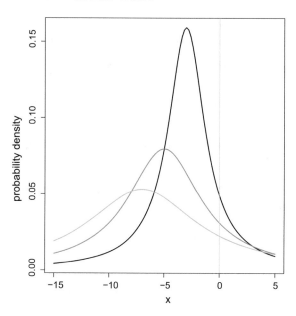

Fig. 4.3 The Cauchy distribution with $l = -3/-5/-7$ and $s = 2/4/6$ (black/grey/light grey), with a reference line at 0

```
> curve(dcauchy(x, location=-5, scale=4), col="grey50", lwd=2, add=T)
> curve(dcauchy(x, location=-7, scale=6), col="grey70", lwd=2, add=T)
> abline(v=0, col="grey")
```

The result (Fig. 4.3) gives a good impression of the effect of both of the parameters. Apparently, the *location* determines where the tip of the curve is, while the *scale* determines the width of the distribution. Since the second and third curve were added with `add=T`, we don't need to provide any `from`/`to` values there. These are simply taken from the first plot.

With `curve`, you can also lay curves on top of a histogram, if these are first produced as a density histogram (with `hist(.,` `freq=F)`). Instead of `curve`, you can also work with the `lines` command or use `plot(., type="l")` to draw lines. An example for this will be shown in the next section for discrete distributions.

4.1.2 Discrete Distributions

For discrete distributions, the function is only defined at certain points (e.g. all positive integers and 0 for the Poisson distribution). As an example, let's look at the hypergeometric distribution.

??hypergeometric leads us to ?dhyper and to the discovery that the hypergeometric distribution has *three* parameters: m, n and k. k is, for example, the number of balls removed from a bucket with m red balls and n blue ones. Drawing a ball is done *without* replacement. There are no default values for the parameters of dhyper.

Let us begin with a simple case: in the bucket there are 10 red and 10 blue balls and we draw a total of 10 balls from the bucket. We would expect that half of the balls would be red and half would be blue, so the most probable value of the distribution would be 5.

```
> x.values <- 0:11
> plot(x.values, dhyper(x.values, m=10, n=10, k=10), type="h", lwd=2, las=1)
```

For the discrete distributions, we can choose a histogram style plot (type = "h"). To use the plot function, we need to provide x and y values. This means that we calculate the probability for the integer values from 0 to 10 in the hypergeometric distribution and then plot these (Fig. 4.4). The lwd and las arguments only specify cosmetic corrections.

We predict that for higher values of m and k, the distribution will shift to the right and that higher n values will shift it to the left. We can test this prediction as follows (Fig. 4.5):

```
> x.values <- 0:11
> plot(x.values-0.2, dhyper(x.values, m=10, n=10, k=15), type="h", lwd=2,
+   las=1, xlab="x-value", ylab="probability density", cex.lab=2)
> points(x.values, dhyper(x.values, m=20, n=10, k=10), type="h", lwd=2,
+   col="grey70")
> points(x.values+.2, dhyper(x.values, m=10, n=20, k=10), type="h",
+   lwd=2, col="grey50")
>
> lines(x.values-.2, dhyper(x.values, m=10, n=10, k=15), lty=2,
+   col="black")
> lines(x.values, dhyper(x.values, m=20, n=10, k=10), lty=2, col="grey70")
> lines(x.values+.2, dhyper(x.values, m=10, n=20, k=10), lty=2,
+   col="grey50")
```

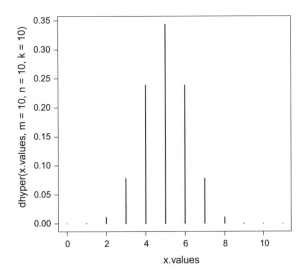

Fig. 4.4 The hypergeometric distribution with $m = 10$, $n = 10$ and $k = 10$

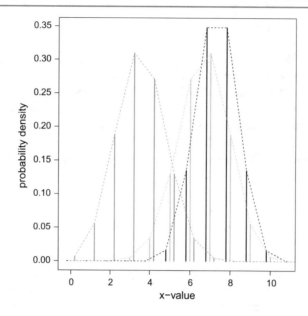

Fig. 4.5 The hypergeometric distribution with $m = 10/20/10$, $n = 10/10/20$ and $k = 15/10/10$ (black/light grey/grey)

In order to keep our lines from being plotted on top of each other, we shifted the first one 0.2 units to the left and the third one 0.2 units to the right. The three `lines` rows of code connect the distribution values with lines for better visibility. Since the distribution is discrete and there are no intermediate values, such lines should really be avoided, since they suggest the existence of intermediate values.

4.2   Calculating the Likelihood of a Data Set

In the previous section we graphically displayed a distribution. Now we want to deal with how to match a measured data set with an assumed distribution. To do this, we first need to calculate the likelihood or the log-likelihood of a data set, as well as the information theoretical values from Sect. 3.5.2.

Here we will use a data set of 1200 fir tree diameters provided by the **FAwR** package (Robinson and Hamann 2011). We load the package[5] and then look at the data set and have a glance at the data structure and variable names:

```
> library(FAwR)
> data(gutten)
> ?gutten   # Opens the help page with a description of the data set
> str(gutten)
```

```
'data.frame':   1200 obs. of  9 variables:
 $ site     : Factor w/ 5 levels "1","2","3","4",..: 1 1 1 1 1 1 1 1 1 1 ...
 $ location: Factor w/ 7 levels "1","2","3","4",..: 1 1 1 1 1 1 1 1 1 1 ...
 $ tree     : int  1 1 1 1 1 1 1 1 1 1 ...
 $ age.base: int  20 30 40 50 60 70 80 90 100 110 ...
 $ height   : num  4.2 9.3 14.9 19.7 23 25.8 27.4 28.8 30 30.9 ...
 $ dbh.cm   : num  4.6 10.2 14.9 18.3 20.7 22.6 24.1 25.5 26.5 27.3 ...
 $ volume   : num  5 38 123 263 400 ...
 $ age.bh   : num  9.67 19.67 29.67 39.67 49.67 ...
 $ tree.ID : Factor w/ 336 levels "1.1","2.1","3.1",..: 1 1 1 1 1 1 1 1 1 ...
```

---

[5]The first time we use the package we need to install it, by entering `install.packages("FAwR")`.

```
> head(gutten)  #  shows first six rows
```

```
  site location tree age.base height dbh.cm volume age.bh tree.ID
2    1        1    1       20    4.2    4.6      5   9.67     1.1
3    1        1    1       30    9.3   10.2     38  19.67     1.1
4    1        1    1       40   14.9   14.9    123  29.67     1.1
5    1        1    1       50   19.7   18.3    263  39.67     1.1
6    1        1    1       60   23.0   20.7    400  49.67     1.1
7    1        1    1       70   25.8   22.6    555  59.67     1.1
```

```
> summary(gutten)  #  provides a summary of each column
```

```
   site       location        tree          age.base          height
 1:231      1:209     Min.   : 1.00   Min.   : 10.00   Min.   : 1.50
 2:376      2: 28     1st Qu.: 7.00   1st Qu.: 40.00   1st Qu.:10.20
 3:242      3:121     Median :12.00   Median : 70.00   Median :17.45
 4:257      4:107     Mean   :16.57   Mean   : 73.08   Mean   :17.47
 5: 94      5:518     3rd Qu.:25.00   3rd Qu.:100.00   3rd Qu.:24.10
            6:102     Max.   :48.00   Max.   :150.00   Max.   :43.50
            7:115
     dbh.cm          volume          age.bh          tree.ID
 Min.   : 0.20   Min.   :   0.10   Min.   :  0.43   1.6    :  15
 1st Qu.:12.90   1st Qu.:  65.75   1st Qu.: 29.81   1.7    :  15
 Median :20.60   Median : 286.00   Median : 59.20   5.25   :  15
 Mean   :20.78   Mean   : 496.71   Mean   : 60.96   5.26   :  15
 3rd Qu.:28.00   3rd Qu.: 697.25   3rd Qu.: 89.38   2.1    :  14
 Max.   :55.40   Max.   :3919.00   Max.   :141.75   2.2    :  14
                                                    (Other):1112
```

We are most interested in the column dbh.cm, the diameter at breast height in cm. In the summary we see that the trees in the mean (and median) have a diameter of 21 cm and the 1st and 3rd quartiles have a similar distance to the mean. This hints at a normally distributed data set. A histogram confirms this suspicion (Fig. 4.6):

```
> hist(gutten$dbh.cm, las=1)
```

However, the frequency distribution is right skewed and naturally cut off on the left, since trees can't have a negative diameter. What we know is that the normal distribution is principally false here, but perhaps it is still *useful* for describing the data.

We fit the normal distribution to these data (using fitdistr from the **MASS** package) and get a maximum likelihood estimate for the mean μ and standard deviation σ of the normal distribution, respectively:

```
> library(MASS)
> fit.norm <- fitdistr(gutten$dbh.cm, "normal")
> fit.norm
```

```
      mean            sd
  20.7820833    10.5987949
 ( 0.3059609)  ( 0.2163470)
```

To calculate the likelihood of these data, we first need to calculate the probability density for each value, and then multiply all of these values together. For the first two points, the likelihoods are:

```
> dnorm(gutten$dbh.cm[1:2], mean=fit.norm$estimate[1], sd=fit.norm$estimate[2])
```

```
[1] 0.01173458 0.02286602
```

The square brackets index the values, i.e. we only grab the first `[1]` or second `[2]` or the first two `[1:2]`. Using the `$` we can access the estimator for the mean and standard deviation stored in the object `fit.norm`. (Try this once with `str(fit.norm)` to see the structure, and with `fit.norm$estimate` to practice accessing an entry within `fit.norm`). The product of these values is:

```
> prod(dnorm(gutten$dbh.cm[1:2], fit.norm$estimate[1], fit.norm$estimate[2]) )
```

```
[1] 0.0002683231
```

Now, if we multiply 1200 values smaller than one, you can imagine that you end up with a number with over 1000 zeros after the decimal.

```
> prod(dnorm(gutten$dbh.cm, mean=fit.norm$estimate[1], sd=fit.norm$estimate[2]))
```

```
[1] 0
```

Not even a computer wants to save all of these zero digits. This is also the reason for calculating the log-likelihood, i.e. the sum of the logarithms of the likelihoods (remember the logarithm rule: $\log(a \cdot b) = \log a + \log b$):

```
> sum(log(dnorm(gutten$dbh.cm, fit.norm$estimate[1], fit.norm$estimate[2])))
```

```
[1] -4535.615
```

(Here we have done a small simplification: instead of entering `mean=` and `sd=` in `dnorm`, we can simply enter the numbers *in the correct order*. R interprets this correctly. This method is a bit more dangerous, but shorter.)

The `d...` functions allow the argument `log=T`, that that we do not have to write a separate section of code to set the `log` for the `d...` function:

```
> sum(dnorm(gutten$dbh.cm, fit.norm$estimate[1], fit.norm$estimate[2], log=T))
```

Since the maximum likelihood has already been used within the `fit.distr` function, we can directly query the log-likelihood from the result (`fit.norm`) and check whether we have reasoned correctly:

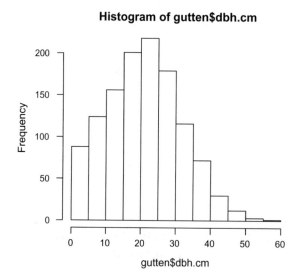

Fig. 4.6 Histogram of 1200 fir tree diameters from Guttenberg (1915, cited in Robinson and Hamann 2011)

```
> logLik(fit.norm)
```

```
'log Lik.' -4535.615 (df=2)
```

That makes our lives much easier! AIC and BIC can be calculated in a similar fashion:

```
> AIC(fit.norm)
```

```
[1] 9075.229
```

```
> BIC(fit.norm)
```

```
[1] 9085.409
```

The correction for small sample sizes is $N/(N - p - 1) = 1200/(1200 - 3) = 1.0025$, where p is the number of fitted parameters. Although this value is so close to 1 in this case, it still has an impact (AIC = 9075, AICc = 9097).

Side Note 4: Newcomb-Benford's Law & Benford's Distribution (from Wikipedia)
"The discovery of Benford's law goes back to 1881, when the American astronomer Simon Newcomb noticed that in logarithm tables the earlier pages (that started with 1) were much more worn than the other pages. Newcomb's published result is the first known instance of this observation and includes a distribution on the second digit, as well. Newcomb proposed a law that the probability of a single number N being the first digit of a number was equal to $\log(N + 1) - \log(N)$.
The phenomenon was again noted in 1938 by the physicist Frank Benford, who tested it on data from 20 different domains and was credited for it. His data set included the surface areas of 335 rivers, the sizes of 3259 US populations, 104 physical constants, 1800 molecular weights, 5000 entries from a mathematical handbook, 308 numbers contained in an issue of *Reader's Digest*, the street addresses of the first 342 persons listed in American Men of Science and 418 death rates. The total number of observations used in the paper was 20,229. [...]
Given a number of decimal numbers that adhere to Newcomb-Benford's law, the number d will appear in the first position with a probability $p(d)$:

$$p(d) = \log_{10}\left(1 + \frac{1}{d}\right) = \log_{10}(d + 1) - \log_{10}(d)$$

for $d \in (1, 2, 3, \ldots, 9)$." [*excerpt from* https://en.wikipedia.org/wiki/Benford%27s_law]
An intuitive interpretation of this "lawfulness" is that most things fit into a lognormal distribution. If you look at a logged string of numbers, then you will see that the range between the first and second position takes up the most space. This is actually quantified by Newcomb-Benford's law.
    Task: Try to implement this law in R and graphically display it. If you fail, R-code is available at http://rwiki.sciviews.org/doku.php?id=tips:stats-distri:benford.

For comparison, we can now fit a log-normal distribution and calculate the AIC (a small trick: if we put the expression in parentheses, then the result is output automatically[6]):

```
> (fit.lognorm <- fitdistr(gutten$dbh.cm, "lognormal"))
```

```
    meanlog         sdlog
  2.82767051    0.77740010
 (0.02244161)  (0.01586861)
```

```
> AIC(fit.lognorm)
```

```
[1] 9591.541
```

The log-normal distribution has a considerably higher (= worse) AIC value and is therefore less appropriate.
    A bright forestry student would tell you at this point that for DBH distributions, the Weibull distribution is usually used. Well, okay then:

```
> (fit.weibull <- fitdistr(gutten$dbh.cm, "weibull"))
```

---

[6]The reason is actually quite simple: the parentheses are interpreted as a `print(x)` command, so that R outputs the object x in the console.

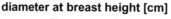

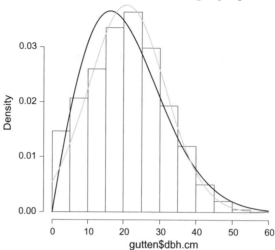

Fig. 4.7 Histogram of 1200 fir tree diameters with a fitted Weibull (black) and normal distribution (grey)

```
      shape           scale
  1.97251051      23.26772749
 ( 0.04664914)   ( 0.35581302)


> AIC(fit.weibull)


[1] 9072.184
```

Indeed, the Weibull distribution is a better fit!

Now we can lay the normal distribution and the Weibull on top of the histogram in order to see if this small difference in AIC (9072– 9075) is at all visible. This code

```
> hist(gutten$dbh.cm, freq=F, las=1, main="diameter at breast height [cm]")
> curve(dweibull(x, fit.weibull$estimate[1], fit.weibull$estimate[2]),
+    add=T, lwd=2)
> curve(dnorm(x, fit.norm$estimate[1], fit.norm$estimate[2]), add=T,
+    lwd=2, col="grey")
```

gives us Fig. 4.7.

The Weibull distribution looks better in the two smallest and the largest size classes, whereas the normal distribution looks to fit better in the middle. But the normal distribution is a good fit, too. It fits the peak actually much better.[7]

4.3 Empirical Cumulative Distribution Function and the Kolmogorov-Smirnov Test

Finally, we want to have a look at how we can use the Kolmogorov-Smirnov test to see whether a data set follows a specific distribution. To do this, it is imperative to display the data as an empirical cumulative distribution function instead of a

---

[7]For practice, you should now draw in the log-normal distribution and the frequency distribution as a density curve (best to do it in another colour)!

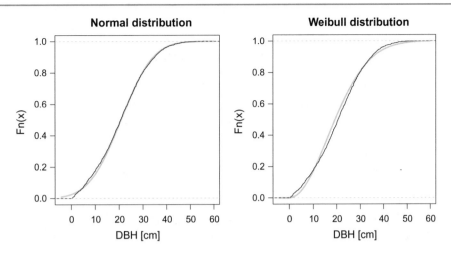

Fig. 4.8 Cumulative frequency distribution (black) of 1200 fir diameters under the fitted normal (left) and Weibull distribution (right) in grey

histogram.[8] This ECDF is then compared to the cumulative distribution function (CDF), and the maximum distance D is measured.

We want to compare the normal distribution and the Weibull distribution with the ECDF of the data. To this end, we plot these on top of each other (with one distribution per graphic; Fig. 4.8):

```
> par(mfrow=c(1,2), mar=c(4,5,4,0.5))
> plot(ecdf(gutten$dbh.cm), pch="", verticals=T, las=1, main="Normal
+   distribution", xlab="DBH [cm]", cex.lab=1.5)
> curve(pnorm(x, mean=fit.norm$estimate[1], sd=fit.norm$estimate[2]), add=T,
+   lwd=3, col="grey")
> lines(ecdf(gutten$dbh.cm), pch="", verticals=T)
> plot(ecdf(gutten$dbh.cm), pch="", verticals=T, las=1, main="Weibull
+   distribution", xlab="DBH [cm]", cex.lab=1.5)
> curve(pweibull(x, fit.weibull$estimate[1], fit.weibull$estimate[2]), add=T,
+   lwd=3, col="grey")
> lines(ecdf(gutten$dbh.cm), pch="", verticals=T)
```

These figures clearly show an overall better fit for the normal distribution. Let's see if the KS-test confirms this. Its probability value indicates whether there is a significant difference between the two compared distributions (i.e. between the ECDF and normal distribution or Weibull distribution). First for the normal distribution:

```
> ks.test(gutten$dbh.cm, "pnorm", fit.norm$estimate[1], fit.norm$estimate[2])


        One-sample Kolmogorov-Smirnov test

data:  gutten$dbh.cm
D = 0.0278, p-value = 0.3105
alternative hypothesis: two-sided

Warning message:
In ks.test(gutten$dbh.cm, "pnorm", mean = fit.norm$estimate[1],  :
  cannot compute correct p-values with ties
```

---

[8]The ECDF is calculated by tabulating and sorting the values. Then, the tabulated values are divided by n, so that the sum is 1. Finally, these relative tabulated values are compared to the measured data. Very confusing to describe in words, but very simple as a chunk of R-Code: `plot(as.numeric(names(table(gutten$dbh.cm))), cumsum(table(gutten$dbh.cm)/length(gutten$dbh.cm)), type='s')` Fortunately, there is a special function in R: `plot(ecdf(.))`.

and for the Weibull distribution:

```
> ks.test(gutten$dbh.cm, "pweibull", fit.weibull$estimate[1],
+     fit.weibull$estimate[2])

        One-sample Kolmogorov-Smirnov test

data:  gutten$dbh.cm
D = 0.0561, p-value = 0.001035
alternative hypothesis: two-sided

Warning message:
In ks.test(gutten$dbh.cm, "pweibull", fit.weibull$estimate[1],
fit.weibull$estimate[2]):  cannot compute correct p-values with ties
```

Here, too, we have a clear result: no difference from a normal distribution was detected (the p-value is bigger than 0.05), however there is a significant difference from the Weibull ($p < 0.01$). The maximum distance between the observed values and the theoretical distribution is twice as large for the Weibull distribution ($D = 0.056$) as for the normal distribution ($D = 0.028$).

The output also returns `alternative hypothesis: two-sided`. Such a two-sided test checks whether the statistic is *larger or smaller* than expected. In the case at hand, it does not matter whether the empirical distribution function is larger or smaller than the analytical one, but it is rather about the absolute difference between the two. If we had the specific hypothesis that the data lie *below* (or above) the analytical distribution, then we would perform a one-sided test.

A word about the warning message: the word *ties* refers to value duplicates—or that the same value comes up multiple times in a data set. If this is the case, we receive a warning message, because `ks.test` can not calculate an exact p-value if there are ties in a data set. Ties in a data set with only 10 data points are much more problematic than if we have 1000 data points. Since we have 1200 data points in this case, we can still trust the "inexact" p-value, especially since the difference in D is so clear.

As a rule, R delivers a warning message, whenever a possible problem comes up, but the function can still be calculated. If the problem is so bad that R stops the calculation, an `Error` is shown instead of a warning.[9]

4.4   Test for Normal Distribution

As mentioned in the side note on the normal distribution test (Side Note 3), formal statistical tests for normal distributions may yield conservative results. The reason is that small data sets rarely fit a distribution nicely, and hence a formal test is likely to be erroneous. Graphic representations, such as those from the previous section, make it easy for the human eye to make comparisons, and is not limited to only normal distributions.

The Komolgorov-Smirnov test is good for all *continuous* distributions, i.e. you can compare both an empirical cumulative frequency distribution (ECDF) with a continuous distribution, or compare two ECDFs with each other.

Lilliefors [1967] questioned the p-value calculated from the KS-test, as the assumption is incorrect that the normal distribution parameters are known. Since the parameters of a distribution are estimated from the data, and are therefore uncertain, the p-values must be corrected. The Lilliefors test offers this potentially substantial correction:

```
> library(nortest)
> lillie.test(gutten$dbh.cm)

        Lilliefors (Kolmogorov-Smirnov) normality test

data:  gutten$dbh.cm
D = 0.0277, p-value = 0.03048
```

---

[9]With `options(warn=2)` you can force R to convert all warnings to error messages, with `warn=-1` they are all ignored: see `?options` for `warn`.

Another test is the Shapiro-Wilk test (Shapiro and Wilk 1965). The derivation of the test is very complex and is outside the scope of this textbook. It is purely a test for "normality".

```
> shapiro.test(gutten$dbh.cm)

    Shapiro-Wilk normality test

data:  gutten$dbh.cm
W = 0.9899, p-value = 2.267e-07
```

The Shapiro-Wilk test has a good reputation for correctly identifying fit, or lack thereof, of data to the normal distribution, based on simulations (Razali and Wah 2011).

Another alternative is the Anderson-Darling goodness-of-fit test (Anderson and Darlink 1952). It is often referred to as the best test for normal distributions, considering all of the limitations that these tests have (Stephens 1974). Like the KS-test, the AD-test can be used for many continuous distributions. The principle behind the test is that an empirical frequency distribution can be transformed into a uniform distribution, if its fits the specified distribution. The deviations of the transformed data can be tested with a distance measurement.

```
> library(ADGofTest)
> ad.test(gutten$dbh.cm, pnorm, mean=fit.norm$estimate[1],
+           sd=fit.norm$estimate[2])

        Anderson-Darling GoF Test

data:  gutten$dbh.cm  and  pnorm
AD = 1.5653, p-value = 0.1616
```

The AD-test is less sensitive to outliers than the Shapiro-Wilk test. For this reason, its use often shows up in ecology literature.

Feel free to discover further tests, such as the Crámer-van-Mises Test and the Jarque-Bera test on your own.[10]

4.5 Exercises

1. Assume that we find a data set scribbled on a piece of paper someone left behind on a tram: (36, 37, 15, 14, 25, 33, 44, 34, 37, 32, 12, 2, 4). Does this look normally distributed? Or perhaps rather Poisson distributed? Fit both of these distributions to the data (`fitdistr`) and compare the fit using the BIC values.
2. Calculate the maximum likelihood of this data set by hand under the assumption of a negative binomial distribution. Fit the data set with the distribution first in order to get the parameter values. Then sum the log-likelihoods of the `dnbinom`-function of the data.
3. Read about the geometric distribution in Wikipedia or a good stats textbook. Plot this distribution for three different parameter values for the (single) parameter p = probability.
4. Read about the triangular distribution. Produce a representation of this distribution similar to those in Sect. 3.4, i.e. with a short description, a plot for three parameter sets, the density function, defined range, first and second moment (mean and variance).
5. Simulate a data set from the logarithmic distribution. Select a set of parameters (for `location` and `scale`; **not** the default values!) and the function `rlogis`. Use 100 for the number of observations. Assign this randomly generated data set to an object (`<-`) and create a histogram of the simulated values. Plot the cumulative distribution of your logarithmic random variables and plot the line of the actual distribution on top (`plogis`) with the actual values chosen by you. What you will see is the difference between a random realisation and the true values from you! So much variability is "normal".

---

[10]Crámer-van-Mises-Test: `cvm.test` also in the **nortest** package; Jarque-Bera-Test: `rjb.test` can be found in the **lawstat** package and as `jarque.test` in the **moments** package.

6. Calculate the maximum likelihood of this data set by hand under the assumption of a logarithmic distribution. Fit the data set to the distribution first to get the parameters of the logarithmic distribution. Then, sum the log-likelihoods of the `dlogis`-function of these data.

7. Use the Crámer-van-Mises test and the Jarque-Bera test to compare the fir tree diameters at breast height[11] and the normal distribution test. Do you find a deviation from the normal distribution (as with the Shapiro-Wilk test) or not (as with the Anderson-Darling test)?

References

1. Anderson, T. W., & Darling, D. A. (1952). Asymptotic theory of certain "goodness-of-fit" criteria based on stochastic processes. *Annals of Mathematical Statistics, 23*, 193–212.
2. Lilliefors, H. (1967). On the Kolmogorov-Smirnov test for normality with mean and variance unknown. *Journal of the American Statistical Association, 62*, 399–402.
3. Razali, N. M., & Wah, Y. B. (2011). Power comparisons of Shapiro-Wilk, Kolmogorov-Smirnov, Lilliefors and Anderson-Darling tests. *Journal of Statistical Modeling and Analytics, 2*, 21–33.
4. Robinson, A. P., & Hamann, J. D. (2011). *Forest analytics with R*. Berlin: Springer.
5. Shapiro, S. S., & Wilk, M. B. (1965). An analysis of variance test for normality (complete samples). *Biometrika, 52*, 591–611.
6. Stephens, M. A. (1974). EDF statistics for goodness of fit and some comparisons. *Journal of the American Statistical Association, 69*, 730–737.

[11] `gutten$dbh.cm`.

Correlation and Association

5

The invalid assumption that correlation implies cause is probably among the two or three most serious and common errors of human reasoning.

—Stephen J. Gould

At the end of this chapter...
... the idea of a correlation will be familiar to you.
... you will be able to use the Pearson's r as a measurement for interpreting the strength of a correlation.
... you will know when you need to use Spearman's ρ or Kendall's τ instead.
... you will have comprehended the principle of association with the help of the χ^2-test.

Up to now, we have dealt with situations where we measured a random variable X, which we assume comes from a distribution. In this chapter, we will be dealing with two variables. Both variables are random variables and for didactic reasons, we will call them X_1 and X_2. Specifically, we are now dealing with a realisation of these random variables, which we denote with a subscript number, that are each a vector[1] of values: $x_1 = (x_{11}, x_{12}, x_{13}, x_{14}, \ldots) = (2, 1.4, 6, 3.9, \ldots)$.

The first part of this chapter will discuss correlation, typically between two continuous variables. In the "association" section of this chapter, we will be looking at the "correlation" between categorical variables (those that are placed in discrete classes, such as feather colour: blue, green, yellow, ...), and specifically, the χ^2-test.[2]

5.1 Correlation

The idea that a stork brings a newborn baby is probably related to a centuries old myth. In any case, there are data from several towns in Germany (Höfer et al. 2004) that documented the simultaneous decline of the stork population and the number of newborns in town. Such data bring up the question of causality. With a correlation, we are only looking at patterns that appear across the two data sets (number of storks and number of babies), whereas with regression, we *assume* that babies appear as a function of the number of storks.[3]

If we have such a data set with two variables, we are often interested in whether the two show the same pattern. Does x_1 increase as x_2 gets bigger? Or does x_1 decrease?

As usual, creating a graphic representation is the most important step of the analysis. As example data we will use the number of spider and ground beetle (carabids) species from 24 different landscapes in Europe (Billeter et al. 2008; Dormann et al. 2008). We want to see if there is a relationship between these two species groups. The data set has two columns, one for beetles and one for spiders, and an additional column used as an ID for the test site (Fig. 5.1, left).

---

[1] Vectors are shown in bold font here.

[2] χ is the Greek letter "chi", pronounced (in English, not in Greek) "ky" (as in "sky"); thus χ^2 is "ky square".

[3] Or maybe it's the other way around: perhaps the wonderful smell of a newborn child attracted more storks, and as the number of births decreased, the storks did not come around any more?

© Springer Nature Switzerland AG 2020
C. Dormann, *Environmental Data Analysis*,
https://doi.org/10.1007/978-3-030-55020-2_5

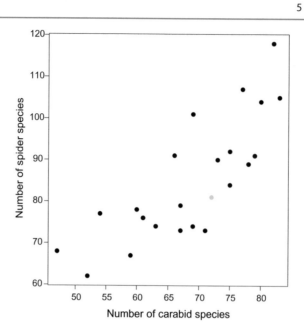

Fig. 5.1 Table and scatter plot of the number of ground beetle and spider species. The grey point shows the data from row 9 (D-QFP/72/81)

Now we can simply plot these values against each other: each site has a combination of ground beetle and spider numbers, which are shown in the figure (Fig. 5.1 on the right). The obvious relationship seen here can be summed up in numbers. The idea is to describe whether the data *covary*: If the value of x_1 increases from one data point to the next and x_2 does as well, then the two positively covary (as in Fig. 5.1). If x_2 decreases whenever x_1 increases, then they negatively covary. In such a case, the points in the scatter plot would then trend from the upper left to the bottom right.

Formally, the **covariance** $s_{x_1 x_2}$ is described as the summed product of the deviation of each value from the respective mean:

$$cov(x_1, x_2) = s_{x_1 x_2} = \frac{1}{n} \sum_{i=1}^{n} (x_{1i} - \bar{x}_1)(x_{2i} - \bar{x}_2) \tag{5.1}$$

This equation looks very similar to that of the variance (Eq. 1.4), only that the respective values from x_1 and x_2 are used here. For our example data, the covariance is $s_{\text{beetles, spiders}} = 108.3$.

As a rule, the greater the covariance value, the stronger the relationship is between x_1 and x_2. The sign of the covariance value tells us the direction of the relationship: a positive value means that the data vary in the same direction, while with a negative value, one data set gets larger as the other gets smaller. A value near 0 tells us that there is no relationship between the two vectors.

The *absolute* value of the covariance depends on the absolute values of x_1 and x_2, i.e for different data (or even different units for the same data) there will be a different covariance value. In order to be able to compare the relationships between different data sets, we need a standardisation so that the value is always between -1 and 1. The result of this standardisation is called **Pearson's correlation coefficient** r:

$$cor(x_1, x_2) = r = \frac{\sum_{i=1}^{n} (x_{i1} - \bar{x}_1)(x_{i2} - \bar{x}_2)}{\sqrt{\sum_{i=1}^{n} (x_{i1} - \bar{x}_1)^2 \sum_{i=1}^{n} (x_{i2} - \bar{x}_2)^2}} \tag{5.2}$$

The correlation coefficient takes values between -1 and 1 ($-1 \leq r \leq 1$). The standard error of Pearson's correlation coefficient, se_r, is calculated as $se_r = \sqrt{\frac{1-r^2}{n-2}}$.

For our example data set, $r = 0.781$ with a standard error of $se_r = 0.133$. Values close to 1 and -1 suggest a very strong correlation (positive and negative, respectively), whereas values around 0 suggest little to no correlation.

Whether a correlation can be considered significant is derived from the ratio of r to se_r. For our data, $r/se_r = 5.86$. This ratio follows the t-distribution (with $n - 2$ degrees of freedom, see Sect. 11.1 on p. 159), which allows the probability of the correlation to be calculated. In this case, the probability that we are dealing with a random correlation is: $p = 6.7 \cdot 10^{-6} \ll 0.05$: beetle and spider species richness are significantly positively correlated.

5.1.1 Non-parametric Correlation

If data display a skewed distribution, then the extreme values have a greater weight. That is to say that a couple of extreme values can dominate the correlation of our data set. If we add the pair of values $(500, 20)$ to our example data, then we get $r = -0.608$, $p = 0.001$. This means that a single (albeit extreme) data point suddenly moved our positive correlation to a negative one! Since Pearson's r assumes normally distributed x_1 and x_2, non-parametric correlations were developed for working with non-normally distributed data.[4] They are robust (less influenced by extreme values), extremely important and often used.[5]

The main idea here is that the data are replaced by their rank and then the correlation is calculated. The rank of a data point is simply its position within the sorted data set (from small to large). From the data $(5, 800, 3, 6)$, we would get a rank transformation of $(2, 4, 1, 3)$. This allows us to "reign in" extreme values.

The most often used non-parametric correlation coefficient is Spearman's ρ,[6] and it simply calculates Pearson's r value for the rank-transformed data. For our data set, $\rho = 0.779$, $p < 0.01$.[7]

An underrated alternative is Kendall's tau.[8] This is calculated as follows: all values of the variables x_1 and x_2 are rank transformed. Then, x_1 is sorted in ascending order. If a value pair (x_{1i}, x_{2i}) is larger than the previous value pair $(x_{1(i-1)}, x_{2(i-1)})$, this is counted as *concordant*, if not, it is counted as *discordant* (ties, i.e. the same rank within one of the variables, are not counted). If $K =$ the number of concordant $-$ number of discordant paired values, then $\tau = \frac{2K}{n(n-1)}$. The big advantage of Kendall's τ is that it is intuitive and interpretable. While a ρ value of 0.4 and 0.8 only shows a stronger correlation for the second comparison, the correlation is actually twice as strong for corresponding values of τ. Unfortunately, Kendall's τ has not yet caught on despite sounder mathematical properties and improved interpretability (Legendre and Legendre 2013, S. 209).

5.1.2 Correlation with and between Discrete Variables

In extreme cases, the data that we want to compare are in different forms, e.g. one continuous and one categorical variable. One example is the correlation between height and sex, the first being rather normally distributed, the latter Bernoulli distributed. Or the correlation between smokers/non-smokers and male/female. For such cases, there is an arsenal of classic association coefficients and tests:

1. continuous-dichotomous (normal-Bernoulli): biserial correlation;
2. continuous-polytomous (normal-multinomial): polyserial correlation;
3. dichotomous-dichotomous (Bernoulli-Bernoulli): χ^2-test;
4. polytomous-polytomous (multinomial-multinomial): contingency table.

Of all of these, the χ^2-test is the most common and the simplest, and is explained in the next section.[9]

---

[4]We can use Pearson's r as an index for the relationship of data from any distribution (e.g. Sokal and Rohlf 1995, P. 560). Pearson's assumption of normality in its derivation is meant to explain why there are also non-parametric correlation coefficients.

[5]For graphical representation, we will keep the raw data form (Fig. 5.1)!

[6]ρ is the Greek letter "rho", pronounced like "row".

[7]A side effect of the rank transformation is that instead of a linear relationship (à la Pearson), we now have any monotonic relationship detected as a correlation (either positive or negative, regardless of how much). For example, a saturation curve has a low Pearson's r, but a high rank-transformed Spearman's ρ.

[8]τ is the Greek letter "tau", pronounced mostly as "tou" ("ou" as in "loud").

[9]Most of these correlations can also be used with parametric methods, in which case they are simply a special case of the GLM (which we will discuss in Chap. 7). For this reason, I find it confusing to unleash the whole armada of disjointed tests (t-test, F-test, binomial test (= Fisher's

At this point, the most important take away message is that for the cases listed above, you can't simply calculate a Pearson's or Spearman's or Kendall's correlation coefficient, but need to use formulas specific to these cases.

5.1.3   Multiple Correlations

If we want to look at multiple variables at the same time, for example not just the number of species from two groups, but from seven, we can put all pairwise correlations into a correlation matrix (often called R):

```
           plants  birds  bees    bugs carabids spiders syrphids
plants      1.000 -0.003 0.197   0.227   -0.007   0.019    0.141
birds      -0.003  1.000 0.229  -0.133    0.055   0.085    0.133
bees        0.197  0.229 1.000   0.616    0.192   0.206    0.204
bugs        0.227 -0.133 0.616   1.000    0.170   0.274    0.580
carabids   -0.007  0.055 0.192   0.170    1.000   0.781    0.140
spiders     0.019  0.085 0.206   0.274    0.781   1.000    0.268
syrphids    0.141  0.133 0.204   0.580    0.140   0.268    1.000
```

Here, the Pearson's r values are given. As it is irrelevant whether we correlate x_1 with x_2, or x_2 with x_1, this matrix is symmetrical. The diagonal, i.e. the correlation of x_i with itself is always 1. The same matrix for Kendall's τ looks like this:

```
           plants  birds  bees    bugs carabids spiders syrphids
plants      1.000  0.093 0.113   0.165    0.066   0.007    0.132
birds       0.093  1.000 0.015  -0.086    0.056   0.011   -0.037
bees        0.113  0.015 1.000   0.610    0.172   0.231    0.158
bugs        0.165 -0.086 0.610   1.000    0.192   0.281    0.359
carabids    0.066  0.056 0.172   0.192    1.000   0.601    0.074
spiders     0.007  0.011 0.231   0.281    0.601   1.000    0.199
syrphids    0.132 -0.037 0.158   0.359    0.074   0.199    1.000
```

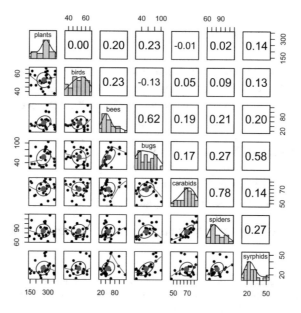

Fig. 5.2 Scatter plot-histogram mixed correlation matrix of species numbers. Pearson's r is displayed above the diagonal, the distributions of the individual variables are shown as a histogram on the diagonal, and the respective scatter plots with the moving average and the total mean are shown below the diagonal

---

sign test), proportionality test, Wilcoxon signed-rank-test (= Mann-Whitney-U-test), ...). If you are interested in using as many different tests as possible, check out the excellent book of Zar [2013].

Of all $7 \cdot (7 - 1)/2 = 21$ pairwise correlations, only three should catch our attention: bugs/bees, bugs/syrphids and spiders/carabids.[10] The bugs/syrphids correlation is a bit weaker for Kendall's τ, but still significant ($p = 0.0158$).

A combination of graphical and numerical displays can be very informative (Fig. 5.2). This dump of information can be hard to interpret for beginners, and for more than 10 variables becomes overwhelming. But with time you will gain experience in which characteristics and deviations from the norm you should pay attention to. The histograms for bees, spiders and syrphids are, for example, are fairly right-skewed. The scatter plot for spiders and beetles, as well as the one for bees and bugs, are non-linear. In the case of non-linearity, non-parametric correlations should be given preference, as long as the relationship is monotonous (i.e. the curve doesn't rise first and then go down, or *vice versa*).

5.2 Test for Association—The χ^2-test

Data consisting of two categorical factors can be summarised using a so-called contingency table.[11] In the following, we want to look at whether the two factors that make up this table are associated with each other.

Our data here result from a survey of a group of retirees looking at the relationship between their biological sex and smoking:

	Smoker	Non-smoker
Men	1	14
Women	12	26

or generally:

	A1	A2
B1	a	b
B2	c	d

Assume that sex and smoking are independent of each other, which allows us to calculate an expected value for each cell. This value is the probability of being a certain sex multiplied by the probability of being a smoker (see Side Note 1 in Sect. 3.2), multiplied by the total number of observations ($N = a + b + c + d$).

	Smoker	Non-smoker
Men	$\frac{(1+14)}{53} \cdot \frac{(1+12)}{53} \cdot 53$	$\frac{(1+14)}{53} \cdot \frac{(14+26)}{53} \cdot 53$
Women	$\frac{(1+12)}{53} \cdot \frac{(12+26)}{53} \cdot 53$	$\frac{(12+26)}{53} \cdot \frac{(14+26)}{53} \cdot 53$

or generally:

	A1	A2
B1	$\frac{(a+b)\cdot(a+c)}{N}$	$\frac{(a+b)\cdot(b+d)}{N}$
B2	$\frac{(a+c)\cdot(c+d)}{N}$	$\frac{(b+d)\cdot(c+d)}{N}$

The χ^2-**test** statistic is then calculated as follows:

$$\chi^2 = \sum \frac{(O - E)^2}{E} \tag{5.3}$$

where O are the observed data, and E are the expected calculated probabilities from the second set of tables. The first cell is therefore $O_1 = a$, $E_1 = \frac{(a+b)\cdot(a+c)}{N}$, and accordingly the first summand is χ_1^2:

$$\chi_1^2 = \left(a - \frac{(a + b) \cdot (a + c)}{N} \right)^2 \bigg/ \frac{(a + b) \cdot (a + c)}{N} \tag{5.4}$$

The corresponding distribution, not surprisingly called the χ^2-distribution, is tabulated in many books (or can be called up in statistics software programs).

---

[10]Here, we will ignore the problem of inevitably finding significance if we compare enough groups. We will look at this problem in detail in Sect. 11.2.5 on p. 171.

[11]Both the χ^2-test and Fisher's exact test are principally used for contingency tables (2×2 tables), but are also defined more generally for $n \times k$ tables.

In the present case, we get a value of $\chi^2 = 2.385$, which provides a p-value of 0.1225. So for this set of data, we do not find any significant relationship between the probability of being a smoker and a person's sex.[12]

A word of caution: The χ^2-test can only be used if the number of *expected* elements of a category is > 5. Otherwise you will need to use the Fisher's test, an expansion of the binomial test (= Fisher's sign test) for testing two proportions (see next chapter for more details).

However, since randomness plays an important role with such small amounts of data, we can expect this test to be very conservative and only recognise an association in extreme cases. Something more sensitive and accurate is Barnard's test (Mehta and Hilton 1993); this would be recommended instead of the Fisher's test, even though it is (currently) not very widely used.

References

1. Billeter, R., Liira, J., Bailey, D., Bugter, R., Arens, P., & Augenstein, I., et al. (2008). Indicators for biodiversity in agricultural landscapes: A pan-European study. *Journal of Applied Ecology, 45*, 141–150.
2. Dormann, C. F., Schweiger, O., Arens, P., Augenstein, I., Aviron, S., & Bailey, D., et al. (2008). Prediction uncertainty of environmental change effects on temperate European biodiversity. *Ecology Letters, 11*, 235–244.
3. Höfer, T., Przyrembel, H., & Verleger, S. (2004). New evidence for the theory of the stork. *Paediatric and Perinatal Epidemiology, 18*, 88–92.
4. Legendre, P., & Legendre, L. (2013). *Numerical ecology*. Amsterdam: Elsevier, 3rd edition.
5. Mehta, C., & Hilton, J. (1993). Exact power of conditional and unconditional tests: Going beyond the 2 x 2 contingency table. *The American Statistician, 47*, 91–98.
6. Sokal, R. R., & Rohlf, F. J. (1995). *Biometry*. New York: Freeman, 3rd edition.
7. Zar, J. H. (2013). *Biostatistical analysis*. Pearson, 5th edition.

---

[12]In most statistical software packages, a different value will be reported, due to a so-called "continuity correction". When Pearson derived the χ^2-test, he made the assumption that the observed frequencies can be approximated by the χ^2-distribution. Frank Yates later showed that this is not correct, and suggested a simple, but somewhat disputed, correction, particularly for small data sets: $\chi^2_{\text{Yates}} = \sum_{i=1}^{N} \frac{(|O_i - E_i| - 0.5)^2}{E_i}$.

Correlation and Association in R

6

> *We used to think that if we knew one, we knew two, because one and one are two. We are finding that we must learn a great deal more about 'and'.*
>
> —Arthur S. Eddington

At the end of this chapter ...
... you will be able to visualise correlations and quantify their strength using different measures.
... you will be able to implement a χ^2-test of association for two categorical variables, and if its assumptions are not met, switch to the Fisher's sign-rank test.

In this chapter, we will first deal with the correlation of two continuous variables, then change gears to look at categorical variables, and end with the χ^2-test of association.

For visualising correlations, a scatter plot is a good option (Fig. 5.1 on p. 72):

```
> gv <- read.csv("speciesrichness.csv", row.names=1)
> names(gv)

[1] "plants"   "birds"    "bees"     "bugs"     "carabids" "spiders"
[7] "syrphids"

> plot(gv$carabids, gv$spiders, las=1, pch=16, cex=1.5, cex.lab=1.5,
+    xlab="Number of carabid species", ylab="Number of spider species")
```

Covariation and correlation can be simply calculated in R using the functions `cov` and `cor`, respectively.[1] You need to provide these functions with a matrix or a purely numerical table (a `data.frame` with numerical values). If we also want to calculate the significance of the correlation at the same time, we can use the function `cor.test`.

```
> cov(gv$spiders, gv$carabids)

[1] 108.2609

> cor(gv$spiders,gv$ carabids)

[1] 0.7807275

> cor.test(gv$spiders, gv$carabids)
```

---

[1] Using `cov2cor` you can quickly and efficiently convert covariance matrices into correlation matrices.

© Springer Nature Switzerland AG 2020
C. Dormann, *Environmental Data Analysis*,
https://doi.org/10.1007/978-3-030-55020-2_6

```
   Pearson's product-moment correlation

data:  gv$spiders and gv$carabids
t = 5.8603, df = 22, p-value = 6.757e-06
alternative hypothesis: true correlation is not equal to 0
95 percent confidence interval:
 0.5508019 0.9005138
sample estimates:
     cor
0.7807275
```

Let's have a closer look at this output: R tells us that it calculated Pearson's *r* correlation and for which variables. The last line gives us the value for the strength of the correlation. Before that, a *t*-test is performed, in order to determine if the correlation is significantly different from 0. We also get the 95% confidence interval for r.[2]

While `cov` and `cor` can also be used for more than two variables and for a matrix, `cor.test` expects exactly two variables.

If a data point is unknown (a cell in the table is blank), then this is designated by R with NA (for not available). The functions `cov` and `cor` also provide a value for NA. Let's see what happens if we remove a piece of data from our ground beetle data set:

```
> gv$carabids[5] <- NA
> cor(gv$carabids, gv$spiders)
```

```
[1] NA
```

The help file for `cor` leads us to a solution. The argument `use` allows a specification for how `cor` should deal with missing data (NAs). The default is `"everything"` and uses all data. If one data point is an NA, the value for the correlation is also returned as NA. The value `"all.obs"` is equivalent, only that an error message is returned instead of NA. More interesting options are `"complete.obs"` and `"pairwise.complete.obs"`. For the correlation of two variables, these options function in the same way: all value pairs where at least one value is an NA are eliminated. For more than two variables `"pairwise.complete.obs"` will not remove all rows of a data record, but only the pairs where there is an NA. `"pairwise.complete.obs"` is therefore the best choice overall.

```
> cor(gv$carabids, gv$spiders, use="pairwise.complete.obs")
```

```
[1] 0.7801731
```

Dealing with missing values is very important in statistical analysis. We may erroneously think that we can simply replace an NA with a 0. This is nearly always wrong, because 0 is information that something has the value of 0, not that it has *no* value!

In R there is an argument for many functions expressed as `na.rm=T` (remove non-available values). This usually excludes any rows in the table that contain an NA for variables used in the function. Further, the option `na.action`, defined in the R options settings `options("na.action")`, determines the behaviour of NAs for many different functions. Here, the most important settings are `na.omit` and `na.exclude`. The latter is preferable if you want to ignore the data points for a calculation, but want to keep the data set intact (e.g. for the calculation of fitted values that fall out when using `na.omit`, but are calculated as an NA with `na.exclude`).

---

[2]Note that this 95% confidence interval is mostly asymmetric, as it has a lower and upper bound. Thus, it cannot be computed, without employing the so-called "Fisher transformation" (which is actually an arctanh-transformation), from the standard error we encountered in the previous chapter as $se_r = \sqrt{\frac{1-r^2}{n-2}}$.

6.1   Non-parametric Correlation

Next to the Pearson's correlation coefficient, the non-parametric variants Spearman and Kendall are the most common.[3] They are all produced in R using the same syntax, but with a specification via the `method`-argument in `cor` or `cor.test`. While `"pearson"` is the default value, `spearman` and `kendall` need to be specified.[4]

```
> gv <- read.csv("speciesrichness.csv", row.names=1)
> cor(gv$carabids, gv$spiders, method="spear")

[1] 0.7792066
```

Here we read in the data once again so that we can restore the data point that we removed in the last example. The option for missing values remains unaffected by the correlation coefficient that we use, so we could also have used `use="pairwise.complete.obs"`. And for Kendall's τ:

```
> cor(gv$carabids, gv$spiders, method="ken")

[1] 0.601476
```

We can also calculate the significance for these non-parametric correlations:

```
> cor.test(gv$carabids, gv$spiders, method="spear")

    Spearman's rank correlation rho

data:  gv$carabids and gv$spiders
S = 507.8248, p-value = 7.235e-06
alternative hypothesis: true rho is not equal to 0
sample estimates:
     rho
0.7792066


Warning message:
In cor.test.default(gv$carabids, gv$spiders, method = "spear") :
  Cannot compute exact p-values with ties
```

(See Section 4.3 for an explanation of the warning message.) And for Kendall's τ:

```
> cor.test(gv$carabids, gv$spiders, method="ken")

    Kendall's rank correlation tau

data:  gv$carabids and gv$spiders
z = 4.0572, p-value = 4.967e-05
alternative hypothesis: true tau is not equal to 0
sample estimates:
     tau
0.601476


Warning message:
```

---

[3] Pearson's correlation is considered "parametric"; the parameters in this case are the mean and standard deviation (see Eq. 5.2). Pearson's correlation coefficient implies that the data are normally distributed (or to be more exact, the data follow a multivariate normal distribution.).

[4] We are allowed to shorten the name as much as we want, but the name needs to remain unique, so here we could reduce it to the first letter. This is known as "partial matching".

```
In cor.test.default(gv$carabids, gv$spiders, method = "ken") :
  Cannot compute exact p-value with ties
```

Both correlation tests output a warning message telling us that because of ties there are multiple values with the same rank, making it impossible to calculate an exact *p*-value.[5]

6.2   Multiple Correlations and the Correlation Matrix

If we want to simultaneously calculate multiple correlations, we can do this by simply providing `cor` with a matrix or a `data.frame`.[6] In this case, these can only contain numerical values!

```
> cor(gv)


              plants        birds       bees       bugs    carabids
plants    1.000000000 -0.003312258 0.1973269  0.2274907 -0.006604945
birds    -0.003312258  1.000000000 0.2285734 -0.1328516  0.054916104
bees      0.197326874  0.228573369 1.0000000  0.6160972  0.191816640
bugs      0.227490668 -0.132851561 0.6160972  1.0000000  0.170281972
carabids -0.006604945  0.054916104 0.1918166  0.1702820  1.000000000
spiders   0.019010181  0.085468814 0.2064895  0.2743538  0.780727493
syrphids  0.140527266  0.132901488 0.2038699  0.5797093  0.139902167
            spiders   syrphids
plants    0.01901018 0.1405273
birds     0.08546881 0.1329015
bees      0.20648952 0.2038699
bugs      0.27435376 0.5797093
carabids  0.78072749 0.1399022
spiders   1.00000000 0.2680577
syrphids  0.26805767 1.0000000
```

This kind of matrix is called a correlation matrix, usually symbolised as R. By defining the method, we can also calculate the correlation matrix for Kendall's τ (R_τ).

Since R is symmetrical, we can also put the information from two correlation matrices into a single matrix: e.g. above the diagonal Pearson's r, below[7] Kendall's *tau*.

```
> R <- cor(gv)
> R_tau <- cor(gv, method="k")
> R[lower.tri(R)] <- R_tau[lower.tri(R_tau)]
> round(R,3)


         plants birds bees   bugs carabids spiders syrphids
plants    1.000 -0.003 0.197  0.227   -0.007   0.019    0.141
birds     0.093  1.000 0.229 -0.133    0.055   0.085    0.133
bees      0.113  0.015 1.000  0.616    0.192   0.206    0.204
bugs      0.165 -0.086 0.610  1.000    0.170   0.274    0.580
```

---

[5] An alternative implementation in other packages such as `Kendall::Kendall` or `pspearman::spearman.test` provides a correction for ties. The notation `pkg::fun` means that the function after the `::` should be called from the package in front of the `::`.

[6] A matrix is the typical object type for mathematical operations in R. Data are usually in tables that sometimes contain non-numerical values (such as observer, region, comments, etc.) In R, data are typically stored as a `data.frame` object. When you read in files (using `read.table` or a similar function), you will end up with a `data.frame`. Convert a (numerical) `data.frame` into a matrix by using `as.matrix`, and use `as.data.frame` for the opposite direction.

[7] The function `lower.tri` indicates an entry below the diagonal of the matrix. Similarly, `upper.tri` is used for entries above the diagonal, and `diag` is used for the diagonal itself.

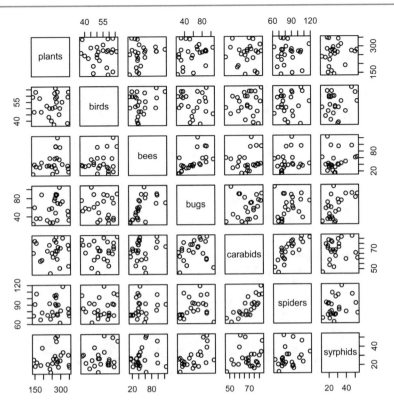

Fig. 6.1 Pairwise scatter plot of the number of species for 7 species groups using `pairs`

```
carabids   0.066   0.056 0.172   0.192      1.000   0.781      0.140
spiders    0.007   0.011 0.231   0.281      0.601   1.000      0.268
syrphids   0.132  -0.037 0.158   0.359      0.074   0.199      1.000
```

A juxtaposition such as this can be very informative, since non-linear relationships have a much stronger effect on Pearson's r than on Kendall's τ. In this case, both correlations are actually quite similar, which suggests that the relationships are predominantly linear.

In the end, we also want to visualise the correlation matrices. The simplest method for doing so is by using the `pairs` command (Fig. 6.1):

```
> pairs(gv, oma=c(2,2,2,2))
```

This plot is essentially the visualisation of the correlation matrix. It is symmetrical, and is really only useful if the number of variables is moderate. For more than a few dozen data points, it is best to use another form of visualisation.[8] If the number of data points is really large, you can create a visualisation with only the density of points instead of all individual points (Fig. 6.2). This can be done as follows:

```
> pairs(gv, panel = function(...) smoothScatter(..., nrpoints=0, add=TRUE),
+    oma=c(2,2,2,2))
```

Caution: density plots are much more complex to calculate and the resulting files are much larger!

An especially elegant looking correlation matrix can be made using the function `pairs.panels` (from the **psych** package; see Fig. 5.2 on p. 75). The code is also very simple:

---

[8]Try `image(cor(gv))`. Caution: the matrix is rotated 90°!

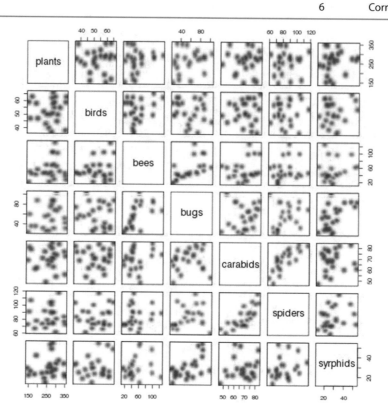

Fig. 6.2 Pairwise smooth scatter plot of the number of species for seven species groups using the combination of `pairs` and `smoothScatter`

```
> library(psych)
> pairs.panels(gv)
```

6.3 Point-Biserial and Point-Polyserial Correlation

If we want to calculate the correlation between a continuous and a categorical variable, we need to use something other than `cor` or the Pearson, Spearman or Kendall approaches. Instead, we can use a point-biserial (for two categories) or point-polyserial (for multiple factor levels) correlation coefficient.

The point-biserial correlation coefficient is defined as:

$$r_{pb} = \frac{\bar{x}_1 - \bar{x}_2}{s_n} \sqrt{\frac{n_1 n_2}{n^2}} = \frac{\bar{x}_1 - \bar{x}_2}{s_{n-1}} \sqrt{\frac{n_1 n_2}{n(n-1)}}$$

where $\bar{x}_1$ and $\bar{x}_2$ are the means of the two groups, n_1 and n_2 are the size of the data set and $n = n_1 + n_2$.

```
> SiSe <- read.csv("SizeSex.csv")
> require(psych)
> biserial(SiSe$size, SiSe$sex)

        [,1]
[1,] 0.6984408
```

The function `hetcor` in the **polycor** package extends the `cor` function for categorical variables in order to calculate normal Pearson correlations for continuous variables, and biserial correlations for mixed continuous/categorical variables.

Non-parametric variants are not available for this R function, but by using rank transformation, we can still calculate the Spearman variant:

```
> biserial(rank(GG$Groesse), GG$Geschlecht)

           [,1]
[1,] 0.700913
```

Polyserial (continuous correlated with multilevel categorical variables) and polychoric correlations (between multiple multilevel categorical variables) can also be executed using `polycor::hetcor` or `psych::polyserial` or `psych::polychoric`, respectively.

I have not seen any of these correlation coefficients used in an ecological or environmental science publication, but they may be useful in some situations. Consider one such situation where the heterogeneous correlation matrix concerns a data set from the sinking of the Titanic. It gives the age, sex, passenger class and the indicator variable "survived" for 1309 passengers. The correlation matrix between these continuous, three-level and binary variables can be calculated using hetcor[9]:

```
> library(effects)
> library(polycor)
> data(TitanicSurvival)
> attach(TitanicSurvival)
> hetcor(TitanicSurvival, use="pairwise.complete.obs")

Two-Step Estimates

Correlations/Type of Correlation:
                survived       sex        age passengerClass
survived               1 Polychoric Polyserial     Polychoric
sex             -0.7453          1 Polyserial     Polychoric
age             -0.07005    0.08149          1     Polyserial
passengerClass  -0.4438     0.1853    -0.4533              1


Standard Errors/Numbers of Observations:
                survived      sex      age passengerClass
survived            1309     1309     1046           1309
sex               0.0252     1309     1046           1309
age              0.03896  0.03939     1046           1046
passengerClass    0.0347  0.04045  0.02715           1309


P-values for Tests of Bivariate Normality:
                survived       sex       age
survived
sex                 <NA>
age            0.0005623 0.0007213
passengerClass    0.6122    0.9433 0.0008194
```

The results are presented in three parts. The uppermost part shows the correlations. The area above the diagonal tells us which form of correlation we are dealing with. In the second part, the standard error and the number of observations used is given. The third part shows the results for testing that the assumptions of polyserial or polychoric correlations are met. We can trust

---

[9]The way a "correlation" between two factors is computed is relatively involved. It uses so-called "latent variables", which are theoretical, unobserved, continuous variables, which are observed by means of one of our factors, e.g. the latent variable "wealth" represented by `passengerClass`. These latent variables are estimated from the factor, and then the latent variables are correlated with each other. So really, polychoric correlation is a correlation of two unobserved variables re-constructed from the observed factor levels. The whole "latent variable" idea is very prominent in the social sciences, where one can hardly ever directly measure what one would really like to know, e.g. wealth, intelligence, social competence and so forth.

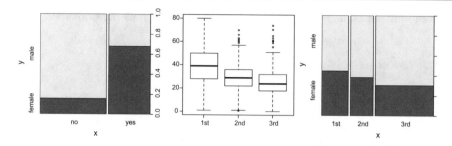

Fig. 6.3 Mosaic plots and box plots of different variables from the `effects::TitanicSurvival` data set. Areas of the mosaic plot (right and left) are proportional to the number of observations in the respective class

the correlations (from the first part) that are *not* shown to be significant in the third part. Note that here all correlations using `age` are violating the test assumptions!

We can create a simple visualisation (Fig. 6.3) with the following code:

```
> par(mfrow=c(1,3), mar=c(4,4,1,1)) # mfrow specifies the number of plots
> plot(survived, sex, las=1)
> plot(passengerClass, age, las=1)
> plot(passengerClass, sex, las=1)
```

Unsurprisingly, we see here that there is a clear correlation between survival and sex ("women and children first!"; Fig. 6.3) and survival and passenger class (lower classes were on lower levels of the ship). Somewhat more surprising is the negative correlation between age and passenger class. Apparently, older passengers could afford a more expensive cabin. The correlation between sex and passenger class also seems clear, even though the strength of the correlation is only 0.185.

6.4   The χ^2-test with R

In order to implement the χ^2-test, the data need to be in the form of a 2×2 matrix. (The option `byrow=FALSE` makes it so that the data are placed by column in the matrix.)

```
> smoking <- matrix(c(1, 12, 14, 26), nrow = 2, byrow = FALSE)
> smoking

     [,1] [,2]
[1,]    1   14
[2,]   12   26
```

Now we can execute the test:

```
> chisq.test(smoking)

    Pearson's Chi-squared test with Yates' continuity correction

data:  smoking
X-squared = 2.3854, df = 1, p-value = 0.1225
```

In this data, there appears to be no relationship between sex and smoking.

With the following command, we can tell R to provide the expected values. These all have to be > 5 for the test to be considered valid.

```
> chisq.test(smoking)$expected
```

```
           [,1]       [,2]
[1,]  3.679245  11.32075
[2,]  9.320755  28.67925
```

We see that the number for the first entry is indeed less than 5. Accordingly, we should not use the χ^2-test here, but rather use the Fisher's Exact test. Fisher's test, as an alternative to the χ^2-test can be implemented in R with the following code:

```
> fisher.test(smoking)

    Fisher's Exact Test for Count Data

data:  smoking
p-value = 0.08013
alternative hypothesis: true odds ratio is not equal to 1
95 percent confidence interval:
 0.003387818 1.292748350
sample estimates:
odds ratio
 0.1590004
```

Using this method, we also find no significant relationship between sex and smoking.

Finally, let's try the Barnard-test:

```
> library(Barnard)
> barnard.test(1, 12, 14, 26)

Barnard's Unconditional Test

            Treatment I Treatment II
Outcome I             1           12
Outcome II           14           26

Null hypothesis: Treatments have no effect on the outcomes
Score statistic = 1.89883
Nuisance parameter = 0.93 (One sided), 0.579 (Two sided)
P-value = 0.0385076 (One sided), 0.0589048 (Two sided)
```

The most important number here is the last one: the P-value for the test for association. The term "Two sided" refers to the fact that we are testing whether there is a positive *or* negative relationship. If we want to specifically test for either a only a positive *or* only a negative relationship, then we could use a one-sided test, which has a significant P-value in this case.

All-in-all, these three tests (χ^2-, Fisher- and Barnard-test) give us the same result: smoking and sex are not associated with each other in this data set.

6.5   Exercises

1. This exercise uses the data set from Bolger et al. (1997, `bolger.txt`). The authors collected data on the presence of rodents (RODENTSP, with the values 0 or 1) in 25 habitat fragments in Californian canyons as well as three environmental parameters.

   Load the data and check for correlation between distance to the next fragment (DISTX) with the length of time isolated (AGE). Then check for correlation of all variables other than RODENTSP using the `pairs.panels` function for Kendall's τ.

2. In the now familiar data set `gutten` (in the **FAwR** package), there are other interesting variables in addition to `dbh.cm`, namely `age.base`, `height` and `volume`.

Make a correlation plot for these four variables. Then calculate the Pearson and Kendall correlation. For which variables are these two correlations drastically different? Why is this (look at your correlation plots)?

3. Ask at least 100 friends or classmates to try to roll their tongue into a tube shape. Record the number of those that can along with the sex of the person into a matrix in R.

Then do a χ^2-test, check the assumptions (expected value > 5 for each cell) and use the Fisher test as a robust alternative.

Reference

1. Bolger, D. T., Alberts, A. C., Sauvajot, R. M., Potenza, P., McCalvin, C., Tran, D., et al. (1997). Response of rodents to habitat fragmentation on coastal southern California. *Ecological Applications, 7*, 552–563.

The purpose of computing is insight, not numbers.
—Richard W. Hamming

By the end of this (long) chapter . . .

. . . the difference between correlation and regression should be obvious. Specifically, you will recognise that with a regression there is always an x and a y, that is, a variable that explains and one that responds. A predictor and an answer—an independent and a dependent variable. The dependent variable is always shown on the y-axis in the figures.

. . . you will be able to explain the logic of the link function and will have learned the corresponding *link* functions for normal, Poisson, and binomial distributions. In addition, you will be able to transform values on the *response* scale and model coefficients on the link scale back and forth.

. . . you will know that a *dummy* can actually be quite clever.

. . . you will learn about how the χ^2-test is an important special case of the GLM.

7.1 Regression

A regression is a specific case of correlation. It implies a direct dependence between the two variables in question: x affects y, but not the other way around. Therefore, we call x the *independent* or *predictor* variable and y the *dependent* or *response* variable. Two typical ways to denote this relationship are $y \sim x$ or $y = f(x)$. Both of these expressions mean that y is a function of x.[1]

7.1.1 Regression: A Distribution Parameter Varies with the Predictor

In Chap. 3 we explored how a distribution can be fit to a random variable. The distribution parameters we looked at were constant for the data set. With regression, we now step into new territory: a distribution parameter that varies with the predictor!

Let's use the following data set as an example (Fig. 7.1). For 25 male collared flycatchers (a species of bird, *Ficedula albicollis*), researchers observed how many pieces of food each individual delivered to their nest per hour. As an additional variable, the symmetry and the sharpness of the feather colouration was evaluated and given an "attractiveness value" (on a scale from 1–5). As per Darwin's theory of sexual selection, female individuals should only select more "attractive" males if

---

[1] The $f(x)$ leaves the exact function unspecified, while $y \sim x$ typically indicates a linear relationship, i.e. $f(x) = ax + b$.

© Springer Nature Switzerland AG 2020
C. Dormann, *Environmental Data Analysis*,
https://doi.org/10.1007/978-3-030-55020-2_7

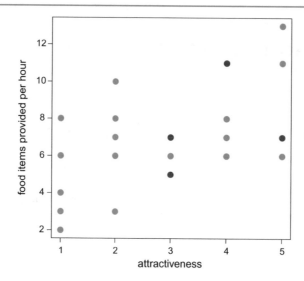

Fig. 7.1 Scatter plot showing food provision depending on the attractiveness of male collared flycatchers. Darker points indicate that there are multiple data points with the same value stacked on top of one another

these males are also better fathers—i.e. they bring more food to the nest to feed their young (therefore increasing the fitness of the mother in the form of a higher survival rate of her offspring).[2] We see from the data that more attractive males do indeed make better fathers.[3]

We can look at the data points for each of the five different attractiveness values as a sample. The mean of each sample increases from left to right: 4.6, 6.8, 6.0, 8.6, 8.8. In other words: the mean of the sample is a function of the independent variable "attractiveness". We assume in this case that the sample is a random variable from a Poisson distribution of all possible food provision behaviours. That is to say, we assume that the observed data come from a Poisson distribution.

We fit a distribution to these data, as we have done before, but we allow the parameter λ of the Poisson distribution to increase with attractiveness: $\lambda = ax + b = a \cdot \textbf{Attractiveness} + b$. Following the description of distributions in Chap. 3, we can write:

$$y \sim \text{Pois}(\lambda = ax + b)$$

or for this specific case:

$$\textbf{Number of food items} \sim \text{Pois}(\lambda = a \cdot \textbf{Attractiveness} + b).$$

You would say: "y is Poisson distributed, whereby λ is a linear function of x."

Of course, this relationship does not necessarily have to be linear, nor does it have to be a Poisson distribution. This is just the simplest possible case. If our dependent y variable came from a negative binomial distribution, for example, and instead of the mean k, the clumping parameter θ was parabolically dependent on x, we would write:

$$y \sim \text{NegBin}(k, \boldsymbol{\theta} = ax + bx^2 + c).$$

Also, our x variable does not necessarily need to be continuous, but can also be a factor. Such a case might be the pH-level of rivers and streams in Germany, Switzerland, and France. The predictor in this case is a factor (country) with three distinct levels (G, CH, F). If the pH-level was normally distributed and the factor "country" only influenced the mean, then we could write[4]:

$$\textbf{pH} \sim N(\boldsymbol{\mu} = a + b \cdot \textbf{G} + c \cdot \textbf{CH} + d \cdot \textbf{F}, \sigma),$$

---

[2]In fact, one can manipulate the attractiveness of males by using coloured leg rings and thus directly test the inverse of this hypothesis: if a male is given the feedback that he is attractive, will he invest more into parenting?

[3]We leave it to the reader to decide how to collect data to test this hypothesis for humans.

[4]This formula is only being shown for didactic purposes. In reality, something else would be fit (see next chapter), because otherwise we would have four parameters for three categories, which is not well-defined. This example is only to show you that y can be a function of categorical predictors.

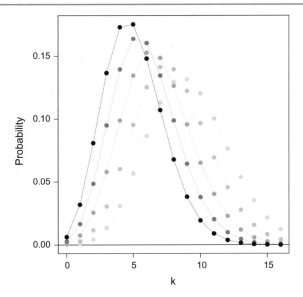

Fig. 7.2 Change of the fitted Poisson distribution with respect to attractiveness. This plot shows the number of food pieces (k) per hour that would be expected by differently attractive individuals (from black = 1 to light grey = 5)

where **G**, **CH** and **F** are three so-called indicators, that are always 0 unless the data point comes from G, CH, or F. Rows from such a dataset might look like this:

Country.CH	Country.G	Country.F	pH
0	1	0	5.6
1	0	0	6.1

The first row of data comes from Germany, the second row from Switzerland. The value of a is then the total mean value of all three countries, while b, c, and d represent the deviation from this mean. If the mean pH-level in France was greater than a, then $d > 0$. Generally, you could say:

A regression quantitatively describes the relationship between one (or more) independent variables (also called predictors) and a dependent variable (response variable). The predictor determines a parameter of the assumed distribution of the response variables.

Apparently, this rather simple idea seems to create substantial conceptual difficulties. The problem may be that it is rather abstract to imagine that every measured data point is actually a random draw from a variable distribution. In the collared flycatcher example, each data point was drawn from a Poisson distribution of the pieces of food whose mean changes, as shown in Fig. 7.2. For an attractiveness value of 1, you would expect 4 or 5 pieces of food an hour, but could get up to 13 pieces per hour. For an attractiveness level of 5 (light grey), the expected value of food items per hour is around 9, and values above 15 would still be plausible.

Before we actually do the calculations for this regression, we need to remember that for an attractiveness of 1, the possible downward deviation is very limited (around 4 or 5, since negative pieces of food is not possible), but the upward deviation is much larger (around 7 or 8). Accordingly, a regression line for these data should not run through the centre of the measured values, but rather run slightly below. For an attractiveness of 5, the Poisson distribution is relatively symmetrical, and therefore the regression line should pass through the middle of the measured values.

7.1.2 Regression and Maximum Likelihood

In Chap. 3, we fit a distribution by using the maximum likelihood. Even if distribution parameters are now determined by another variable, we can use the same approach.[5] That is, we try to find the parameters of the distribution *and* the model, which make the observed data most likely. To do so, we need to take two steps: first, we need to decide on the distribution

---

[5]At this point in most statistics textbooks, the least squares method is introduced. Since this implicitly assumes a normal distribution, we'll just forget about it here. Our logical framework here includes all distributions and maximum likelihood.

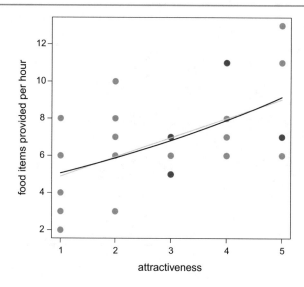

Fig. 7.3 The flycatcher data set with a fitted line (grey) and exponential function (black)

and thereby specify the likelihood-function for our model; second, we need to use some technical approach to find the best parameters (an optimisation technique). Here, we don't need to bother about *how* a computer does the optimisation; there are many books about it. Let us thus tackle the first step.

As the data are count data, we assume (for now) that they are Poisson-distributed. The parameter of the Poisson distribution is called λ. In a regression, each observed value y_i is linked to the value of the predictor x_i. However, the value y_i itself is a random draw from a Poisson distribution, and it is in fact the Poisson distribution's parameter λ which is affected by x_i: $\lambda_i = ax_i + b$.

To compute the likelihood of observing y_i given our model $ax_i + b$ is computed with the Poisson distribution function (Eq. 3.3.3 on p. 41). We thus set $k = y_i$ and then $\lambda_i = ax_i + b$ and we get:

$$L(y_i) = \frac{\lambda_i^{y_i}}{y_i! e^{\lambda_i}} = \frac{(ax_i + b)^{y_i}}{y_i! e^{(ax_i+b)}}$$

This is the expression that computes the probability of observation y_i, given parameter values a and b for the assumed Poisson distribution. At the moment we don't know the values of a and b, so let us choose haphazardly 1 and 4, respectively.

We can now compute the likelihood of the first collared flycatcher data point: attractiveness = 1, pieces of food = 3. We get a likelihood of

$$P(k = y_i|\lambda_i = a \cdot x_i + b) = P(k = 3|\lambda_i = 1 \cdot 1 + 4) = \frac{(1 \cdot 1 + 4)^3}{3! e^{1 \cdot 1+4}} = \frac{125}{3 \cdot 148.4} = 0.140$$

This means that the observed value has a likelihood of 0.14. For the log-likelihood of the full data set, we can use the same equation for all observed pairs of data (y_i, x_i), take the natural log and and sum them up to get the log-likelihood (as we did in Sect. 3.3.1). Next, we can maximise this by trying out different values for a and b.

Instead of just one parameter (λ), we now have to optimise two parameters (a and b). An analytical approach may not always exist (see below), but we can solve the problem numerically. Any self-respecting statistical software provides an optimisation algorithm or has already implemented it for this type of problem.

For the example here, the computer finds the best parameters as $a = 3.864$ and $b = 1.032$ (see next chapter for the technical details). With these values, we can lay a maximum likelihood-curve directly through the data points (Fig. 7.3).

7.1.3 The Other Scale and the *Link* Function

The grey line in Fig. 7.3 has a distinct flaw: at $x = -1.032/3.864 = -0.267$ it becomes negative.[6] In our case, where attractiveness can't be negative, this is no problem. However, if we have a predictor that can take values below the fitted range, the fitted y-values slide below 0. This is impossible in a Poisson distribution. When fitting for values less than -3.74, we would obtain undefined values—and an error message from our software.

Therefore, it was proposed that each distribution has a corresponding *link* function, $g(y)$. This link function ensures that the predicted $\hat{y}$-values conform to the distribution. The typical link function of a distribution (practically the default setting for analysis) is called the "canonical" link function.

So what does the link function do exactly? In short, it changes the form of the relationship between the predictor and the response variable (see Fig. 7.4). From a linear relationship ($y = ax + b$), the log-link creates an exponential relationship, the logit-link creates a sigmoidal relationship, and the inverse-link creates a hyperbolic relationship (see Table 7.1).

The link function determines the way that both the mean and the variance are modeled. The distributions shown in Table 7.1 all belong to the so-called exponential distribution family (with the exception of the negative binomial distribution). In this family, the variance is a function of the mean (see the last column).

Four different link functions are commonly used (Table 7.1) and are described here. The forms of these functions are displayed in Fig. 7.4. The following description is a bit ... clumsy. This is due to the fact that we essentially want to say: the observed values are modeled on the *link scale*. This, unfortunately, can only really be understood if one already knows it! This is why the description of the "phenomenon" of the link function proceeds in such an awkward manner. Hopefully it will lead to a better understanding in the end!

The normal distribution is the simplest case. Here, data are fitted just as they were collected.

For distributions that are defined in the interval $[0, \infty)$ (Poisson, negative binomial, log normal), the log is the usual link function.

For the binomial distribution (and additionally, the special case of the Bernoulli distribution), values must be between 0 and 1. The logit-function ensures that this is the case: for each value from $f(x) \in (-\infty, \infty)$, then $g(f(x)) \in (0, 1)$. In fact, $g(x)$ only reaches 0 and 1 asymptotically.

For the γ-distribution, the inversion can often cause confusion. In this case, large y-values suddenly become small y'-values. It's all a matter of just getting used to it!

You should not think of the link functions as transformations of the y-variable, but rather as a transformation of the relationship. If our data are, for example, 0/1 distributed (perhaps: living/dead; male/female; red/blue), you can't logically transform these variables to values between $-\infty$ and ∞. Instead, we fit the inverse function. For Bernoulli data, we would write:

$$y \sim \text{Bern}\left(\mathbf{p} = \frac{e^{ax+b}}{1 + e^{ax+b}}\right)$$

Table 7.1 Distributions and their typical (canonical) link functions. Instead of using the function for fitting (e.g. $f(x) = ax + b$), the inverse or "link function" is used: $g(f(x))$

Distribution	Canonical link function	Inverse function	Variance function
Normal	Identity ($y' = y$)	$g(f(x)) = f(x)$	1
Poisson	log ($y' = \ln(y)$)	$g(f(x)) = e^{f(x)}$	$\hat{y}$
Binomial	logit ($y' = \ln\left(\frac{y}{1-y}\right)$)	$g(f(x)) = \frac{e^{f(x)}}{1+e^{f(x)}}$	$\hat{y}(1 - \hat{y})$
γ	inverse ($y' = 1/y$)	$g(f(x)) = 1/f(x)$	$\hat{y}^2$
Negative binomial	log ($y' = \ln(y)$)	$g(f(x)) = e^{f(x)}$	$\hat{y}/(1 - \hat{y})^2$

---

[6]To remind you: the intersection with the x axis can be found when we set $y = 0$: $y = ax + b = 0 \Leftrightarrow x = -b/a$.

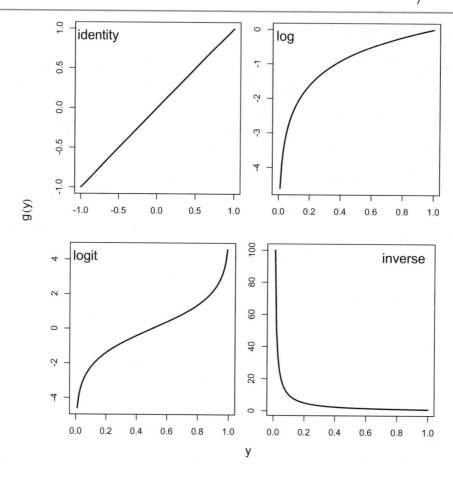

Fig. 7.4 Common link functions for GLMs. **a** *identity link*, **b** *log link*, **c** *logit link* and **d** *inverse link*. The value on the *x*-axis represents the measured values, the values on the *y*-axis are those on the link scale

You would say: We fit the model on the link scale.

Accordingly, the values for a and b are also fit on the link scale.

And if we use such a model to calculate predictions $\hat{y}$ for new values of x, then we have to back-transform them onto the *response scale* afterwards.[7]

When we fit our data for the collared flycatchers, we used the identity as the link function, instead of the "canonical" logarithm. Let's analyse the data again with the "common" approach:

$$\textbf{Pieces of food} \sim \text{Pois}\left(\lambda = e^{a \cdot \textbf{Attractiveness} + b}\right).$$

The result is an exponential function (Fig. 7.3, black line), with $a = 1.47$ and $b = 0.148$. These parameter estimates are on the link scale! If we transform this to the response scale, the difference between the linear and the exponential function is (in this case) minimal.

It has become customary to produce figures on the link scale, in this case with a logarithmic *y*-axis. Here, the fitted function x becomes linear (see Fig. 7.5). The response scale seems more natural, because it is on this scale that we collected our data. If you want the emphasis to be placed on the information gained from the model, however, the link scale may be more appropriate, because it more clearly shows the form of the fitted linear model. "Linear", in the statistical sense, means that the model is a linear combination of the predictors, that is to say, it only consists of additive terms. If, for example, in addition to an x, an x^2 is also entered in the model ($y \sim ax + bx^2 + c$), then the resulting function is not mathematically linear, but the statistical model is still "linear in its parameters".

---

[7] Yes, this *is* a transformation!.

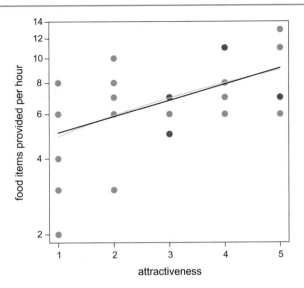

Fig. 7.5 The flycatcher data set with a fitted line (grey) and exponential function (black), this time with a logged *y*-axis (so it is on the *link scale*). This turns the (identity-link) line into a curve and the (log-link) curve into a straight line!

A regression where you can choose the distribution and that is linear on the link scale is called a *generalised linear model* (GLM).[8] Many statistics textbooks traditionally use the term "regression" only for regressions involving normally distributed data. These books call GLMs for Poisson distributed data "Poisson regressions" (or a log-linear model, which takes the link function into account), while binomial regression models are referred to as "logistic regression".

7.2 Categorical Predictors

Up to now, we have acted as if we don't care at all what the predictor looks like. The term *regression* usually implies that the predictor is continuous. If our explanatory variable is something like male/female, day/night or even the day of the week, we are dealing with a categorical predictor. Fundamentally, this changes nothing about the principles of the regression we have just learned about. In the interpretation, however, there are some considerable differences.

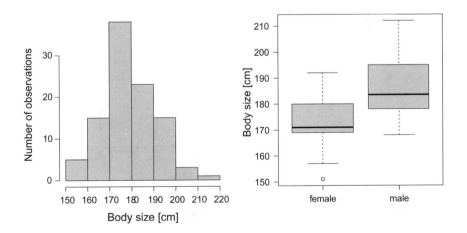

Fig. 7.6 Histogram showing the height of 50 men and 50 women: no height difference between the sexes? In the box plot, it looks as if men are slightly taller

---

[8]The "generalised" part comes from the fact that different distributions can be used.

7.2.1   A Categorical Predictor with Two Levels

Let's begin with a categorical predictor, also called a *factor*, with two possible values: male and female. For this data set, the height of 50 men and 50 women was measured (Fig. 7.6). If we were to do a regression at this point, we would get a *y*-intercept and a slope. The *y*-intercept in the case of a categorical predictor refers to the mean of one of the groups, and the slope shows the difference between the groups. Usually, factors are listed in alphabetical order, so that the first factor level (here: Female) is used as the reference group.

```
Coefficients:
            Estimate Std. Error t value Pr(>|t|)
(Intercept)  173.140      1.363 127.036  < 2e-16 ***
SexMale       12.900      1.927   6.693 1.37e-09 ***
---
Signif. codes:  0 `***' 0.001 `**' 0.01 `*' 0.05 `.' 0.1 ` ' 1
```

We see here that females in our sample are on average 173 cm tall, and males in our sample are 13 cm taller. In this simple case (with only two factor levels), the significance test for the "increase" in SexMale is also the test for the difference between the two sexes.

So how is this a regression? Where is the slope, and if it does exist, what is the *x*-value for the factor level Female or Male?

Simply put, every statistical software recodes factors internally. A column with values F M F M M M F F M is internally read as 0 1 0 1 1 1 0 0 1, in alphabetical order, with the additional information (= attribute) that 0=F and 1=M. In our example, all females are assigned the *x*-value 0 and all males are assigned the *x*-value 1.[9] Now it finally becomes clear that we can calculate a regression (Fig. 7.7). The slope corresponds to the factor that is multiplied by the *x*-value ($y = ax + b$). The *y*-value for factor level M is therefore 173.1 (=*y*-intercept) + 12.9 · 1 = 186.0.

This calculation is the simple comparison of two normally distributed samples. We came to this calculation by means of a regression, which we fitted to our data using maximum likelihood. Traditionally, this test is called a *t*-test and has a different motivation. Since the *t*-test shows up again and again, and forms the basis of many significance tests, it has its own section in this book (Sect. 11.1 on p. 159). However, it is really nothing more than a special form of regression.

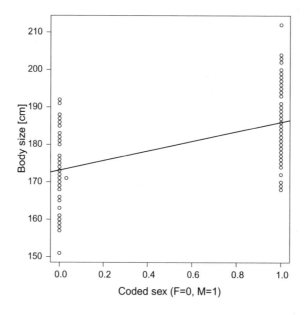

Fig. 7.7 The heights of men and women as a scatter plot with regression lines. The *y*-intercept corresponds to the mean for the level F=0

---

[9]This shouldn't be taken as sexism. You could also rename the categories here or do whatever you like. It's actually a bit confusing, because these "factor levels" are often internally named 1 and 2, and then are recoded *again* when calculations are done. Phew!

7.2.2   A Categorical Predictor with More that Two Levels

If we have a predictor with more than two levels, not much changes from what we learned in the previous section. Having said that, we use a sort of trick to make a single variable with multiple levels into multiple variables with just 2 levels. These new variables are called *dummies*.

From the data set

```
                                  Colour
                                   red
                                   yellow
                                   blue
                                   blue
                                   red
                                   ...
```

we get

```
                    Colourred Colouryellow Colourblue
                        1          0            0
                        0          1            0
                        0          0            1
                        0          0            1
                        1          0            0
              ...
```

with the dummy-variables Colourred, Colouryellow and Colourblue, which are ordered alphabetically. Instead of a single response variable, we now have three![10]

For three colour variations of a single species of butterfly kept in an aviary, we measured how long it took for birds in the aviary to notice the butterflies and devour them. Apparently, yellow was the most conspicuous colour (Fig. 7.8). Waiting times usually have a very skewed distribution and such data are often analysed using a γ-distribution. The typical link function is then the inverse (see Tab. 7.1). On the other hand, if we apply a logarithm to the data, they look rather normally distributed (Fig. 7.8, right panel), which means you could also go this route. First, the GLM with the γ-distribution:

```
Coefficients:
Estimate Std. Error t value Pr(>|t|)
(Intercept)    0.106587    0.020837    5.115 3.82e-06 ***
Colourred     -0.009871    0.028136   -0.351    0.7270
Colouryellow   0.072080    0.040671    1.772    0.0817 .
---
Signif. codes:  0 `***' 0.001 `**' 0.01 `*' 0.05 `.' 0.1 ` ' 1

    Null deviance: 43.838  on 59  degrees of freedom
Residual deviance: 39.767  on 57  degrees of freedom
AIC: 372.25
```

As expected, the alphabetically first level (Colourblue) is the y-intercept, and the other two levels are shown relative to this. The difference in colour is apparently not significant, although you can see a slight preference for Colouryellow. That the slope for yellow is larger than the reference value for blue means that the values for yellow are *smaller*, since the link function is the reciprocal: Blue = $1/0.1066 = 9.4$, Yellow = $1/(0.1066 + 0.072) = 5.6$.

For the log-normal model, we get:

```
Coefficients:
Estimate Std. Error t value Pr(>|t|)
```

---

[10] All this happens behind the scenes of the statistical software: we do not have to do any dummy-coding ourselves.

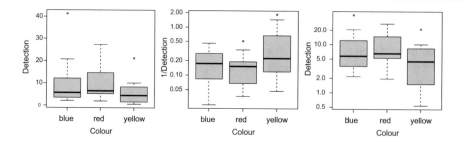

Fig. 7.8 Time until detection of three colour variations of a butterfly species by birds (in seconds) shown on the normal (left), inverse (middle) and log scale (right)

```
(Intercept)     1.9188      0.1917   10.008 3.65e-14 ***
Colourred       0.1574      0.2712    0.580   0.5639
Colouryellow   -0.6111      0.2712   -2.254   0.0281 *
---
Signif. codes:  0 `***' 0.001 `**' 0.01 `*' 0.05 `.' 0.1 ` ' 1


    Null deviance: 48.501  on 59  degrees of freedom
Residual deviance: 41.909  on 57  degrees of freedom
AIC: 156.74
```

With this model, `yellow` is detected earlier as well, this time even significantly earlier. The difference between `blue` and `red` remains insignificant.

In order to help us decide which of the two models we should trust, we can roughly estimate how well a fitted model will match the data. Just as with comparing two distributions, here we also want to choose the better, more appropriate model. In Chap. 3, we used either the log-likelihood or the AIC. Unfortunately, we are **not** able to use the AIC of the respective models in this case, because the models are based on different data (one using the measured data, and one using the logarithmic data).[11]

Instead, we compare the difference between the null deviance and residual deviance with the null deviance. This corresponds to the amount of deviance explained by the model: the more explained, the better the fit.

In this case with the butterflies, the log-normal model performed a bit better (14% instead of 9% explained deviance), and therefore, we would choose this model.

7.3   A Couple of Examples

What we just tiresomely learned is known as the generalised linear model (GLM) in the literature. It is linear, because our regression function is derived from linear (= additive) elements (i.e. $ax + b$ or $ax^4 + bx^2 + c$ or something similar), and it is generalised, because we don't have to use the normal distribution, but can choose from many options.[12] Today, the GLM is the standard solution for regression problems.

7.3.1   Height and Sex—A GLM Without The G

For 50 males and females respectively, height was measured. A histogram of this data does not indicate that there are differences in the sexes in this sample (Fig. 7.6, left panel).

---

[11]For this reason, the null deviance is also completely different. The null deviance denotes the variability in the data, whereas the residual deviance is the remaining variability *after* the regression. We'll learn much more about this in detail in the next chapter.

[12]The first regressions could only be done for normally distributed data because an analytical solution exists. The definitive publication by McCullough and Nelder (1989) expanded this to include many distributions and is accessible even to non-statisticians.

We now fit a model to these data using maximum likelihood assuming a normal distribution,[13] first without considering the sex of the subjects:

$$y \sim N(\mu, \sigma).$$

We get $\mu = 179.6$ and $\sigma = 11.52$. These data, when taking the parameter values as given, have an $AIC = 776.5$.

Now we have our regression model, in which we use sex as a predictor for height, by making the mean μ a function of sex:

$$y \sim N(\mu = aF + bM, \sigma),$$

where F and M are indicators for the sex "female" and "male", respectively. We get $a = 173.1$ and $b = 186.0$, $\sigma = 9.64$; and $AIC = 740.9$ (see output below). The distinction by sex leads to a much better fit (difference in AICs is Δ AIC $= 35.6$), and is also highly significant ($p \ll 0.05$). You can regard these two fits as two separate regressions (see box plot in Fig. 7.6). In the first case, we only fit a horizontal line ($=$ a grand mean), in the second, we fit a sex-specific mean value. A typical result output for the first fit would look something like this (here slightly edited output from R):

```
Coefficients:
            Estimate Std. Error t value Pr(>|t|)
(Intercept)  179.590      1.157   155.2   <2e-16 ***
---
Signif. codes:  0 `***' 0.001 `**' 0.01 `*' 0.05 `.' 0.1 ` ' 1


(Dispersion parameter for gaussian family taken to be 133.9615)


Residual deviance: 13262  on 99  degrees of freedom
AIC: 776.54
```

The grand mean is shown as the y-intercept in this case, the dispersion parameter for normally distributed data (called *gaussian* in R), is the variance. Therefore, σ is $\sqrt{133.96} = 12.9$.

The output for the model with separate values for females and males looks like this:

```
Coefficients:
            Estimate Std. Error t value Pr(>|t|)
(Intercept)  173.140      1.363 127.036  < 2e-16 ***
SexMale       12.900      1.927   6.693 1.37e-09 ***
---
Signif. codes:  0 `***' 0.001 `**' 0.01 `*' 0.05 `.' 0.1 ` ' 1


(Dispersion parameter for gaussian family taken to be 92.87694)


    Null deviance: 13262.2  on 99  degrees of freedom
Residual deviance:  9101.9  on 98  degrees of freedom
AIC: 740.89
```

The intercept value here shows the value for females (because the factor level Female comes before Male alphabetically). The value for Male is the difference from the intercept. The fitted mean for males is therefore $173.1 + 12.9 = 186.0$. The fitted standard deviation is now $\sqrt{92.88} = 9.64$.

With regression models, it is common to compare the fitted model with an "intercept-only" model (also called "null model"). The goal here is to reduce the remaining noise, the residual deviance, as much as possible with the fewest explanatory variables or factor levels. In this case, we only estimated one parameter for our intercept-only model (for this reason there are $100-1 = 99$ degrees of freedom left over), and for the second model, only two parameters were estimated (one mean for each sex, accordingly, there are 98 degrees of freedom left). With this one additional parameter, we reduce the residual deviance from over 13,000 to just above 9,000.

---

[13] A linear model; for mathematical formulas and analytical solutions, see Sect. 15.5 on p. 248.

For this example, we could actually go back to the historically older (and faster to calculate) linear model (LM). This would change nothing with regard to the results; however, it would allow us to get rid of the G in GLM in the results section of our study.

7.3.2   Smokers and Sex—The χ^2-test as a Binomial GLM

In Sect. 5.2 on p. 75 we analysed a dataset that we could also analyse with the help of a GLM. We wanted to explore the relationship between smokers and sex. To do this here, we need to reformulate the four numbers from the χ^2-test into the *long format*:

```
Sex         Smoker
Male          1
Male          0
... (14 times)
Female        1
... (12 times)
Female        0
... (26 times)
```

Now we can fit a binomial model (more precise: a Bernoulli-distribution), with sex as the predictor and smoker (yes or no) as the response variable.

Similar to the height-sex model from the last section, here we get:

```
Coefficients:
Estimate Std. Error z value Pr(>|z|)
(Intercept)  -0.7732     0.3490   -2.215    0.0267 *
sexMale      -1.8659     1.0923   -1.708    0.0876 .
---
Signif. codes:  0 `***' 0.001 `**' 0.01 `*' 0.05 `.' 0.1 ` ' 1

(Dispersion parameter for binomial family taken to be 1)

    Null deviance: 59.052  on 52  degrees of freedom
Residual deviance: 54.746  on 51  degrees of freedom
AIC: 58.746
```

The difference between the sexes is not significant (just as we found in Sect. 5.2 on p. 75). However, there are three things that differentiate this binomial GLM from the LM in the previous section:

1. The coefficients are on the link scale. We need to back-transform them in order to understand and interpret them. The y-intercept corresponds to the estimated probability that females smoke. According to Table 7.1, the inverse function is $e^{f(x)}/(1 + e^{f(x)})$. Since we only want to transform one value, $f(x) = x = -0.7732$. Thus, the back-transformed value is $e^{-0.7732}/(1 + e^{-0.7732}) = 0.316$. 12 of 38 females in our data set were smokers, which correlates to a proportion of 0.316. Aha! So for the males in this study, the value is equal to that of the females plus the deviation: $e^{-0.7732-1.8659}/(1 + e^{-0.7732-1.8659}) = 0.067$. The value is considerably smaller, but due to the small sample size of males, the difference is not significant.

2. The deviance is *not* the same as the variance![14]
3. Whereas with a normal distribution a dispersion parameter (the variance) is fitted, for *all other distributions* it is **assumed**. We now need to check whether a dispersion of 1 is plausible. To do this, we divide the residual deviance by the residual degrees of freedom, in this case 54.7/51. If this value is less than 1, then the dispersion parameter of 1 is fine.[15]

For more examples and explanations, see the next chapter on implementation in R.

Reference

1. McCullough, P., & Nelder, J. A. (1989). *Generalized linear models*. London: Chapman & Hall, 2nd edition.

---

[14] In a GLM, no variances are computed, but rather deviances. However, deviance does correspond to the sum of difference squares in ANOVA/regression, and is thus identical to these for normally distributed data (McCullough and Nelder 1989). For other distributions, the deviance is a bit more complicated, and for understanding further methods, being able to understand the exact mathematical definition is not important. I'll define it here anyway ($\hat{y}$ is the predicted value):

Normal distribution:	$\sum(y_i - \hat{y})^2$
Poisson-distribution:	$2\sum(y_i \log(y_i/\hat{y}) - (y_i - \hat{y}))$
Binomial-distribution:	$2\sum(y_i \log(y_i/\hat{y}) + (n_i - y_i)\log((n_i - y_i)/(n_i - \hat{y})))$
Gamma-distribution:	$2\sum(-log(y_i/\hat{y}) + (y_i - \hat{y})/\hat{y})$
Inverse normal distribution (inverse Gaussian):	$\sum(y_i - \hat{y})^2/(\hat{y}^2 y_i),$

where n_i in a binomial distribution is the number of trials for the respective value (i.e. 1 for 0/1 = Bernoulli-distributed data). As you can see from this formula, the deviance is derived from the log-likelihoods of the respective distributions, or more precisely, from the difference between the actual model and the maximal model.

[15] There is no dispersion parameter for simple Bernoulli-models.

He uses statistics as a drunken man uses lamp-posts—for support rather than illumination.

Andrew Lang

At the end of this chapter...
... a GLM with a single predictor will flow easily from your fingertips.
... you will be able to interpret the output of the `summary(glm(.))` function.
... you will be capable of drawing the regression line and confidence interval in a plot of data points from a `glm`-analysis.
... you will be able to calculate a GLM using an optimisation approach or `vglm`, when necessary.

8.1 Regression Using GLM

A regression using maximum likelihood is called a *Generalised Linear Model* (GLM). The R function has the same name: `glm`. Within this function, you can specify the independent and the dependent variable, the distribution, and a number of optional arguments. In order to show the regression relationship, we use the formula syntax: $y \sim x$. Using this notation, we say that y is a function of x.

Let's have a look at an example. The volume of a tree depends on its height and its width. To estimate width, it is common to only use the diameter at breast height (DBH, in the old days referred to as "girth"). For each tree species, we need a calibration curve in order to convert girth to cubic meters of wood. An appropriate dataset for 31 black cherry trees can be found in R under the name `trees`. We load the data and have a look at it in a scatterplot (Fig. 8.1)[1]: trees

```
> data(trees)
> ?trees
> par(mar=c(4,5,1,1))
> plot(Volume ~ Girth, data=trees, las=1, cex.lab=1.5, pch=16, cex=1.5,
+    xlab="Girth [in]", ylab="Volume [cubic ft]")
```

With `data` we specify where the variables can be found; `las` rotates the axis labels; `cex.lab` defines the size of the axis labels; `pch` defines the type of symbol, `cex` defines the size of the symbol, and `xlab`/`ylab` specify the axis labels. Graphical elements are explained in the help page for `par` (`?par`).

The relationship here looks very linear, so we will fit a straight line. Or, more precisely, we model the mean of the normal distribution, from which the measured points are a random draw, as a linear function of the girth at breast height: **Volume** $\sim N(\mu = a \cdot$ **Girth** $+ b, \sigma)$. In R the formula looks like this: `Volume ~ Girth, family=gaussian`. We do not need to specify the y-intercept, because it is included automatically (in order to remove it, we would need to include a `-1`

---

[1]A "proper" metric version of this data set is available as "tree.metric.csv" among the data for this book.

© Springer Nature Switzerland AG 2020
C. Dormann, *Environmental Data Analysis*,
https://doi.org/10.1007/978-3-030-55020-2_8

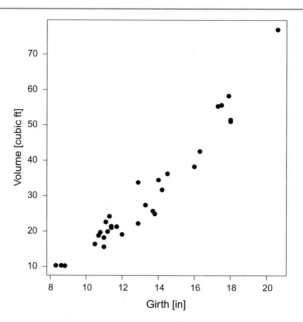

Fig. 8.1 Volume of black cherry trees as a function of girth

or a +0 after `Girth` in the formula above). In the `glm` function of R, the normal distribution is called `gaussian`, and instead of distribution, we write `family`. Only a few distributions are available for use in the `glm` function in R (see `?family`): `gaussian`, `poisson`, `binomial` and `gamma`.

We fit the line as follows:

```
> fm <- glm(Volume ~ Girth, data=trees, family=gaussian)
> summary(fm)

Call:
glm(formula = Volume ~ Girth, family = gaussian, data = trees)

Deviance Residuals:
   Min      1Q   Median      3Q      Max
-8.065   -3.107    0.152   3.495    9.587

Coefficients:
            Estimate Std. Error t value Pr(>|t|)
(Intercept) -36.9435     3.3651  -10.98 7.62e-12 ***
Girth         5.0659     0.2474   20.48  < 2e-16 ***
---
Signif. codes:  0 `***' 0.001 `**' 0.01 `*' 0.05 `.' 0.1 ` ' 1

(Dispersion parameter for gaussian family taken to be 18.0794)

    Null deviance: 8106.1  on 30  degrees of freedom
Residual deviance:  524.3  on 29  degrees of freedom
AIC: 181.64

Number of Fisher Scoring iterations: 2
```

In the next chapter, we will more intensively explore whether the assumptions are met and what the numbers under `Deviance Residuals` tell us. For the time being, however, the `estimate` for `Intercept`, slope of the effect of girth at breast

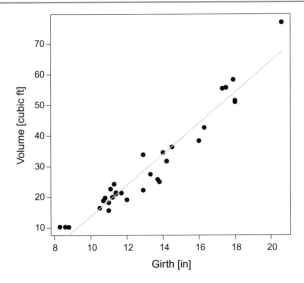

Fig. 8.2 Volume of black cherry trees as a function of girth with a fitted calibration curve

height (`Girth`), and the estimated standard deviation (ultimately, we also fit σ as part of our model!) are of interest. The `dispersion parameter` of 18.08 corresponds to the estimated value for the variance of the underlying normal distribution, i.e. σ^2.[2]

Each estimate of the model, i.e. the estimate for the intercept and that for the slope, is compared to its standard error and tested for significance using a t-test. From this, you may deduce that the estimated values appear to be normally distributed, and this also happens to be the case:

The parameters of the generalised linear model (a and b in this case) are normal distributed on the link scale.

Since the canonical link function for the normal distribution is the identity function, the estimates on the link and response scales are the same. What the t-tests tell us here is that both the intercept and the slope are significantly different from 0. In other words, the relationship is highly significant. The comparison of null and residual deviance shows that we can explain $(8106 - 524)/8106 = 0.94$ or 94% of the variance in the data. Now we can place the fitted line in the plot (Fig. 8.2).

The simplest way to draw this line would be with the command `abline(fm)`. Using this command, however, the line would be drawn all the way to the edge of the plot window. This is neither common practice, nor correct. We only want to make a statement about the range for which we have values, not beyond it!

The method described below is more complicated, but is (unfortunately) the normal procedure in R. It involves three steps:

1. We first define a data sequence (`seq`) for which we want to calculate the points of the calibration function and store these numbers in an object (here: `Girthnew`). In this case, two data points would have been sufficient, but with curved lines you would want at least 50. The sequence goes from the minimum value to the maximum value of the explanatory variable.[3]
2. For these new values, we can now calculate the model predictions using the function `predict`. This function is applied to the previously stored GLM object (here: `fm`)[4] The `newdata` argument is then used to create a `data.frame` containing the values f or which the model predictions are to be calculated. The variable names must be entered exactly as they appear in the GLM formula (i.e. `Girth` not `Girthnew`). Of course, we could also calculate this step by hand. For the lowest value, 8.3, the model gives a volume of $-36.9 + 8.3 \cdot 5.07 = 5.18$. We can then do the same calculation for the highest value: $-36.9 + 20.6 \cdot 5.07 = 67.54$. We can now connect these two points, (8.3, 5.18) and (20.6, 67.54), with a line.
3. Lastly, we draw a line using the command `lines`. The first argument of this command is the x value, then the y value, and finally, we can enter parameters to spruce up the thickness and colour of our line.

---

[2]Caution: This is not the variance of the sample, i.e. $s^2 = \frac{\sum (x-\bar{x})^2}{n-1}$, but rather the population variance: $\sigma^2 = \frac{\sum (x-\mu)^2}{n}$.

[3]If there are any NAs here, then you need to expand the argument in R for `min` and `max` to `min(trees$Girth, na.rm=T)`.

[4]`fm` stands for *fitted model* and is a typical innocent name for model objects. You could also give it any other name that you could think of but avoid repetitions and using names already defining an R-function (such as `c, C, D, F, I, q, t, T`)!

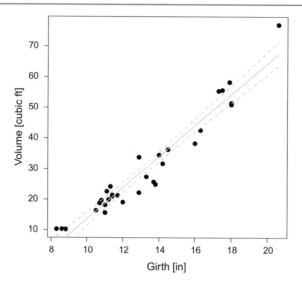

Fig. 8.3 Volume of cherry trees as a function of girth with a fitted calibration curve and its 95% confidence interval

```
> Girthnew <- seq(min(trees$Girth), max(trees$Girth), len=101)
> preds <- predict(fm, newdata=data.frame("Girth"=Girthnew))
> plot(Volume ~ Girth, data=trees, las=1, cex.lab=1.5, pch=16, cex=1.5)
> lines(Girthnew, preds, lwd=2, col="grey")
```

Ideally, we would show an error interval in which the line likely lies, i.e. graphically show the standard errors of the estimates. To do this, we must remember that the 95% confidence interval of a sample estimator lies ± 2 standard deviations around the estimate (Chap. 1). Now we tell R to output not only the value for the calibration curve of the estimated values, but also the corresponding standard deviation. We then double these values and draw them in our plot above and below the curve (Fig. 8.3):

```
> preds2 <- predict(fm, newdata=data.frame("Girth"=Girthnew), se.fit=T)
> str(preds2)

List of 3
 $ fit           : Named num [1:101] 5.1 5.73 6.35 6.97 7.6 ...
  ..- attr(*, "names")= chr [1:101] "1" "2" "3" "4" ...
 $ se.fit        : Named num [1:101] 1.44 1.42 1.39 1.37 1.34 ...
  ..- attr(*, "names")= chr [1:101] "1" "2" "3" "4" ...
 $ residual.scale: num 4.25

> lines(Girthnew, preds2$fit +2*preds2$se.fit, lwd=2, lty=2, col="grey")
> lines(Girthnew, preds2$fit -2*preds2$se.fit, lwd=2, lty=2, col="grey")
```

In 95% of cases will the true regression line lie between these narrow lines.[5] This does not mean that each new measured value will also lie within these boundaries! In reality, some data points will be outside of these bounds. The confidence interval just plotted refers to the fitted line, not to the measured values. We should make a distinction between a confidence interval (shown here) and a prediction interval for new data points (which is much wider). For the latter we must add the normal distribution with the standard deviation specified in the model to the regression error.

---

[5]This is the shortest formulation! The full version of this sentence is substantially longer, and more confusing. Here it goes: "This 95%-confidence interval will be overlapping with the truth in 95% of the cases, if we were to repeatedly draw samples from the underlying true, but unknown distribution." The second half of this sentence is often abbreviated to "under repeated sampling". The confidence interval was introduced by Jerzy Neyman and remains a source of great confusion and misinterpretation. Note, for example, that the 95%-CI is a statement about the regression line, not about the underlying truth which we typically are interested in.

Note that we are making predictions on the link scale. If the link function is not the identity function, then we need to back-transform our predicted values. We will do just that in the next example.

Our next example for a regression comes from ecotoxicology. In this field, 'bioassays' are often used to test the toxicity of a substance for test organisms. For such experiments, a series of dilutions of the substance in question is made, and then a number of organisms (e.g. water fleas) are added to the solution. After a certain amount of time (e.g. 2 h), the number of organisms still alive are counted.

Data from a bioassay contain three variables: 1. The total number of organisms put into the solution (**totals**); 2. The number of surviving organisms after 2 hour (**y**); 3. The concentration of the substance in the solution (**S**).

We assume that these data follow a binomial distribution (Sect. 3.4 on p. 42):

$$\mathbf{y} \sim \text{Binom}(n = \textbf{totals}, \, p = \text{logit}^{-1}(a\mathbf{S}+b)) = \text{Binom}\left(n = \textbf{totals}, \, p = \frac{e^{a\mathbf{S}+b}}{1+e^{a\mathbf{S}+b}}\right)$$

The number of survivors depends on the number of organisms put into the solution, as well as on the survival probability, which is modeled as a function of the substance concentration (connected by the logit-link).

In R, the analysis of binomial data is performed as follows (see ?glm, section "Details"): First we need to express our dependent variable (the one to the left of the tilde $\sim$ in the formula above) as a two column matrix of *successes* and *failures*:

```
> binodat <- read.csv("binomialdata.csv")
> combi.y <- cbind(succ=binodat$alive, fail=binodat$total-binodat$alive)
> combi.y
```

```
       succ fail
 [1,]   10    0
 [2,]   18    0
 [3,]   20    0
 [4,]   11    1
 [5,]   11    0
 [6,]   10    0
 [7,]    8    3
 [8,]   15    2
 [9,]    5    5
[10,]    3    7
[11,]    3   11
[12,]    4   14
[13,]    1   13
[14,]    1   13
[15,]    1   11
```

Now we plot these data by plotting the proportion of survivors against the concentration of the substance (Fig. 8.4). We already have an expectation of what should come out of the analysis: a negative coefficient for *S*.

```
> fm <- glm(combi.y ~ S, binomial, data=binodat)
> summary(fm)
```

```
Call:
glm(formula = combi.y ~ S, family = binomial, data = binodat)

Deviance Residuals:
    Min      1Q   Median      3Q      Max
-1.0611  -0.6261   0.2903  0.8255   1.4012

Coefficients:
```

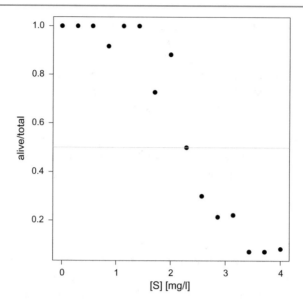

Fig. 8.4 Proportion of survivors as a function of substance concentration S (abbreviated, in chemistry, as "[S]"). This figure hides the fact that the number of individuals placed in each substance varies and that the data points are therefore differently variable. The grey line shows a survival proportion of 50%

```
            Estimate Std. Error z value Pr(>|z|)
(Intercept)   5.7031     0.8062   7.074 1.51e-12 ***
S            -2.3135     0.3132  -7.387 1.50e-13 ***
---
Signif. codes:  0 `***' 0.001 `**' 0.01 `*' 0.05 `.' 0.1 ` ' 1

(Dispersion parameter for binomial family taken to be 1)

    Null deviance: 157.141  on 14  degrees of freedom
Residual deviance:  10.734  on 13  degrees of freedom
AIC: 38.803

Number of Fisher Scoring iterations: 5
```

And our expectations are confirmed. From these estimates, we learn very little at first. We only see that a higher concentration of the substance has a significantly negative effect on the survival of the organisms.

A typical value of interest for toxicologists is the substance concentration at which half of the organisms survive (LD_{50}: lethal dose for 50% of the individuals). In R we can calculate this value (or other survival proportions) with the function dose.p (Package **MASS**)[6]:

```
> dose.p(fm, p=0.5)

            Dose        SE
p = 0.5: 2.465162 0.09896017
```

This means that at a concentration of 2.47 ± 0.099 mg/l of the substance in water, half of the organisms survive.

Now we also want to plot the fitted regression with a 95%-confidence interval. To do this, we first predict the expected values for 100 new values in the range of S, together with standard errors:

---

[6]The argument p=0.5 specifies that we are interested in the 50% survival.

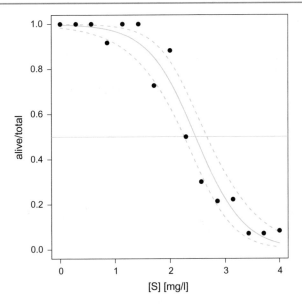

Fig. 8.5 Proportion of survivors as a function of substance concentration *S*. Fitted model (solid grey line) with 95% confidence interval (dashed lines)

```
> newS <- seq(min(binodat$S), max(binodat$S), len=100)
> preds <- predict(fm, newdata=data.frame(S=newS), se.fit=T)
```

This forms the basis for our plot. To make our plot, we first create an empty plot (`type="n"`) and draw a midline and 95%-confidence interval lines. Only after this do we draw our points, so that they are drawn on top of the lines (Fig. 8.5):

```
> par(mar=c(5,5,1,1))
> plot(alive/total ~ S, type="n", data=binodat, las=1, cex.lab=1.5,
+    ylim=c(0,1), xlab="[S] [mg/l]")
> lines(newS, plogis(preds$fit), lwd=2, col="grey")
> lines(newS, plogis(preds$fit + 2*preds$se.fit), lwd=2, col="grey", lty=2)
> lines(newS, plogis(preds$fit - 2*preds$se.fit), lwd=2, col="grey", lty=2)
> abline(h=0.5, col="grey")
> points(alive/total ~ S, data=binodat, pch=16, cex=1.5)
```

The function `plogis` back-transforms values from the link- (here logit) to the response-scale.

8.2 Regression: Maximum Likelihood by Hand

In the previous section, we implemented a regression with a GLM. The GLM saves us the nitty-gritty work of specifying the model to be fitted and then optimising the parameters with maximum likelihood. Indeed, it is not usually necessary to do these steps ourselves (that is, we don't need to calculate a GLM "by hand"). Sometimes, however, we have no choice. This could be because the distribution is not covered by the function `glm`, or because the relationship to be modelled is non-linear. In the following, we will therefore go through an easier and then a more complex example of calculating a GLM by hand. In the process, we will learn some new skills with R, which may be challenging at first, like learning the vocabulary and grammar rules of a new language. Whenever a new function appears, we can look up the syntax and arguments using the help ("?") function.

8.2.1 Poisson Model by Hand

Here we will use the data set that we became familiar with in Sect. 7.1 on p. 89: the Poisson-distributed number of food pieces delivered to the nest by male collared flycatchers as a function of their attractiveness.

For a Poisson-distribution, the likelihood function takes the following form:

$$\mathscr{L}(\lambda|\mathbf{y}) = \frac{\lambda^{y_1}}{y_1!e^\lambda} \cdot \frac{\lambda^{y_2}}{y_2!e^\lambda} \cdots \frac{\lambda^{y_n}}{y_n!e^\lambda} = \frac{\lambda^{\sum y_n}}{y_1!\cdots y_n!e^{n\lambda}} \tag{8.1}$$

We take the log to get the *log-likelihood*:

$$\ln \mathscr{L} = \ell = \sum_{i=1}^{n} \left(-\lambda + (\ln \lambda) \cdot y_i\right) - \ln\left(\prod_{i=1}^{n} y_i!\right)$$

The next step is the formulation of the regression equation on the link scale: $y' = \beta_0 + \beta_1 x$. Since the log is the canonical link function for the Poisson-distribution (Table 7.1 on page 85), we calculate the actual value using the inverse function:

$$y \sim g^{-1}(y') = g^{-1}(\beta_0 + \beta_1 x) = e^{\beta_0 + \beta_1 x}.$$

We see that we need to calculate two parameters simultaneously (the two βs). Let us assume, for the sake of simplicity, that we knew that the y-intercept had the value $\beta_0 = 1$, and the slope had the value $\beta_1 = 0.1$. Knowing this, we can predict an attractiveness value for each observed y-value (`pieces`):

```
> fc <- read.table("flycatcher.txt")
> attach(fc)
> pieces

 [1]   3   6   8   4   2   7   6   8  10   3   5   7   6   7   5   6   7  11   8  11  13  11   7   7   6

> exp(1 + attract * 0.1)

 [1] 3.004166 3.004166 3.004166 3.004166 3.004166 3.320117 3.320117
 [8] 3.320117 3.320117 3.320117 3.669297 3.669297 3.669297 3.669297
[15] 3.669297 4.055200 4.055200 4.055200 4.055200 4.055200 4.481689
[22] 4.481689 4.481689 4.481689 4.481689
```

These values are our model predictions $\hat{y}$ for the coefficient values $\beta_0 = 1$ and $\beta_1 = 0.1$. In order to calculate the likelihood of these values, we put the values into Eq. 8.1. That is, the $\hat{y}$ corresponds to the λ, since our expected value varies with attractiveness.

Since this is such an important concept, here it is again in other words: For each data point y_i, we can now calculate the probability that it originates from a Poisson-distribution with a certain mean value λ_i. This is calculated in R as `dpois`(y_i, λ_i). We then take the log of these and sum them. We have now obtained our log-likelihood. Note that we can link the observed values to the model with the log link function!

```
> sum(dpois(pieces, exp(1 + 0.1*attract), log=T))

[1] -84.40658
```

Using the argument `log=T` in the d...-functions, we can directly log-transform the probability density.

Now we vary the values for β_0 and β_1 a little to see what happens with the log-likelihood:

```
> sum(dpois(pieces, exp(1.1 + 0.2*attract), log=T))

[1] -59.43922
```

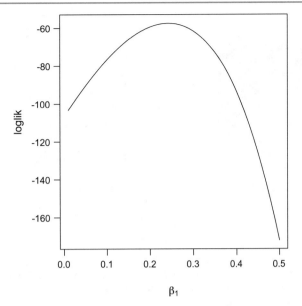

Fig. 8.6 Log-likelihood of the different values for β_1 given that $\beta_0 = 1.1$. A figure such as this is called a likelihood-profile

It gets better (i.e. less negative).

Now we will repeat this process for a series of values for β_1, from 0.01 to 0.5, and plot the log-likelihood sums to produce Fig. 8.6.

```
> loglik <- 1:50 # a vector where the results will be inserted
> beta1 <- seq(0.01, 0.5, len = 50)
> for (i in 1:50){
+    loglik[i] <- sum(dpois(pieces, exp(1.1 + attract * beta1[i]),
+    log=T))
+ }
> par(mar=c(5,5,1,1)) # reduces the whitespace of the border
> plot(beta1, loglik, type = "l", xlab = expression(beta[1]), cex.lab = 1.5,
+    las=1)
```

This chunk of R-code contains two elements that are new for us: in the first row, we created an object named `loglik`, in which we first stored the values 1 to 50. This will later take the values of the log-likelihoods, but needs to exist first. In rows 3–6, we use a `for`-loop, in order to calculate the log-likelihood for all values of `beta1`.

According to the result, depicted in Fig. 8.6, the maximum log-likelihood (for a given $\beta_0 = 1.1$) with $\beta_1 \approx 0.23$ is around -58. Now we must try to find the optimal value for β_0 as well. Since β_0 and β_1 are likely dependent on each other, we need to optimise both values at the same time.

Now we will do this systematically in two different ways. First we will vary both coefficients along an equidistant grid of values. The result is a three-dimensional surface that should resemble a mountain. Secondly, we use an optimisation method to pinpoint the best values.

Let's start by choosing 100 values for the y-intercept and slope from 0.1 to 2 and 0.01 to 0.5, respectively, and then calculating the Poisson-likelihood for each combination of these values.[7] Then we can plot the result (Fig. 8.7): first three-dimensionally using `persp`, then using a clearer two-dimensional visualisation with `contour`.[8]

```
> beta0 <- seq(0.1, 2, length = 100)
> beta1 <- seq(0.01, 0.5, length = 100)
```

---

[7]This would be possible using less code with `outer`, but the `for`-loops make more sense didactically.

[8]There is also another variation of this, `filled.contour`, which uses bright color gradients instead of contours.

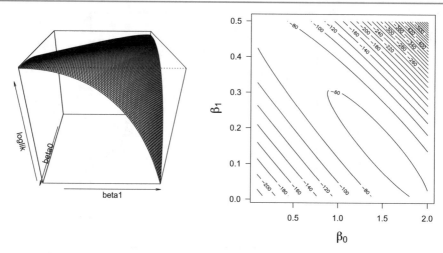

Fig. 8.7 3D- and contour graphics of the log-likelihood calculation for the flycatcher data. We can clearly see that the coefficients are not independent of one another: the higher the value of one, the lower the value is for the other. For the correlation between axis intercept and slope, this is practically always the case

```
> llfun <- function(parms) {
+       # calculates LL for given values of parms
+       sum(dpois(pieces, exp(parms[1] + parms[2]*attract), log=T))
+ }
> loglik.m <- matrix(ncol = 100, nrow = 100)
> for (i in 1:100) {
+     for (j in 1:100) {
+         loglik.m[i, j] <- llfun(c(beta0[i], beta1[j]))
+     }
+ }
> par(mfrow = c(1, 2))
> persp(beta0, beta1, loglik.m, phi = 30, theta = 90, xlab = "beta0",
+     ylab = "beta1", zlab = "loglik")
> contour(beta0, beta1, loglik.m, nlevels = 30, xlab = expression(beta[0]),
+     ylab = expression(beta[1]), cex.lab = 1.5)
```

Another bit of new R-grammar for us: writing a function ourselves! In row three we define a new function (using `function`), which we then use in the double-nested `for`-loop. In this case, our function `llfun` is defined in dependence of a parameter vector `parms`, whose first and second value are placed in the body of the function definition for β_0 and β_1. **`llfun` is the heart of our approach. This is where the model is defined!** We can give this function a vector with two arbitrary values, and it calculates the log-likelihood of this combination.[9]

Now, we would like to have the maximum likelihood values for β_0 and β_1. In order to extract these values, we use the `which` function:[10]

```
> loglik.m[which(loglik.m) == max(loglik.m)]
```

```
[1] -55.71565
```

```
> (llmax <- which(loglik.m == max(loglik.m), arr.ind = T))
```

---

[9]Go on, try it out! `llfun(c(-1, 0))`.

[10]With `which.max(loglik.m)` (and its analog `which.min`) we can directly obtain the point of the maximum value. Unfortunately, these commands do not have the option `arr.ind`, which is necessary for our purposes here.

```
     row col
[1,]  73  28
```

The argument `arr.ind=T` returns the position of the desired values in the form of rows and columns. Without `arr.ind=T`, we obtain the position only as a "running number" of the matrix. When we have both elements of `llmax` (i.e. the row number and the column number of the maximum value) we can move to the next step:

```
> beta0[llmax[1]]
```

```
[1] 1.481818
```

```
> beta1[llmax[2]]
```

```
[1] 0.1436364
```

The numbers we want are then $\hat{\beta}_0 = 1.48$ and $\hat{\beta}_0 = 0.14$.[11] (The accuracy depends on the density of the grid, for which we calculated the log-likelihood. In our case here, more than two decimal places are unnecessary.) Thus, we identified the maximum likelihood estimates for β_0 and β_1 as $\hat{\beta}_0 = 1.48$ and $\hat{\beta}_1 = 0.14$, respectively. The downside of this approach is that we have to be reasonably certain, in which interval to find the optimal value (for we use it to define the sequence of values for `beta0` and `beta1`). Also, for higher accuracy, we need *far* finer grids, making the algorithm very inefficient (since it also computes a lot of values far away from the optimum). Such an approach is called "greedy" (because it uses up a lot of computing time), or "brute force" (since there is no smart way about it, only a simple grid search).

Now we come to the second approach, where instead of the regular sequence of values, we use an optimisation procedure. The nice thing about this approach is that it jumps in large steps towards the optimum, and as it approaches it, the steps become finer. However, the optimisation also requires multiple steps:

1. We need a **function to be optimised** (score function or goal function). This must take the parameters to be optimised as an input vector and output a single number as a result of the calculation. So, the skeleton of this function looks like this:

   ```
   optfun <- function(parms, ...){
        res <- do.something(parms, ...)
        return(res)
   }
   ```

   The three dots (`...`, or "ellipsis") stand for arguments that are passed on to internal functions. These are typically data to be used in the function. The internal function `do.something` is used here as a placeholder for a density function, for example. What is important here is that `optfun` is written so that for all possible values of `parms`, a real value emerges as a result. Otherwise, `optim` may try a value that delivers an `NA` and then aborts the optimisation altogether.
2. We need to identify **starting values for the parameter to be optimised**, `parms`. You can use any number for this, as long as it delivers a real value with `optfun`. It is perhaps best to test whether the numbers will work before the optimisation is started: for example, `optfun(c(1,2,3))`, if three parameters are to be optimised.
3. Select the **optimisation process and the optimisation conditions**. There are dozens of optimisation algorithms, and five are provided by `optim`. It would be a bit much to explain all five of them here in detail, but it may be useful to know that only one of them, `method="L-BFGS-B"`, allows us to set a minimum and maximum for the optimisation area. All other algorithms assume that all real values for `parms` are fair game. The default setting is `method="Nelder-Mead"`, whereas `method="BFGS"` is a faster process (see Side Note 5).
   Finally, within `optim` you can define whether you want to minimise (default setting) or maximise (in the argument option: `fnscale`), set the maximum number of iterations to do (`maxit`) and define how exact the solution should be (`reltol`).

If you consider how complex the optimisation is internally, the code syntax here is surprisingly simple:

```
> optim(par=c(1, 0.1), fn=llfun, control=list(fnscale=-1))
```

---

[11] The "hat" above a parameter (e.g. $\hat{\beta}$) indicates that it has been estimated.

```
$par
[1] 1.4745978 0.1479505

$value
[1] -55.70837

$counts
function gradient
      61       NA

$convergence
[1] 0

$message
NULL
```

We only need to provide `optim` with three arguments: 1. Starting values for the function (`par`); this can be anything, but has to deliver a value (no NA). 2. The function to be maximised (`fn`); this **must** be defined in such a way that a vector with parameters is passed as the first argument of the function. If we wanted to provide data to the function, this would have to be the second argument of the function, not the first! 3. A declaration as to whether we should maximise or minimise. The default setting is minimisation, but we want to maximise. Therefore, we simply multiply the result of `fn` with −1. This is coded by changing the optimisation control option `fnscale`.[12]

Side note 5: Comparing the speed of two algorithms

R also has a built-in stopwatch that can be called with the command `system.time`. We can use it to measure the amount of time it takes to complete the above calculation:
```
> system.time(o1 <- optim(par=c(1, 0.1), fn=llfun, control=list(fnscale=-1),
+ hessian=T))
user system elapsed
0.007 0.000 0.008
```
This calculation required 0.008 seconds. With the BFGS-algorithm, we can do it even faster:
```
> system.time(o2 <- optim(par=c(1, 0.1), fn=llfun, method="BFGS",
+ control=list(fnscale=-1), hessian=T))
user system elapsed
0.002 0.001 0.002
```

The output here might be confusing, since the naming goes against some conventions. First, under `$par`, we get the optimal parameter values (i.e. $\hat{\beta}_0$ and $\hat{\beta}_1$). These should now be familiar to us by now. Under `$value`, we get the optimal function value, or our log-likelihood of −55.7. Under `$counts` we see that the algorithm required 61 function calls to find the optimum (information that doesn't really help us in any way). Under `$convergence`, `optim` tells us whether the algorithm converged, that is to say, that it found a final value. *Everything other than 0 tells us that we have a problem!* It is annoying that a "0" tells us that everything is OK; a "1" (= TRUE) would really be more logical.[13] Finally, any error messages are shown under `$message`. Here, we are happy if we see the word NULL.

---

[12]It is much more common to see a minus sign in the target function, such as `res <- - do.something(.)`. I find this to be confusing without a proper explanation, and therefore prefer the more cumbersome `fnscale=-1`.

[13]Simply changing this in the next release of R is not an option! Imagine all code that uses `optim` and checks for convergence would all a sudden break and have to be re-written. Doable, but a nightmare for the R Core Team. Personally, I'd love it.

Side note 6: Quantifying the standard errors of the parameters to be optimised.

The argument `hessian=T` gives us the data that we need in order to compute the standard errors of the parameters that we want to estimate. Without going into to much detail, the Hessian is the matrix of the second-order partial derivatives (named after a mathematician with the German-most name possible: Ludwig Otto Hesse). If you invert the inverse, then the variances of the parameter estimators are on the diagonal. (If we carried out a maximisation, the negative variances are on the diagonal.) We can then take the root of these values to get the standard deviation (which for estimates is called a standard error). These should correspond to the output of the GLM.

(Here, understanding isn't the main point, but rather the craft of it: *how* can we get the standard error, not *why* do we do it this way. For those who want a more technical explanation: for the maximum of a function, its first derivative is 0 and its second is negative. The second derivative tells us how steep it is around the optimum, or how steep the gradient is. The steeper it is, the smaller the possible error: it is easy to find the peak. During calculation, the matrix is inverted so that the Hesse matrix is the inverse of the gradient derivative.)

   Let's try it out:

```
> sqrt(diag(solve(-o1$hessian)))
[1] 0.19442347 0.05436619
```

And with the BFGS-algorithm:

```
> sqrt(diag(solve(-o2$hessian)))
[1] 0.19442645 0.05436748
```

The standard error according to the GLM is:

```
> summary(glm(pieces ~attract, poisson))$coefficients
            Estimate Std. Error z value  Pr(>|z|)
(Intercept) 1.4745872 0.19442732 7.584259 3.343919e-14
attract     0.1479366 0.05436782 2.721033 6.507822e-03
```

These values are identical to the fifth decimal place. The reason that they vary after that has to do with the choice of the optimisation algorithm.

8.2.2   Non-linear Regression by Hand

The last few pages were drier than toast. Why do we need all that, if the GLM simply produces a result for us? Because not everything works with a GLM. There are two cases where the standard implementation of a GLM cannot be handled in R: (1) non-standard distributions (in reality, there are "only" four available distributions for `glm`: normal, Poisson, binomial and γ);[14] and (2) non-linear functional relationships between response and predictor.

   Therefore, we will now take a look at an example of a non-linear regression. An often used non-linear regression is the Michaelis-Menten regression from the field of enzyme kinetics. In this case, the function starts at some low value (often 0), increases and approaches an asymptote. The Michaelis-Menten-kinetic is represented by the following formula:

$$\mathbf{y} = f(x) = \frac{v_{\max}\mathbf{x}}{\mathbf{x} + k_M} + B,$$

where $\mathbf{y}$ is the concentration of the product, $\mathbf{x}$ is the concentration of the substrate, $v_{\max}$ is the maximum rate of turnover (the asymptote) and k_M is the concentration at which half of $v_{\max}$ is reached (the 50% saturation concentration). B is a constant that represents the turnover without substrate. According to Selwyn [1995], B is an important and essential component, which is unfortunately left out in many textbooks.

   Now let's fit the Michaelis-Menten-kenetics for an example dataset. We will assume that the turnover rate, $\mathbf{v}$, comes from a normal distribution, whose mean is a function of the substance concentration, $\mathbf{S}$. Or, in formal notation:

$$\mathbf{v} \sim N(\mu = \frac{v_{\max}\mathbf{x}}{\mathbf{x} + k_M} + B, \sigma).$$

---

[14]This shortcoming is resolved with the use of additional packages, especially **VGAM** see Sect. 8.3 on page 111.

We now follow the three optimisation steps that we just learned in order to fit this dataset with maximum likelihood after we load the data:

```
> library(nlstools)
> data(vmkm) # Example data for the Michaelis-Menten kinetics
> attach(vmkm)
> head(vmkm)

    S    v
1 0.3 0.17
2 0.3 0.15
3 0.4 0.21
4 0.4 0.23
5 0.5 0.26
6 0.5 0.23


> # 1. Optimisation function
> llfun <- function(parms){
+   mu <- parms["vmax"]*S/(parms["Km"] + S) + parms["B"]
+   res <- sum(dnorm(v, mean=mu, sd=exp(parms["sigma"]), log=T))
+   return(res)
+ }
> # 2. Starting value:
> initial <- c("vmax"=0.7, "Km"=0.2, "B"=0, "sigma"=1)
```

Note that we set the estimate of σ into the exponent, to ensure that the resulting value is always positive (i.e. just like a link-function). `initial` must be a named vector, since we can work with it only if it has a name (e.g. `parms["vmax"]`)!

```
> # 3. Optimisation settings:
> optim(par=initial, fn=llfun, method="BFGS", control=list(fnscale=-1),
+   hessian=T)

$par
      vmax         Km          B       sigma
2.12493217  4.70371567  0.04994101  -3.75884639

$value
[1] 51.47793

$counts
function gradient
108       35

$convergence
[1] 0

$message
NULL

$hessian
              vmax            Km             B         sigma
vmax  -1.500191e+03   517.99083563  -7.102810e+03   0.003146408
Km     5.179908e+02  -179.93477544   2.531017e+03  -0.001141080
B     -7.102810e+03  2531.01718529  -4.048696e+04   0.012418786
sigma  3.146408e-03    -0.00114108   1.241879e-02  -44.000213673
```

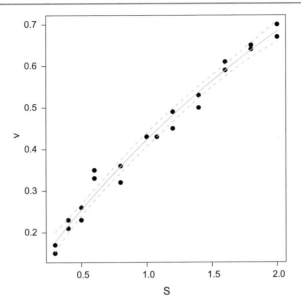

Fig. 8.8 Michaelis-Menten-kinetics for the vmkm data set from **nlstools**. The solid line is drawn with the parameters of the maximum likelihood estimate, the dashed lines are 95% confidence intervals

The parameters for our Michaelis-Menten-kinetic are now readily extractable, but we have to back-transform the estimate for σ as $\hat{\sigma} = e^{-3.76} = 0.023$.

Next we calculate the asymptotic parameter standard error (see Side Note 6)

```
> sqrt(diag(solve(-op$hessian)))
```

```
     vmax          Km          B       sigma
0.68545903  2.34513041  0.02944576  0.15075531
```

From this, we can use a t-test to test for significance:

```
> op$par/sqrt(diag(solve(-op$hessian)))
```

```
     vmax          Km          B       sigma
 3.100013     2.005737    1.696034  -24.933427
```

The t-values are significant if the absolute value lies above qt(0.05, 18, lower.tail=F) $= -1.734$.[15] This is the case for $v_{\max}$ and K_m, but not for B.

And finally we plot the data and the fit (Fig. 8.8):

```
> par(mar=c(5,5,1,1))
> plot(v ~ S, data=vmkm, las=1, pch=16, cex=1.5, cex.lab=1.5)
> parms <- op$par # we assign the optimal parameters to parms
> curve(parms["vmax"]*x/(parms["Km"] + x) + parms["B"], add=T, lwd=2,
+    col="grey")
```

Computing confidence intervals for non-linear regression fits, particularly those gained from applying the optim function, is not a trivial task. In fact, we need to deploy the so-called "Delta method", which is based on a Taylor-series approximation of the asymptotically normally distributed difference between an estimator and its true value.[16] The key equation we need

---

[15]There are 22 data points and we fit four parameters, therefore we have 18 degrees of freedom left over. lower.tail=F forces the specification of the right part of the distribution so that values are larger than x.

[16]See, e.g., https://en.wikipedia.org/wiki/Delta_method.

to know is that, at the link scale, the variance of an estimated value $\hat{y}$ depends on the function modelled, $f(x, \boldsymbol{\beta})$, and the variance-covariance of the estimates for the function's parameters, $\text{cov}(\hat{\boldsymbol{\beta}})$, in the following way:

$$\text{var}(\hat{y}) = f'(x, \hat{\boldsymbol{\beta}})^T \text{cov}(\hat{\boldsymbol{\beta}}) f'(x, \hat{\boldsymbol{\beta}}), \tag{8.2}$$

where f' are the first (partial) derivative with respect to each model parameter (i.e. for each element of $\mathbf{fi}$). Thus, we first have to find the derivatives for each parameter, and evaluate it at the value of x that we are interested in. In our case, $f = \frac{v_{\max}x}{x+k_M} + B$ and hence $f'_{v_{\max}} = \frac{x}{x+k_M}$, $f'_{k_M} = -\frac{v_{\max}x}{(x+k_M)^2}$ and $f'_B = 1$.[17] We get the variance-covariance matrix as the inverse of the Hesse matrix (the "Hessian"), after stripping the column/row for the estimate of σ (which does not concern us here).

Let's try to estimate the variance for $x = 1$, using the above estimates for the parameters: $v_{\max} = 2.12$, $k_M = 4.70$, and $B = 0.050$. $f' = (f'_{v_{\max}}(x),\ f'_{k_M}(x),\ f'_B(x)) = (1/5.70,\ -2.12/(5.70^2),\ 1)^T = (0.175,\ -0.065,\ 1)^T$. We invert the Hessian (`solve(-op$hessian[-4,-4])`) to get

$$\begin{pmatrix} 0.466\ 1.587\ 0.018 \\ 1.587\ 5.450\ 0.062 \\ 0.018\ 0.062\ 0.001 \end{pmatrix}.$$

Thus, for $x = 1$ our variance estimate is

$$\begin{pmatrix} 0.175 \\ -0.065 \\ 1 \end{pmatrix} \begin{pmatrix} 0.466\ 1.587\ 0.018 \\ 1.587\ 5.450\ 0.062 \\ 0.018\ 0.062\ 0.001 \end{pmatrix} (0.175\ -0.065\ 1) = 6.13 \cdot 10^{-5}.$$

The R-code is a bit negligent at the transposition of the first term, which is taken care of automatically:

```
> c(0.175, -0.065, 1) %*% solve(-op$hessian[-4,-4]) %*% c(0.175, -0.065, 1)

         [,1]
[1,] 6.129486e-05
```

The standard error of an estimate is the square-root of the variance, so if we want to draw 95% confidence intervals, we compute $\pm 2 \cdot \sqrt{6.13 \cdot 10^{-5}} = \pm 0.0157$.

To make the computation available for not only one point but many, we have to define functions for the derivative, compute the values for many points, etc. Below is the code required for standard error computations and plotting the resulting 95% confidence interval (see Fig. 8.8), but it is for future reference, rather than to understand right now.

```
> micmenSE <- function(x, op=op){
+     # for a general value of x:
+     fvmax.prime <- function(x) x/(x + op$par[2])
+     fkM.prime <- function(x) -(op$par[1]*x)/((x + op$par[2])^2)
+     fB.prime <- function(x) return(1)
+     pars <- c(fvmax.prime(x), fkM.prime(x), fB.prime(x))
+     varest <- pars
+     #varest
+     return(sqrt(varest))
+ }
> # micmenSE(2, op) # test of the function!
> Svals <- seq(0.3, 2, len=100)
> preds <- op$par["vmax"]*Svals/(op$par["Km"] + Svals) + op$par["B"]
> SEs <- sapply(Svals, micmenSE, op=op) # compute SE for each value
```

---

[17]Remember that x is now a constant. For the derivative with respect to k_M, we can re-write the problem, after substituting $a = v_{\max}x$, as $a \cdot (k_M + x)^{-1}$. Then, using the chain rule, the outer derivative is $-\frac{a}{(k_M+x)^{-2}}$, while the inner is simply 1. Multiplying them yields our partial derivative.

```
> # plotting:
> par(mar=c(5,5,1,1))
> plot(v ~ S, data=vmkm, las=1, pch=16, cex=1.5, cex.lab=1.5)
> parms <- op$par
> lines(Svals, preds, lwd=2, col="grey")
> lines(Svals, preds + 2*SEs, lwd=2, col="grey", lty=2)
> lines(Svals, preds - 2*SEs, lwd=2, col="grey", lty=2)
```

8.3   GLM with VGAM

The standard function `glm` works for four distributions. With `glm.nb` (Package **MASS**), the negative binomial distribution is also implemented and works in the same way as `glm`. The log-normal distribution can be implemented by simply taking the log of the response variable. But how can we estimate the parameters of a β-distribution or a Weibull distribution?

The powerful package **VGAM** is our best friend in this case. Yee and Wild [1996] implemented over 50 distributions and apply them to a GLM. The central function of the package is called `vglm`, and the distributions are implemented in so-called "family functions" (see `?"vglmff-class"`).

VGAM offers a number of functions and options, all of which I could not possibly cover here (see Yee 2008 for an introduction). Here we will just calculate a negative binomial model and compare it to the result we get with **MASS**. For this, we will simulate data just to be sure that the data are, in fact, negative-binomially distributed.

The model that we want to fit looks like this: $\mathbf{y} \sim \text{NegBin}(\mathbf{k}, \theta)$, where $\mathbf{k} = e^{ax+b}$. We simulate a dataset where $a = 1$, $b = 3$ and $\theta = e^1$:

```
> set.seed(101) # fixes the random generator to a specific starting value
> ndata <- data.frame(x = runif(500))
> ydata <- rnbinom(500, mu = exp(3 + ndata$x), size = exp(1))
> ndata <- cbind(ndata, "y"=ydata)
> # now we fit a negbin-GLM with vglm:
> library(VGAM)
> fit1 <- vglm(y ~ x, negbinomial, ndata)
> summary(fit1)

Call:
vglm(formula = y1 ~ x, family = negbinomial, data = ndata)


Pearson residuals:
Min      1Q  Median    3Q     Max
loge(mu)    -1.486 -0.7556 -0.1465 0.4853 4.0352
loge(size) -6.431 -0.3877  0.3812 0.6678 0.7439


Coefficients:
Estimate Std. Error z value Pr(>|z|)
(Intercept):1  3.05990    0.06153  49.733   <2e-16 ***
(Intercept):2  0.85336    0.06555  13.019   <2e-16 ***
x              0.87469    0.10485   8.342   <2e-16 ***
---
Signif. codes:  0 `***' 0.001 `**' 0.01 `*' 0.05 `.' 0.1 ` ' 1


Number of linear predictors:  2

Names of linear predictors: loge(mu), loge(size)
```

```
Log-likelihood: -2185.716 on 997 degrees of freedom

Number of iterations: 4
```

Now let's try the same thing with `glm.nb`:

```
> library(MASS)
> summary(glm.nb(y ~ x, ndata))

Call:
glm.nb(formula = y1 ~ x, data = ndata, init.theta = 2.347522506,
link = log)

Deviance Residuals:
    Min       1Q    Median       3Q      Max
-3.6480   -0.9357   -0.1515    0.4418    2.5113

Coefficients:
Estimate Std. Error z value Pr(>|z|)
(Intercept)   3.05990    0.06152   49.735   <2e-16 ***
x             0.87469    0.10484    8.343   <2e-16 ***
---
Signif. codes:  0 `***' 0.001 `**' 0.01 `*' 0.05 `.' 0.1 ` ' 1


(Dispersion parameter for Negative Binomial(2.3475) family taken to be 1)

Null deviance: 607.56  on 499  degrees of freedom
Residual deviance: 536.08  on 498  degrees of freedom
AIC: 4377.4

Number of Fisher Scoring iterations: 1


Theta:  2.348
Std. Err.:  0.154


2 x log-likelihood:  -4371.432
```

Let's start at the bottom: `glm.nb` provides an output that is the same as that of `glm`. In addition to the parameters of the linear equation (on the link scale: $a = 0.875, b = 3.060$), it also returns an estimate for the dispersion parameter $\theta = 2.3475$).

The `vglm` delivers the same information, but presents it differently: a is easy to find, b is the first of the two intercepts. The second intercept shows the estimate for θ on the link-scale (shown at: `names of linear predictors`). If we do a back-transformation, $\theta = e^{0.853} = 2.3475$. As we know (since we simulated the data), the actual value of $b = 3$ and $a = 1$. The differences are simply due to the random generation of the data.

8.4   Modelling Distribution Parameters (Other Than Just the Mean)

An important class of statistical models for the normal distribution are so-called *Generalised Least Square*-models (GLS), where not only the mean of a normal distribution is a function of some predictors, but also the variance. This allows the user to have variances increase with the measured value, which is certainly very reasonable when weighing mammals from mice to elephants: the heavier the animal, the larger the absolute variability of the measurements.

The use of GLS-models is definitely an advanced topic, which won't be covered in detail here. However, it is based on the fact that we can model not only the mean value, but also the variance. This is actually quite simple also for non-normal distributions, because we only need to provide the linear model with a different distribution parameter: instead of $\mathbf{y} \sim \text{Gamma}(\text{shape} = \frac{1}{a\mathbf{x}+b}, \text{scale} = c)$ we simply model $\mathbf{y} \sim \text{Gamma}(\text{shape} = c, \text{scale} = \frac{1}{a\mathbf{x}+b})$. Or we can customise different functions for both distribution parameters. This can be more complicated, because the parameters we want to determine are often correlated with each other, but with a few fancy tricks, this can be done as well (Yee 2014).[18]

A simple example should help to make this concept clear. In behavioural ecology, sexual selection plays an important role. Females select the best males to sire their offspring in order to maximise their reproductive fitness. For various reasons, the ratio of males to females in the population is nevertheless often 50:50 (Krebs and Davies 1993). However, if, for example, 200 males and 200 females produce 400 offspring (so an average of 2 offspring per female), not all males and all females will likely have contributed to the next generation evenly. Some may have produced more offspring, and some fewer (or none at all). In other words, the distribution of the number of offspring is likely different between males and females even though the average is the same for both sexes.

In this sample data set (Stunken and Logen 2012) we counted the number of offspring for 200 male and female spotted hyenas that were still living in the pack after one year. Now let's load this data set and look at a histogram based on sex:

```
> offspring <- read.csv("offspring.csv")
> summary(offspring)
```

```
       M                 F
Min.    : 0.00   Min.    :0
1st Qu.: 0.00   1st Qu.:1
Median : 1.00   Median :2
Mean    : 2.00   Mean    :2
3rd Qu.: 2.25   3rd Qu.:3
Max.    :24.00  Max.    :8
```

```
> par(mfrow=1:2, mar=c(4,5,1,1))
> hist(offspring$M, col="grey", las=1, xlab="offspring per male", main="",
+   ylab="frequency", cex.lab=1.5)
> hist(offspring$F, col="grey", las=1, xlab="offspring per female", main="",
+   ylab="frequency", cex.lab=1.5)
```

This produces Fig. 8.9. In the summary, we can also see that males have up to 24 offspring, whereas females have a maximum of eight cubs per litter.

Now we will fit a negative binomial model, where we will model the clumping parameter r (instead of the mean μ) depending on the sex[19]:

$$\mathbf{y} \sim \text{NegBin}(\mu, \mathbf{r} = e^{a+b\mathbf{G}}),$$

where $\mathbf{G}$ is an indicator function for sex (with 0 for male and 1 for female). We once again follow the three steps of optimisation after we transform the data:

```
> offstack <- stack(offspring)
> colnames(offstack) <- c("offspring", "parentalsex")
> # 1. Goal function:
> llfun <- function(parms){
+   size.est <- exp(parms[2] + parms[3]*(as.numeric(offstack$parentalsex)
+     -1))
+   res <- sum(dnbinom(offstack$offspring, mu=exp(parms[1]),
```

---

[18]The reason that this causes so many problems for analysts is that many textbooks don't introduce the GLM along with distributions, but rather use the case of the normal distribution to generalise about other distributions. Due to this structure, it is hard to keep track of the central meaning of the distribution parameter. While only μ determines the mean in the normal distribution, for many other distributions, both parameters are included in the mean (as depicted in the mean formulas in Sect. 3.4).

[19]This is an example, not necessarily a reasonable model. Of course we should also model the mean here!

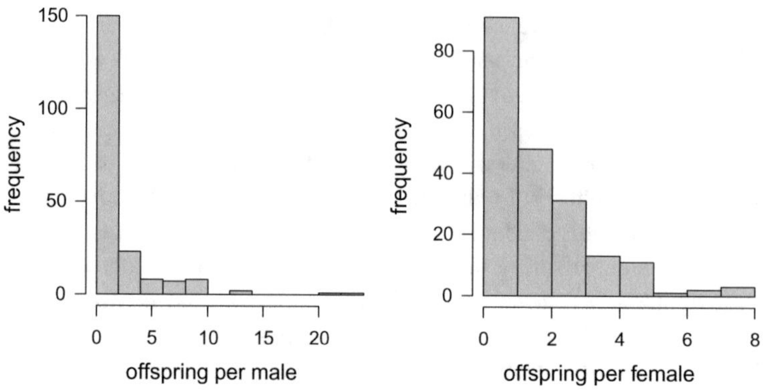

Fig. 8.9 Histogram of the number of offspring per individual for male (left) and female (right) spotted hyenas

```
+           size=size.est, log=T))
+     res
+ }
> # 2. Starting values:
> initial <- c(1,1,1)
> # 3. Optimisation:
> (op <- optim(initial, llfun, control=list(fnscale=-1), method="BFGS", hessian=T))

$par
[1]   0.6931472 -0.7672723  2.6128038

$value
[1] -726.084

$counts
function gradient
      39       13

$convergence
[1] 0

$message
NULL

$hessian
              [,1]            [,2]            [,3]
[1,] -3.793397e+02 -2.447109e-05 -2.597744e-05
[2,] -2.447109e-05 -4.959455e+01 -5.606904e+00
[3,] -2.597744e-05 -5.606904e+00 -5.606904e+00
```

We get an estimate for the mean ($\hat{c} = e^{0.693} = 2.00$), one for the intercept $\hat{a} = e^{-0.767} = 0.464$ and one for the effect of "female" ($\hat{b} = e^{2.613} = 13.64$). The error for these estimates is:

```
> (se <- sqrt(diag(solve(-op$hessian))))

[1] 0.05134354 0.15077684 0.44842524
```

on the link-scale. We then transform these values and realise that they become asymmetrical:

```
> upperCI <- exp(op$par+2*se)
> lowerCI <- exp(op$par-2*se)
> cbind(lowerCI, mean=exp(op$par), upperCI)
```

```
         lowerCI         mean      upperCI
[1,]   1.8048186   2.0000000   2.2162892
[2,]   0.3434114   0.4642777   0.6276839
[3,]   5.5619751  13.6372330  33.4367052
```

The 95% confidence interval for the mean is around 10% of the estimated value, for a it is around 50% and for b around -50 and $+150\%$. Nevertheless, all of the estimators are significantly different from 0 (as shown by the t-test value on the link scale[20]:)

```
> op$par/sqrt(diag(solve(-op$hessian)))
```

```
[1] 13.500182 -5.088794  5.826621
```

```
> qt(0.05, 200-3, lower.tail=F) # computes the critical t-value
```

```
[1] 1.652625
```

All t values are clearly greater than 1.65.

So with respect to hyena offspring, we can say that the reproduction rate is much more evenly distributed amongst individuals for females than it is for males (clumping parameter of 0.46 for males and 13.6 for females, where a larger value indicates less clumping). Now we only need to calculate how many male and female individuals had any offspring at all. To do this, we simply tabulate the number of offspring per individual:

```
> table(offspring$F)
```

```
 0  1  2  3  4  5  6  7  8
31 60 48 31 13 11  1  2  3
```

```
> table(offspring$M)
```

```
 0  1  2  3  4  5  6  7  8  9 10 13 14 22 24
91 37 22 14  9  5  3  5  2  6  2  1  1  1  1
```

So while 85% of all females reproduced (169 out of 200), only 55% of males reproduced (109 out of 200).

So what have we learned from this? We can fit not only the mean using maximum likelihood, but also the variance, scale parameters and other distribution parameters. Of course, we can also fit functions for all parameters of a distribution, such as: $y \sim N(\mu = a + b\mathbf{x}, \sigma = c + d\mathbf{x} + d\mathbf{x}^2)$. The approach just introduced above is only limited by the number of data points and the imagination of the analyst.

8.5   Exercises

1. Make a regression to see whether or not it makes sense to fertilise. The data set is called `cornnit` and can be found in the **faraway** pacakge. The response variable is `yield` (in bushels per acre), the predictor is `nitrogen` (in lbs per acre). Always plot first, then fit, then draw in points! Try a linear regression first. How much better is the quadratic polynomial? Or even a cubic polynomial (`yield ~ poly(nitrogen, 3)`)?

---

[20]The sign is unimportant here, it only tells us whether a value is above or below 0.

2. A medication to reduce blood pressure is given to rabbits. Does it work? Look at the `Rabbit` data set from the **MASS** package: response variable `BPchange`, predictor `Treatment`. This is a "regression" with a categorical variable. Here, we want to know if the `Treatment` effect is significant.

3. Jaw length was measured in jackals. Do male and female jackals have different jaw lengths? (`jackal` data set from the **permute** package.)

4. Make a regression for the height of pine trees as a function of their age. Load the `Loblolly` data set (load the data via `data(Loblolly)`, put the two variables `age` and `height` in the right spots in the model and determine the parameters of the regression lines. Now attempt to calculate the regression "by hand": write a likelihood function, enter starting values, put into `optim`.

5. If you were able to do that, let's try a more challenging problem. In the dataset `ChickWeights` (`data(ChickWeights)`), not only does the weight of the chicks vary with time, but the standard deviation also increases. This means that we only do a regression with `glm` to get starting values for calculation "by hand". Using optimisation, we let both the mean and the standard deviation of the normal distribution vary with the predictor `time` (respectively independent). This means that we need to fit 4 parameters for the likelihood function: intercept and slope for the mean (so: *a* and *b*), and the intercept and slope for the standard deviation (so: *c* and *d*). This problem is not easy. Don't get frustrated if it isn't working! For example, if the optimisation returns an error message, just start over with different start values.

6. These exercises deal with the data set from Bolger et al. (1997, `bolger.txt`). In it, the authors collected data on the presence of rodents (`RODENTSP`, with values 0 and 1) in 25 habitat fragments in canyons in California, as well as three environmental parameters.

    (a) Plot the presence of rodents (variable name `RODENTSP`) as a function of shrub cover (`PERSHRUB`). Fit a GLM (which distribution?) and draw the regression line and 95% confidence intervals.

    (b) Calculate the shrub cover for when the presence probability of rodents is at 80%.

7. In a further study, the same authors looked at the abundance of individual rodents in connection with different predictors. Here, we will look at whether the regression for the native species `RMEGAL` differs from the regression for the non-native species `MMUS`. The predictor that we choose describes how long a fragment has been isolated and is called `AGE`.

    (a) Read in the data set `bolger4.txt` (you could use `read.delim`) and assign it to an object.

    (b) Perform a Poisson regression for `RMEGAL`. Repeat this with a negative binomial regression (`glm.nb` in **MASS**). Compare the AIC values: which model seems more appropriate? Compare the significance of the `AGE` effect in both models. What happens when we change from a Poisson to a negative binomial model?

    (c) Now extend this analysis to include an evaluation of the occurrence (not the abundance). For this, we will define a variable `RMEGAL>0` and use is as the response variable in a binomial GLM.

    (d) Repeat these three analyses for the non-native species *Mus musculus*. Does the pattern look any different?

    (e) Plot the `AGE` effect for `RMEGAL` and `MMUS` from the negative binomial model in the same plot.

    (f) If you're ecological interest is piqued, do the same for `RRATTUS` (non-native) and `PCALIF` (native) as well.

References

1. Bolger, D. T., Alberts, A. C., Sauvajot, R. M., Potenza, P., McCalvin, C., & Tran, D., et al. (1997). Response of rodents to habitat fragmentation on coastal southern California. *Ecological Applications, 7*, 552–563.

2. Krebs, J. R., & Davies, N. B. (1993). *An introduction to behavioural ecology.* Oxford, UK: Blackwell Scientific Publications, 3rd edition.

3. Selwyn, M. J. (1995). Michaelis-Menten data: Misleading examples. *Biochemical Education, 23*, 138–141.

4. Stunken, R., & Logen, R. (2012). Pretend-analysis of hyaena reproductive successes to demonstrate modelling of variance. *Collection of Didactical Examples in Ecology and Evolution, 1*, 111–113.

5. Yee, T. W. (2008). *The VGAM package. R News, 8*, 28–39.

6. Yee, T. W. (2014). Reduced-rank vector generalized linear models with two linear predictors. *Computational Statistics and Data Analysis, 71*, 889–902.

7. Yee, T. W., & Wild, C. J. (1996). Vector generalized additive models. *Journal of Royal Statistical Society, Series B, 58*, 481–493.

Regression—Part II

The most misleading assumptions are the ones you don't even know you're making.

Douglas Adams

At the end of this chapter ...
... you will have learned the different steps that are necessary to do *after* formulating the model.
... you will know how predictors should be distributed.
... you will know that there are more than one type of residual, and that these can provide important information about the model fit.
... you will be able to test for "outliers" and influential data points.
... you will know that you should test for alternate curve forms.

Every statistical analysis has some underlying assumptions. How can we check whether these assumptions have been met for our analysis? How can we measure the quality of our regression? What metrics will inform us that our model is "sick", i.e that our model gives us a result that doesn't match our expectations after taking multiple plots and figures into account?

The word "sick" is actually quite suitable here, because the topic we will be dealing with in this chapter is model diagnostics. Just because we have learned how to fit a GLM does not mean that everything is peachy and our analysis is finished.

The core message of this chapter is that statistics (such as the R^2 or t-test of the estimator) alone do not constitute model diagnostics. The visualisation of the data and the residuals is extremely important. To illustrate this point, I'll use the data set from Anscombe (1973, Fig. 9.1). This data set is constructed so that all four regressions have the same slope and R^2 value. At first glance without running any model diagnostics, we would maybe have accepted all of these models!

In reality, only the figure in the first panel (upper left) shows an acceptable regression. In the second (upper right), we see a non-linear, or to be more exact, a quadratic relationship. The fit of the line is bad for pretty much the whole range. Here, we would need to include x^2, in addition to x, as a predictor in our model. The outlier in the third panel has a strong influence on the otherwise perfect looking relationship. In the last panel, we have a regression even though there is only one value that is actually different from the rest. The relationship there is questionable at best.

We can only do model diagnostics *after* we have fit the model. A test for normal distribution on the raw data is pointless if the mean of the normal distribution varies with a predictor. We don't want to test $y \sim N(\mu, \sigma)$, but rather $y \sim N(\mu = ax + b, \sigma)$. Or in words: the data are not taken from a distribution with *fixed* parameters, but are taken from a distribution with *variable* parameters. Accordingly, we need to take this parameter variability into account when we test for distribution assumptions.

The most important step is the visualisation of the fitted regression: comparing data and regression lines visually can not be replaced by any numerical test!

© Springer Nature Switzerland AG 2020
C. Dormann, *Environmental Data Analysis*,
https://doi.org/10.1007/978-3-030-55020-2_9

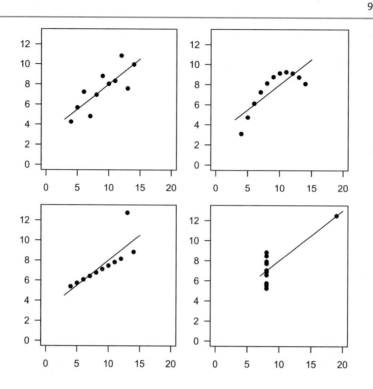

Fig. 9.1 The Anscombe quartet: Four different data sets that have the following similarities: Regression line: $y \sim N(\mu = 3.0 + 0.5x, \sigma)$, number of data points = 11, $R^2 = 0.67$, deviation squares = 27.5 and t-test of the slope = 4.24, $P = 0.002$ (we will learn the meaning of these numbers in later chapters). (Data from Anscombe 1973)

9.1   Model Diagnostics

Numerical model diagnostics depends on the type of distribution, but in general should include the following five points:

1. Analysis of the predictor distribution;
2. Analysis of influential points;
3. Analysis of the dispersion;
4. Analysis of the residuals;
5. Analysis of the functional relationship between y and x.

9.1.1   Analysis of the Predictors

While we need to assume a distribution for the response variable, up until now we haven't paid much attention to the predictors. Why should we?

Figure 9.2 shows us the problem. The two data points with x values over 30 turn the clearly negative relationship (in light grey) into a positive one (dark grey). Since these points lie so far from the rest, they have a higher weight when calculating the regression. If we accept the light grey line, then the two extreme right points fit terribly in the regression. But with the dark grey line, all of the points are at least somewhat close to the line.

In short: extreme x values dominate a regression. Ideally, the x values would be distributed in a uniform fashion: the same data density throughout the range.

The important point here is that the distribution of x can play a significant role. We don't make any assumptions about it (like we do for the response variable), but outliers on the x-axis can influence the result, in some cases quite strongly (see Fig. 9.1 lower left). Non-uniform x values are the norm (due to the central limit theorem, they are often normally distributed) and a transformation may be indicated. Important transformations for capturing such "outliers" include the log and square-root transformations (e.g. for distances or areas). If the regression result looks the same after transforming the x values, then the regression was robust. However, it is often the case that significance is lost as soon as an extreme skew is transformed away.

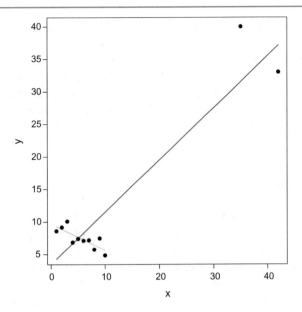

Fig. 9.2 A simple regression?

9.1.2 Analysis of Influential Points

There is no justification for eliminating a data point just because it is an "outlier"! If there was a data entry or measuring error, then of course you can eliminate the point, but it is not OK to "fix" your data after the fact, just to make it so the results "work"!

Some programs define outliers as values that differ from the expected value by more than 3 times the standard deviation.[1] There are, however, more subtle methods for identifying influential data points. Typically, each data point is left out one at a time and the regression is then recalculated to calculate the change in the estimators or the residual variance. The most important metric in this case is Cook's distance, which clearly shows influential data points. Most other methods have the drawback that they can detect a single extreme value, but not two (as there were in our example).

Cook's distances for the 12 data points from Fig. 9.2 are found in the following table under cook.d.

```
      dfb.1_     dfb.x    dffit  cov.r   cook.d    hat inf
1    0.39187  -0.24321   0.39230 1.160 7.71e-02 0.1354
2    0.35068  -0.20420   0.35235 1.175 6.29e-02 0.1255
3    0.34138  -0.18453   0.34536 1.154 6.02e-02 0.1166
4    0.01288  -0.00637   0.01317 1.385 9.63e-05 0.1088
5   -0.00526   0.00234  -0.00546 1.375 1.66e-05 0.1021
6   -0.07680   0.02993  -0.08145 1.347 3.66e-03 0.0963
7   -0.11835   0.03888  -0.12902 1.311 9.08e-03 0.0917
8   -0.25319   0.06594  -0.28587 1.131 4.15e-02 0.0880
9   -0.18160   0.03343  -0.21418 1.214 2.42e-02 0.0854
10  -0.38797   0.03801  -0.48251 0.819 1.01e-01 0.0839
11   0.54919  -1.57864  -1.70513 1.956 1.31e+00 0.5833   *
12  -0.57682   2.23560   2.52739 0.436 1.66e+00 0.3830   *
```

Whatever the other measures are, not all of them would consider both 11 *and* 12 (the values in the upper right of Fig. 9.2) as influential (the following do: dfb.x, cook.d, hat). For Cook's distance, values greater than 1 (Cook and Weisberg 1982) or greater than $4/n$ (n = number of data points) are designated as influential (Bollen and Jackman 1990). For our example, these would be 1 or 0.33. Both of these would identify points 11 and 12 as influential.

---

[1] Another approach (with perhaps limited usefulness) is based on the distance of a data point from the multi-dimensional mean calculated using the Mahalanobis distance. Unfortunately, this can sometimes lead to data points that actually fit well in the regression to be falsely identified as an outlier. The methods introduced here are more useful, because they quantify the weight of the data point *for the regression model*.

What we learn from this is that it can be difficult to quantify something that is simple to observe. Cook's distance, however, is one measure that we can trust.

While we are not allowed to simply remove extreme values from our data set, we can do another analysis without influential data points and compare the results and test for robustness of the results. This always needs to be done *in addition* to running the analysis with all data points, not instead of it. If regression results are robust, i.e. don't change after omission of influential data points, then we can report the full data set and state that "omission of influential data points had no substantial effect on the results". If, however, the results change, we should report *both* analyses!

9.1.3 Analysis of the Dispersion

For many distributions, the expected value (kind of the mean of a distribution) and variance have a certain relationship (see Sect. 3.4 on p. 54). For example, the expected value for the Poisson distribution is *equal to* the variance. Or for a binomial distribution, the expected value is $\mu = np$ and the variance is $s^2 = np(1-p) = \mu(1-p)$.

Dispersion quantifies how much more (or less) the variability increases with the expected value than implied by the distribution. For the normal distribution, the standard deviation (as a measure of the variability) is independent from the mean and the dispersion is fitted (it is the variance in this case). For the Poisson distribution, the variance is supposed to increase 1:1 with the mean, making the dispersion = 1. If we now fit a GLM with a Poisson distribution, the variance in the data set has to increase exactly by a factor of 1 with the mean. If it increases more slowly, then the data are underdispersed, if it increases more quickly (which is more often the case), then the data are overdispersed. Under- and overdispersion can also occur in other distributions, such as the binomial or negative binomial distribution, but not for the γ or normal distributions, since the dispersion is an explicit parameter in these cases. There is also no under- or overdispersion for the Bernoulli distribution (e.g. Gelman and Hill 2007, p. 302).

Once we execute a GLM regression, we have to do a rough check for overdispersion. To do this, we divide the residual deviance of the model by the residual degrees of freedom. This value should be around 1. If it is much bigger (>2) or smaller (< 0.6), then we either need to use a different distribution, or use quasi-likelihood, a modified maximum likelihood optimisation where the dispersion parameter is fit alongside the model.

9.1.4 Analysis of the Residuals

Residuals are the difference between the observed ($\mathbf{y}$) and fitted ($\hat{\mathbf{y}}$) data.[2] In the case of a regression with normally distributed data (so, $\mathbf{y} \sim N(\mu = a\mathbf{x} + b, \sigma)$), the residuals are simply "observed minus expected": $\mathbf{y} - \hat{\mathbf{y}}$. These residuals are called the response residuals.

For other distributions (Poisson, binomial, ...), such response residuals are not really useful, since the variance changes with the mean. Accordingly, the response residuals for large values of $\mathbf{y}$ are higher. Therefore, for standardisation, the response residuals are divided by the square root of the variance[3]:

$$\mathbf{r}_P = \frac{\mathbf{y} - \hat{\mathbf{y}}}{\sqrt{\mathrm{var}(\hat{\mathbf{y}})}} \tag{9.1}$$

$\mathbf{r}_P$ are called the *Pearson residuals* or standardised residuals. Pearson constructed them in such a way that $\chi^2 = \sum \mathbf{r}_P^2$, i.e. that the sum of the residual squares results in the χ^2-value of the model, with df =number of fit parameters. In this way, $\sum \mathbf{r}_P^2$ is also a goodness-of-fit measure, with which you can test the fit of the data (for details, see Faraway 2006, p. 121)

Alternatively, you can also scale $\mathbf{y} - \hat{\mathbf{y}}$ so that the sum of the residuals is equal to the deviance. These *deviance residuals* are defined as:

$$\mathbf{r}_D = \mathrm{sign}(\mathbf{y} - \hat{\mathbf{y}})\sqrt{d^2} \tag{9.2}$$

---

[2]Note that, from a statistical perspective, a fitted value is the same as a model prediction, i.e. the expectation for this specific value of x_i. In either case, we would use a hat to indicate that these are model predictions: $E(y_i|x_i) = \hat{y}_i$.

[3]Remember here that the variance has the "wrong" dimensions, while the square root of the variance (= standard deviation) has the same dimensions as the mean. This means that the standardised residuals are dimensionless, as the units cancel.

with **d** (the deviance vector of the data set as mentioned in Sect. 7.3.2.[4] For binary residuals, d_i^2 would be calculated as, for example, $d_i^2 = -2(\log(|\hat{y}_i - y^c|)$, where ^{c} is the complement (so 1 if $y = 0$ and 0 if $y = 1$).

Finally there are also the *studentised residuals*, where the deviance residuals are corrected for the estimated dispersion parameter $\hat{\phi}$ (e.g. variance, *scale*) and for the *leverage* (the hat value from the analysis of influential points):

$$\mathbf{r}_{SD} = \frac{\mathbf{r}_D}{\sqrt{\hat{\phi}(1 - \mathbf{h})}}. \tag{9.3}$$

Deviance and Pearson residuals are similar to each other and have much different values than both the response residuals and the studentised residuals.[5]

Now that we know how to calculate different types of residuals, what are we going to do with this knowledge? The most useful plot we can make is one showing the residuals (on the y-axis) plotted against the fitted value (on the x-axis).

As an example, we will use a **Poisson** data set also used by Faraway [2006] to illustrate a similar point: the number of plant species on the 30 Galapagos islands explained by the area of the island, distance to Santa Cruz, maximum meters above sea level, distance to next island and distance to the next island.[6]

Figure 9.3 shows the three types of residuals we just discussed on both the response and link scale. As we can see, the residuals on the link scale are corrected for the variance increasing with the mean. The lower plots in this figure should show no real pattern and look a bit like "stars in a clear night sky", as Crawley [2002] so elegantly put it. The response scale is not very informative in this case.

When diagnosing residuals, we have to pay particular attention to two phenomena: is there a trend in the residuals, and is the dispersion the same everywhere (homogeneous)? Fig. 9.4 illustrates the four combinations of these two possible problems. Ideally, the residuals should look like the upper left plot in Fig. 9.4. A trend, e.g. a quadratic correlation of the residuals with the expected value, as in the upper right plot of Fig. 9.4 shows that our model was incorrectly specified (see Sect. 9.1.5). The

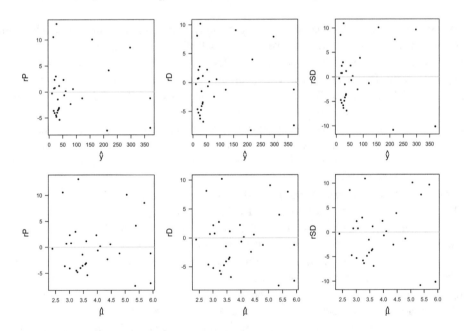

Fig. 9.3 Residuals for a GLM of the number of plant species on the Galapagos Islands. The upper plots are on the response scale, and show the difference between observed and fitted data ($y - \hat{y}$). The lower plots are on the link scale (the scale of the linear predictor), so in this case, they show the difference between $\log y$ and $\log \hat{y}$. The first column shows Pearson, the second deviance, and the third studentised residuals. (For the studentised residuals, the islands of Isabela and San Cristobal have the same area and residual, the bottom-right most point.)

---

[4]"sign" only tells us if it is positive or negative, but not by how much. From -3.2 and 1.7 we would get -1 and 1.

[5]Hardin and Hilbe (2007, S. 44 ff) also provide formulas for a number of other residuals, but recommend using deviance residuals. Therefore, these other types of alternative residuals are not discussed here.

[6]We will ignore the problem that many of these variables are correlated (such as altitude and area), and also ignore the fact that we should transform some of the variables to capture extreme values.

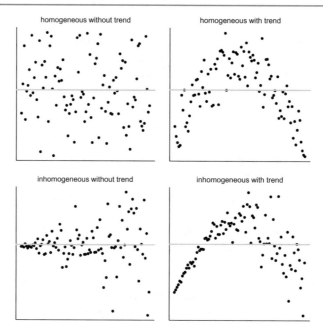

Fig. 9.4 Residual plots with the expected value on the y-axis. The grey line shows the mean of the residuals (typically 0). The goal of a good model structure is a plot like the one in the upper left corner

variability increases with the expected value in the plot on the bottom left, a clear violation of the required homogeneity of variance.

For **normally distributed** data, the residual analysis is especially important. The same principles apply here, but they have special names. Since mean and variance are independent of each other in a normal distribution, and only the mean value is modelled in a regression, the variances should be the same for all values of y. This requirement is called homogeneity of variance or *homoscedasticity*. The opposite, i.e. with a variance that is not constant, is called variance heterogeneity or *heteroscedasticity*. This is a violation of the GLM assumptions for normally distributed data, as we fit the regression with a constant standard deviation ($y \sim N(\mu = ax + b, \sigma = \text{constant})$).

For a regression of normally distributed data, the response and the link scale are identical (since the link function is the identity). Accordingly, the residual plots here should look exactly the same as the studentised residuals on the link scale for Poisson data (Fig. 9.3, lower right). These studentised residuals on the link scale are the generally recommended form of residuals for this type of model diagnostics.

For **Bernoulli distributed** data (0/1 data), this residual plot is not very helpful (Fig. 9.5). Here, the observed data has a value of either 0 or 1 and the expected values vary between 0 and 1. In this case, where there is a 1 in the data, the model predicts an expected value of 0.6, giving the response residual a value of 0.4. A value of 0, on the other hand, has an expected value of 0.7, and therefore a response residual of -0.3. Since the fitted curve is continuous, we see that it also varies continuously. Since the discrete values 0 and 1 cannot be transformed into continuous values on the link scale, this pattern is retained.

What do we learn here? For some distributions (normal, Poisson, γ), residual-fitted plots are informative and useful for checking whether the assumptions with regard to the relationship between the variance and the mean are met. For other distributions (Bernoulli), such plots are not useful. In such cases, residual analysis is not available as a part of model diagnostics. In the next chapter we will encounter a way to do model diagnostics also for binary data, based on simulating data from the fitted model, but this approach requires a bit of coding, so we postpone until Sect. 10.1.4, Side Note 7.

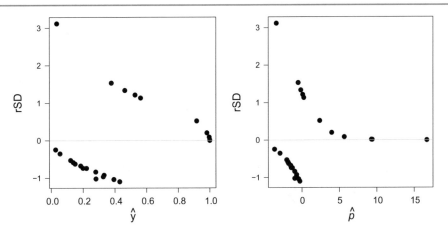

Fig. 9.5 Residuals for a GLM of the probability that a mouse species (PCALIF) occurs in a habitat fragment (Bolger et al. 1997). Depicted here are the (linear) studentised residuals on the response (left) and link scales (right; note that the parameter of the binomial distribution is called p). Pearson and deviance residuals would result in very similar plots. Note that only the response residuals lie between -1 and 1, whereas the calculation of the studentised residuals depicted here can also give us larger values

9.1.5 Analysis of the Functional Relationship Between y and x

When we fit a regression, we assume a linear relationship (for a GLM, one on the link scale). However, there might not be a linear relationship. Whether an assumed functional relationship is suitable or not can generally be found out only by comparison with another model.

Assuming an incorrect relationship is the most common and arguably most serious form of model misspecification. It influences the shape of the residuals, the AIC, extrapolation, etc. Therefore, it is **very important** that we do not simply fit whatever we want and call it good, but that we compare alternative models. But which ones?

Every "standard" (mathematically: continuously differentiable) function can be approximated by a polynomial ($f(x) = a + bx + bx^2 + cx^3 + dx^4 + \ldots$), a variation on the Taylor series. Since our data are often "simply" non-linear (not like a sine wave or a quartic function that goes up and down), it is usually sufficient to just fit the linear model ($f(x) = a + bx$) and a quadratic model ($f(x) = a + bx + cx^2$) and compare the two. In doing so, we advance our simple regression into one with two predictors: x and x^2. The actual shape of this curve depends strongly on the link function.

Here is our approach: We assume that the response variable y is dependent on multiple predictors $\mathbf{x}_1, \mathbf{x}_2, \mathbf{x}_3, \ldots$. In our model, however, we allow for a flexible form of this relationship. For a Poisson-distributed random variable, we get for example:

$$\text{Pois}(\lambda = b_0 + b_1\mathbf{x}_1 + b_2\mathbf{x}_1^2 + b_3\mathbf{x}_2 + b_4\mathbf{x}_2^2 + b_5\mathbf{x}_3 + b_6\mathbf{x}_3^2 + \ldots).$$

If, say, b_6 does not deviate much from 0, then this is a sign that for the variable $\mathbf{x}_3$, we only need to assume a linear relationship.

In fact, we can, and should, also allow so-called "interactions" between variables (see Sect. 15.2 on page 207 for details). In this case, we would include as well the product of the predictors in the model:

$$\text{Pois}(\lambda = b_0 + b_1\mathbf{x}_1 + b_2\mathbf{x}_1^2 + b_3\mathbf{x}_2 + b_4\mathbf{x}_2^2 + b_5\mathbf{x}_1\mathbf{x}_2 + \ldots).$$

Often we can detect non-linear relationships in small data sets by plotting the data. With large data sets and multiple predictors, this is rarely possible because the patterns of the different predictors overlap. For this reason, it is very important that we compare alternative model structures.

Let's have another look at the Galapagos data set. Since the log is the link function for the Poisson distribution, we put log(number of plant species) on the y-axis (but of course analyse the number of plant species themselves). The area of the islands is very right skewed, so we transform this as well. Then we focus on the explanatory variable, area, and compare the two models: (1) log(area) as a single predictor; and (2) log(area) + log(area)^2 as predictors.

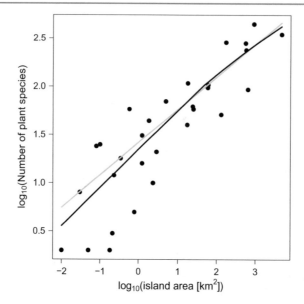

Fig. 9.6 Number of species as a function of area: both variables are $\log_{10}$ transformed. The light grey line is the linear model (1), the dark grey includes a quadratic term (model 2)

We can first visually compare these models (Fig. 9.6). The two curves are only different for small islands, but there is no clear deviation from a linear pattern.

Next, we can look at the AICc values of these two models. Model 1 has an AICc value of 816.9, model 2 has a value of 808.9. From this, it seems like the quadratic model (2) is actually better (lower AICc value). For the BIC, the conclusions are the same ($BIC_1 = 819.3$, $BIC_2 = 812.2$).

We can also carry out the model comparison statistically, by using a χ^2-test to test the difference in the log-likelihood (or the ratio of the likelihoods; the *likelihood-ratio test*). The idea behind this is that we have two "competing" models. The comparison of their likelihoods shows us which one is better. In fact, the logarithm of the quotient of the likelihood, or the difference of the log-likelihoods, is asymptotically χ^2-distributed.[7] The likelihood-ratio-test is valid only under certain conditions. Most importantly, we should apply it only if the two compared models are "nested", meaning that one is a simplified form of the other.[8] In our example, model 1 is a nested simplification of model 2, which results from leaving out the quadratic term.

The result looks like this:

```
Analysis of Deviance Table

Model 1: Species ~ log10(Area)
Model 2: Species ~ log10(Area) + log10(Area)^2
  Resid. Df Resid. Dev Df Deviance P(>|Chi|)
1        28     651.67
2        27     641.19  1   10.478   0.001208 **
---
Signif. codes:  0 `***' 0.001 `**' 0.01 `*' 0.05 `.' 0.1 ` ' 1
```

---

[7]The word "asymptotically" is often used in a way resembling "approximately", but it has a specific statistical meaning: as sample size increases towards infinity, the likelihood ratio *is* χ^2-distributed. Since we are far away from infinity, we hope that the ratio still is approximately χ^2-distributed. Or, in other words: the logic of the test is sound for infinite sample sizes; we *assume* it also works in our case.

[8]You can think of these models as Russian matryoshka dolls. Although the name we use is *nested*, mathematically speaking, it would be more accurate to say that the simpler model can be derived from the other by linear parameter transformation. In our example, this is the multiplication of the square term by 0.

The difference here is significant. The additional parameter for the quadratic term provides 10.5 more units of explained deviance, and this difference is significant according to the χ^2-test ($P < 0.01$). We would thus choose to fit the relationship between the number of plant species and the area with a quadratic polynomial.

In addition to the overall comparison of the models, a look at the estimated model parameters also allows us to see whether these two models differ. If the additional quadratic term in model 2 has an estimated value close to 0, then the effect is also very low. In this case, the difference between the linear model 1 and the quadratic model 2 would also be very small (the estimate for the quadratic effect is only -0.0099, even if we remember that it is included in the quadratic).

To be sure, we can visualise the data by plotting the observed data against the fit (Fig. 9.7). The lines correspond to model 1 (linear, grey) and 2 (quadratic, black). The difference is very small.

We should also look at how this affects the residuals. For this, we calculate the studentised residuals of both models and then plot these against the fit (Fig. 9.8).

Statistically, the quadratic model is better, but the differences are so small that a linear model also seems justifiable.[9] The differences become relevant if we make a prediction beyond the range of values, i.e. extrapolate to islands that are smaller or larger than the observed ones. A larger island in the Galapagos archipelago is not expected, but there are hundreds of rocks that rise out of the water like an island. We could now make a prediction for such a rock with both models. If these differ (which does not have to be the case!), then a new observation could help decide between models 1 and 2.

The smallest island in this data set has an area of 0.01 km^2, or 10000 m^2. Now we extrapolate for an island that is a bit smaller, with an area of only 1000 m^2 (i.e. a value of -3 on the x-axis in Fig. 9.6). Model 1 gives us an expected value of

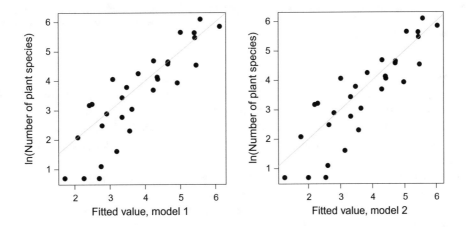

Fig. 9.7 Observed data plotted against the fitted data for model 1 (linear) and model 2 (quadratic). Minimal differences can be seen in the left half of the value range. The grey line shows the ideal 1:1 equivalence

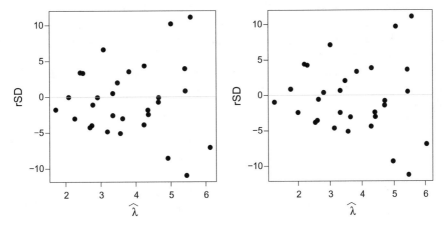

Fig. 9.8 Studentised residuals plotted against the fit on the link scale for model 1 (linear) and model 2 (quadratic). There are slight differences in the left half of the plot. Note that since we are looking at a Poisson-GLM, the predicted value is $\widehat{\lambda}$

---

[9]Nothing prevents us from trying higher polynomials now!

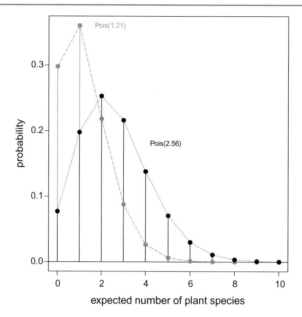

Fig. 9.9 Poisson distribution for the number of plant species on a 1000 m² island according to model 1 (grey) and model 2 (black). In fact, both parameters still have an error, so that the distributions are practically indistinguishable in the worst case (lower confidence interval of model 1: 2.15 compared to the upper confidence interval of model 2: 2.02)

$\lambda = 2.56$ [2.15, 3.06] (95% confidence interval) species, whereas model 2 gives us $\lambda = 1.21$ [0.72 2.02] species.[10] This does not mean that we would only expect either one or two species on this island, but that the number of species on a 0.1 ha island in the Galapagos is a Poisson-distributed random variable with $\lambda = 2.56$ and 1.21, respectively. In fact, even here the differences are still small. In both cases, two species of plants are expected to be found on a good 20% of such 1000 m² islands, but even 1 or no plant species are compatible with these values of a Poisson distribution (Fig. 9.9). We better check quite a few small islands then!

This example shall teach us that although statistically there may be a clearly better fitting model, this does not necessarily make very different ecological predictions than a simpler one. What we really need here is a sense of proportion: is the rigorous testing of one (or more) hypotheses the goal, or do we just want to roughly compare the data statistically? In the first case we should use the quadratic term if demanded by the hypothesis, in the latter we can drop it in favour of a simplified view.

At this point, you may be thinking to yourself that statistics are arbitrary. It's not like that. Statistics are (mostly) unambiguous, but the interpretation of them rarely is. Since we are not primarily statisticians but rather ecologists, biologists, environmental researchers or the like, statistics merely provides the basis of our interpretation. This should be solid and technically correct. But if, as in the present case, our analyses show that the linear model *for all practical purposes* does not differ from the quadratic one, then we can stick to a linear interpretation.

Later (more precisely in Chap. 15) we will deal with the systematic comparison of many models with several explanatory variables. At the moment, it is especially important to recognise that we finally have a method in our toolbox to investigate non-linearity: the polynomial.

References

1. Anscombe, F. J. (1973). Graphs in statistical analysis. *American Statistician, 27,* 17–21.
2. Bolger, D. T., Alberts, A. C., Sauvajot, R. M., Potenza, P., McCalvin, C., & Tran, D., et al. (1997). Response of rodents to habitat fragmentation on coastal southern California. *Ecological Applications, 7,* 552–563.

---

[10]Here it is not a question of where these values come from, but rather that the models are hardly distinguishable, despite the apparently very different predictions, due to their uncertainty. The prediction for model 1 ($\hat{y}_1$) results from the estimated parameters and the island size of -3 (corresponds to 0.001 km²) as follows: $\hat{y}_1 = e^{3.273+0.78\cdot(-3)} = e^{0.938} = 2.56$. We calculate the confidence interval ([2.15,3.06]) more easily with a statistics program (see next chapter). The analogous calculation for model 2 is: $\hat{y}_2 = e^{3.323+0.902\cdot(-3)-0.037\cdot(-3)^2} = e^{0.191} = 1.21$.

3. Bollen, K. A., & Jackman, R. (1990). Regression diagnostics: An expository treatment of outliers and influential cases. In J. Fox & J. S. Long (Eds.), *Modern methods of data analysis* (pp. 257–291). Newbury Park: Sage.

4. Cook, R. D., & Weisberg, S. (1982). *Residuals and influence in regression*. New York: Chapman & Hall.

5. Crawley, M. J. (2002). *Statistical computing. An introduction to data analysis using S-Plus*. Chichester: Wiley.

6. Faraway, J. J. (2006). *Extending the linear model with R*. Boca Raton: Chapman & Hall/CRC.

7. Gelman, A., & Hill, J. (2007). *Data analysis using regression and multilevel/hierarchical models*. Cambridge, UK: Cambridge University Press.

8. Hardin, J. W., & Hilbe, J. M. (2007). *Generalized linear models and extensions*. College Station, Texas, USA: Stata Press, 2nd edition.

Regression in R—Part II

10

At the end of this chapter...
... you will be able to execute the most important steps of model diagnostics.
... you will no longer be impressed simply by a good-looking fit: the residuals and results of alternative model structures are also important.
... you will see that the graphic representation of the relationship is just as important as plotting the data itself.
... you will understand why a linear model is more popular than a generalised linear model (GLM): it makes model diagnostics easier.

10.1 Model Diagnostics

Let's go through the five points of model diagnostics mentioned in the last chapter by using an example.

1. Analysis of the predictor distribution;
2. Analysis of influential points;
3. Analysis of the dispersion;
4. Analysis of the residuals;
5. Analysis of the functional relationship between y and x.

As an example, we will use the data set from Chap. 9, Fig. 9.2:

```
> exa <- read.csv("regressionsexample.csv")
> exa
```

```
    x   y
1   1   9
2   2   9
3   3  10
4   4   7
5   5   7
6   6   7
7   7   7
8   8   6
9   9   7
```

© Springer Nature Switzerland AG 2020
C. Dormann, *Environmental Data Analysis*,
https://doi.org/10.1007/978-3-030-55020-2_10

```
10 10   5
11 42  33
12 35  40
```

We will first calculate a GLM under the assumption that the data are Poisson distributed: $\mathbf{y} \sim \text{Pois}(\lambda = a\mathbf{x} + b)$:

```
> fm <- glm(y ~ x, family=poisson, data=exa)
> summary(fm)

Call:
glm(formula = y ~ x, family = poisson, data = exa)

Deviance Residuals:
    Min       1Q    Median       3Q      Max
-1.5929  -0.7669  -0.2627   0.9187   1.8961

Coefficients:
            Estimate Std. Error z value Pr(>|z|)
(Intercept) 1.795581   0.133607  13.439   <2e-16 ***
x           0.045078   0.004893   9.213   <2e-16 ***
---
Signif. codes:  0 `***' 0.001 `**' 0.01 `*' 0.05 `.' 0.1 ` ' 1

(Dispersion parameter for poisson family taken to be 1)

    Null deviance: 88.216  on 11  degrees of freedom
Residual deviance: 12.226  on 10  degrees of freedom
AIC: 65.536

Number of Fisher Scoring iterations: 4
```

So we have our model. Now for the diagnostics.

10.1.1 Analysis of the Predictors

To analyse our predictor x with respect to its distribution, we will have a look at our trusty histogram. Code:

```
> hist(exa$x, cex.lab=1.5, col="grey")
```

We end up with Fig. 10.1, left plot. As we can see, the distribution is very right skewed and far from a uniform distribution. What should bother us is the large gap between the two value clusters (between 10 and 35). Are we sampling two totally different systems here? Is it appropriate to draw a straight line between these two clusters of points?

Let's try to reduce the gap by distributing the x values more evenly, which we achieve by transforming the data. One transformation often used in such a case (few outliers on the upper end) is the logarithm. After taking the logarithm, the histogram already looks much better (Fig. 10.1, right), and the gaps are closed.

It is good to see that in this case, the uneven spread of x did not affect the results much: the regression remains significant even after log-transforming the data (although obviously the estimate for the effect of x is different, as it is now measured in log-units):

```
> fm2 <- glm(y ~ log(x), family=poisson, data=exa)
> summary(fm2)
```

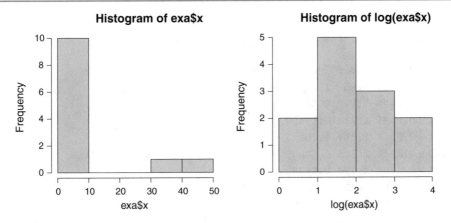

Fig. 10.1 Histogram of the predictor: raw data (left) and log-transformed data (right)

```
Call:
glm(formula = y ~ log(x), family = poisson, data = exa)

Deviance Residuals:
    Min       1Q   Median       3Q      Max
-2.5688  -1.2953  -0.3945   1.4002   2.5892

Coefficients:
            Estimate Std. Error z value Pr(>|z|)
(Intercept)  1.19004    0.21738   5.474 4.39e-08 ***
log(x)       0.60126    0.08006   7.510 5.92e-14 ***
---
Signif. codes:  0 `***' 0.001 `**' 0.01 `*' 0.05 `.' 0.1 ` ' 1

(Dispersion parameter for poisson family taken to be 1)

    Null deviance: 88.216  on 11  degrees of freedom
Residual deviance: 30.895  on 10  degrees of freedom
AIC: 84.205

Number of Fisher Scoring iterations: 5
```

The explained *deviance* has decreased significantly (from $88 - 12 = 76$ to $88 - 31 = 57$ units). This is the result of changing the influence of the two outliers.

As the result of this first step, we will only look at the model with the logged x values for further diagnostics (fm2).

10.1.2 Analysis of Influential Points

There are different measures for the influence of a data point. The two most important are Cook's distance and the so-called "hat" values.[1] In R, the command influence.measures calculates six measures including the two that were just mentioned:

```
> influence.measures(fm2)
```

---

[1] The hat matrix $\mathbf{H}$ analytically measures the influence of a data point on the fitted value $\hat{y}$: $\mathbf{H} = \mathbf{X}\left(\mathbf{X}^{\top}\mathbf{X}\right)^{-1}\mathbf{X}^{\top}$, where $\mathbf{X}$ is the matrix of the predictors.

```
Influence measures of
     glm(formula = y ~ log(x), family = poisson, data = exa) :

    dfb.1_ dfb.lg..   dffit cov.r  cook.d    hat inf
1    0.756  -0.6999   0.7564 0.807 1.08084 0.1553   *
2    0.395  -0.3465   0.3982 1.168 0.30478 0.1397
3    0.291  -0.2421   0.2995 1.232 0.16927 0.1248
4   -0.040   0.0314  -0.0426 1.387 0.00303 0.1129
5   -0.102   0.0749  -0.1136 1.345 0.02043 0.1040
6   -0.143   0.0978  -0.1706 1.290 0.04382 0.0978
7   -0.171   0.1061  -0.2202 1.231 0.06957 0.0938
8   -0.241   0.1319  -0.3405 1.070 0.14553 0.0918
9   -0.198   0.0913  -0.3125 1.108 0.12753 0.0916
10  -0.297   0.1063  -0.5331 0.796 0.28421 0.0930
11  -0.124   0.2040   0.2665 2.487 0.12344 0.5111   *
12  -0.527   0.9575   1.3457 1.146 2.66982 0.3841   *
```

(For the result with the untransformed x values, see the output.)

Cook's distance and the hat values can also be directly queried:

```
> cooks.distance(fm2)

         1          2          3          4          5          6
1.08083548 0.30477668 0.16927193 0.00303057 0.02043060 0.04382066
         7          8          9         10         11         12
0.06956570 0.14553078 0.12752602 0.28421163 0.12343615 2.66981692


> hatvalues(fm2)

         1          2          3          4          5          6
0.15533992 0.13969170 0.12479343 0.11293215 0.10404443 0.09778637
         7          8          9         10         11         12
0.09381782 0.09184772 0.09163454 0.09297833 0.51105239 0.38408119
```

Not only are the two high values (as in the untransformed analysis) noteworthy, but now data point 1 is as well. The transformation of x has significantly reduced the range of *Cook's distance* values, i.e. the data points now have a more equal effect on the regression estimators:

```
> range(cooks.distance(fm)); range(cooks.distance(fm2))

[1] 0.0003973005 3.9191471566
[1] 0.00303057 2.66981692
```

As described in the previous chapter, values greater than the thresholds of 1 or $4/12 = 0.33$ should raise a flag. Both limit values would identify points 1 and 12 as influential (in contrast to 11 and 12 for untransformed x values!). What are we going to do with this knowledge now? Since we usually have no reason to delete the data as being unsubstantiated, we can only check whether the relationship found exists even if the influential points are excluded. In our example, without points 1 and 12, we may expect the positive regression to tilt. Let's try it out [2]:

```
> fm3 <- glm(y ~ log(x), family=poisson, data=exa[-c(1,12),])
> summary(fm3)
```

---

[2]To do this we simply eliminate these two rows of data for the analysis of our data set by adding a negative index: `exa[-c (1,12),]`. This means: delete rows 1 and 12, but keep all columns (blank space behind the comma) of `exa`.

```
Call:
glm(formula = y ~ log(x), family = poisson, data = exa[-c(1,
    12), ])


Deviance Residuals:
    Min      1Q    Median       3Q       Max
-2.0576  -1.0261   -0.2989   1.0868    1.9504


Coefficients:
             Estimate Std. Error z value Pr(>|z|)
(Intercept)    1.0638     0.2709   3.927 8.59e-05 ***
log(x)         0.5836     0.1085   5.380 7.43e-08 ***
---
Signif. codes:  0 `***' 0.001 `**' 0.01 `*' 0.05 `.' 0.1 ` ' 1


(Dispersion parameter for poisson family taken to be 1)


    Null deviance: 42.833  on 9  degrees of freedom
Residual deviance: 16.623  on 8  degrees of freedom
AIC: 60.349


Number of Fisher Scoring iterations: 4
```

This is not the case: even if we get rid of points 1 and 12 the significant positive relationship remains.[3] If we were writing a report, after presenting the model for all data points, we would write, "Data points 1 and 12 were identified as conspicuous according to Cook's distance; their exclusion, however, did not qualitatively change the result (estimator for $\log(x) = 0.58 \pm 0.109$, $p < 0.001$)."

If the result does change (see footnote), then we would write, "After excluding the influential data points 11 and 12 (Cook's distance > 1: Cook and Weisberg 1982), the relationship between x and y was no longer significant ($p = 0.158$)."

10.1.3   Analysis of the Dispersion

With a GLM, a certain relationship between mean value and variance is assumed for each distribution (family). Dispersion refers to the factor by which the variance is greater than assumed. For the normal distribution, mean and variance are independent of each other, meaning that the dispersion here describes "only" the variance itself. With the Poisson distribution, the variance is equal to the mean (see Sect. 3.4.4; here, the dispersion parameter describes whether the dispersion is stronger than the mean value (*overdispersed*), or weaker (*underdispersed*).

While there are R-functions for individual distributions that calculate and test these dispersion parameters,[4] a rough estimate is good enough for us. To do this, we divide the residual deviance, provided in the summary output for a glm, by the residual degrees of freedom (found directly after the residual deviance). In our model, fm2 this is $30.9 : 10 = 3.09$, which is clearly overdispersed.

---

[3]This would be different with the untransformed x values, where 11 and 12 were identified as conspicuous. If we omit them, the relationship is no longer significant:

```
> summary(glm(y ~ x, family=poisson, data=exa[-c(11,12),]))


Coefficients:
             Estimate Std. Error z value Pr(>|z|)
(Intercept)   2.30488    0.23581   9.774  <2e-16 ***
x            -0.05765    0.04081  -1.413   0.158
```

We also see here that the first step of model diagnostics can influence the second step!.

[4]Such as dispersiontest in **AER** for Poisson GLMs.

The consequences of overdispersed data is an underestimated standard error of the model parameters (McCullough and Nelder 1989). So if we observe a clear deviation from the assumed value of 1 ("Disperson parameter for the poisson family taken to be 1"), we have to change the GLM and take this deviation into account. We can do this in two ways. Either we choose a distribution that has a different ratio between mean and variance, and therefore may not be falsely dispersed for our data set. Or we adapt the relationship between mean and variance by fitting it as additional parameter; we thus give up the assumption that the data come from exactly this distribution. Let's take a quick look at both options for our example.

10.1.3.1 Dealing with Overdispersion 1: Assume a Different Distribution

Instead of the Poisson distribution, we can fit a negative binomial distribution to our data (see Ver Hoef and Boveng 2007). It also describes count data, but has a second parameter that allows for clumping of data (it has a changing variance: see Sect. 3.4.5). A negative binomial GLM can be fit most easily with the function glm.nb from the **MASS** package[5]:

```
> library(MASS)
> fm5 <- glm.nb(y ~ log(x), data=exa)
> summary(fm5)

Call:
glm.nb(formula = y ~ log(x), data = exa, init.theta = 6.618929543,
    link = log)

Deviance Residuals:
    Min       1Q    Median        3Q       Max
-1.6045   -0.8696   -0.4380    0.6885    1.4020

Coefficients:
            Estimate Std. Error z value Pr(>|z|)
(Intercept)   1.4900     0.3234    4.607 4.08e-06 ***
log(x)        0.4633     0.1417    3.270  0.00108 **
---
Signif. codes:  0 `***' 0.001 `**' 0.01 `*' 0.05 `.' 0.1 ` ' 1

(Dispersion parameter for Negative Binomial(6.6189) family taken to be 1)

    Null deviance: 26.963  on 11  degrees of freedom
Residual deviance: 11.833  on 10  degrees of freedom
AIC: 78.513

Number of Fisher Scoring iterations: 1

          Theta:  6.62
      Std. Err.:  4.24

 2 x log-likelihood:  -72.513
```

Compared to our Poisson model (fm2), the parameter estimators have changed: the effect of log(x) is now weaker (0.46 compared to 0.60; we can directly compare these values since they both use the log as the *link* function). Moreover, the standard error of these variables has nearly doubled (from 0.08 to 0.14). As a result, the model has lost some significance.

With this approach, we also have to check for overdispersion: $11.8 : 10 = 1.18$, this looks good. The clumping parameter of the negative binomial distribution (in this textbook referred to as "r": Sect. 3.4.5), called θ (theta) by glm.nb, is estimated at 6.62 (Theta: 6.62 in the output). For $\theta \to \infty$ we get the Poisson distribution; so the smaller the value for θ, the less appropriate it is to use a Poisson GLM.

---

[5]Or alternatively with vglm from **VGAM**: summary(vglm(y log(x), data=bsp, family=negbinomial)).

By using the negative binomial distribution, we have now found an acceptable model with respect to dispersion.

10.1.3.2 Dealing with Overdispersion 2: Quasi-Distributions

The second way to deal with overdispersion is to modify the Poisson distribution to a quasi-Poisson distribution. The ratio of mean and variance is also taken into account and fitted. However, this is not possible for the Poisson distribution itself, and hence a similar quasi-likelihood is defined and fitted. There are specific quasi-likelihoods for Poisson and binomial distributions (`quasipoisson` and `quasibinomial`), but we can also more generally refer to another distribution, whose mean-to-variance function is specified by `quasi`. We can fit a quasi-Poisson distribution in the `glm` as follows:

```
> fm6 <- glm(y ~ log(x), family=quasipoisson, data=exa)
> summary(fm6)

Call:
glm(formula = y ~ log(x), family = quasipoisson, data = exa)

Deviance Residuals:
    Min      1Q    Median      3Q      Max
-2.5688  -1.2953  -0.3945   1.4002   2.5892

Coefficients:
             Estimate Std. Error t value Pr(>|t|)
(Intercept)    1.1900     0.3941   3.019   0.0129 *
log(x)         0.6013     0.1452   4.142   0.0020 **
---
Signif. codes:  0 `***' 0.001 `**' 0.01 `*' 0.05 `.' 0.1 ` ' 1

(Dispersion parameter for quasipoisson family taken to be 3.286975)

    Null deviance: 88.216  on 11  degrees of freedom
Residual deviance: 30.895  on 10  degrees of freedom
AIC: NA

Number of Fisher Scoring iterations: 5
```

The dispersion parameter of 3.27 is basically the ratio of residual deviance to residual degrees of freedom. The estimators for y-intercept and `log(x)` slope are identical to model `fm2`, but the standard errors are about twice the size, similar to those in the negative binomial model.

Since the fit is based on a quasi-likelihood, no AIC value can be calculated. So we are poorly equipped to compare models here.[6]

We find out that both corrections for overdispersion have produced weaker significance, but in both cases a clear, positive relationship remains. Ver Hoef and Boveng [2007] recommend method 1 as long as it is available and does not have issues with overdispersion itself.

10.1.4 Analysis of the Residuals

To analyse the residuals, we must first calculate them. We remember that there were three useful variants of residuals: Pearson, deviance and studentised. For the negative binomial model `fm5` that we have determined as our favourite, these are calculated as follows, together with the fit values at the link scale ($\hat{\mu}$, the parameter of the negative binomial distribution that changes with x) and at the level of the response variables ($\hat{y}$):

---

[6]qAIC-values also exist (e.g. `qaic` in the package **bbmle**), but stray a bit from the topic here.

```
> rP <- residuals(fm5, type="pearson")
> rD <- residuals(fm5) #  type="deviance" = default
> rSD <- rstudent(fm5)
> muhat <- predict(fm5, type="link")
> yhat <- predict(fm5, type="response")
```

We can now plot all six variants (Fig. 10.2), even if they are so similar in this case that we only need to look at the studentised residuals on the link-scale (bottom right):

```
> par(mfrow=c(2,3), mar=c(4,5,1,1))
> plot(rP ~ yhat, pch=16, las=1, cex.lab=1.5, xlab=expression(hat(y)))
>     abline(h=0, col="grey")
> plot(rD ~ yhat, pch=16, las=1, cex.lab=1.5, xlab=expression(hat(y)))
>     abline(h=0, col="grey")
> plot(rSD ~ yhat, pch=16, las=1, cex.lab=1.5, xlab=expression(hat(y)))
>     abline(h=0, col="grey")
> # link scale
> plot(rP ~ muhat, pch=16, las=1, cex.lab=1.5, xlab=expression(hat(mu)))
>     abline(h=0, col="grey")
> plot(rD ~ muhat, pch=16, las=1, cex.lab=1.5, xlab=expression(hat(mu)))
>     abline(h=0, col="grey")
> plot(rSD ~ muhat, pch=16, las=1, cex.lab=1.5, xlab=expression(hat(mu)))
>     abline(h=0, col="grey")
```

The pattern shown by the first ten residuals is very noticeable. Here we have to be suspicious: such a straight line, when these residuals should not have an obvious pattern at all?

So we are suspicious. What now?

The goal of model diagnostics is to detect possible errors and weaknesses in our analysis. That's what happened right here. We question our entire model! Apparently, $\log(x)$ is not enough to describe the pattern of the data well. The opposing tendencies of the first 10 data points compared to the trend across all data points are not captured. We obviously need a more appropriate functional relationship between y and x!

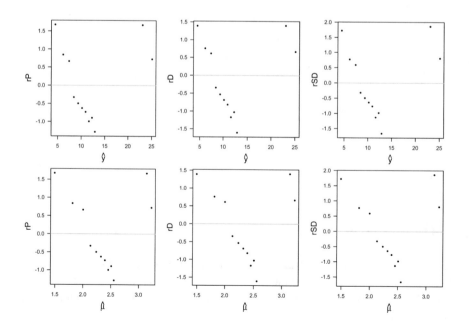

Fig. 10.2 The six variants of the residual plots: on the response scale (upper) and on the link scale (lower), with Pearson, deviance and studentised residuals (right to left)

10.1.5 Analysis of the Functional Relationship Between y and x

It is best to draw the modeled functional relationship as a line through the data, at least as a visual aid for yourself (sometimes it doesn't make sense to include the data in the background, especially if we are looking at multiple predictors simultaneously). To do this, we simply depict the fit relationship as a function. Again, here is the model that we fit:

$$\mathbf{y} \sim \text{NegBin}(\mu = e^{a \cdot \log(\mathbf{x}) + b}, r = \theta)$$

For a line that only plots the expected values, r (or θ) is irrelevant; it only describes the spread around the line. So we only need to plot the function $e^{a \cdot \log(x) + b} = e^{0.4633 \log(x) + 1.49}$. For the sake of completeness, we can also include our earlier model (fm2) as well (Fig. 10.3):

```
> plot(exa$x, exa$y, pch=16, cex=1.5, las=1, ylab="y", xlab="log(x)")
> curve(exp(coef(fm2)[2]*log(x)+coef(fm2)[1]), from=1, to=42, add=T,
+    col="lightgrey", lwd=3)
> curve(exp(0.4633*log(x)+1.49), from=1, to=42, add=T, col="darkgrey", lwd=3)
```

First, we notice that the linear relationship $a \log(x) + b$ on the response scale has become a non-linear relationship (since the function $e^{a \log(x) + b}$ is modeled here). If we take the inverse log of the x-axis, we get a completely different picture!

Depicting the 95% confidence intervals around the regression curve is not possible with the curve function. To do this, we need to go a more cumbersome route: define new x values, calculate the expected values of the model and the 95%-confidence intervals (so mean ± 2 standard deviations), and plot the latter (se.fit=T; Fig. 10.4):

```
> newx <- seq(min(exa$x), max(exa$x), len=100)
> nbpreds <- predict(fm5, newdata=list(x=newx), se.fit=T)
> qppreds <- predict(fm6, newdata=list(x=newx), se.fit=T)
> #Data points:
> plot(exa$x, exa$y, pch=16, cex=1.5, las=1, ylab="y", xlab="x", cex.lab=1.5)
> #neg.bin. curves:
> matlines(newx, exp(cbind(nbpreds$fit, nbpreds$fit-2*nbpreds$se.fit,
+    nbpreds$fit+2*nbpreds$se.fit)), lwd=2, col="grey30", lty=c(1,2,2))
> #quasipois. curves:
> matlines(newx, exp(cbind(qppreds$fit, qppreds$fit-2*qppreds$se.fit,
+    qppreds$fit+2*qppreds$se.fit)), lwd=2, col="grey70", lty=c(1,2,2))
```

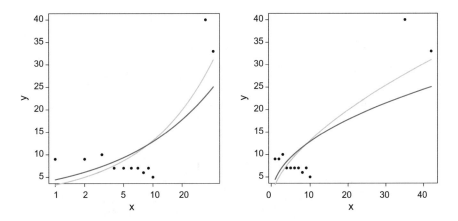

Fig. 10.3 Data and fit of our model: Poisson (light grey) and negative binomial (dark grey). Poisson and quasi-Poisson models make the same predictions with regard to the expected values (shown here) and differ only in their confidence intervals. Note that the x-axis is on the log scale on the left and is linear on the right. A logged y-axis on the left would turn the curves into straight lines

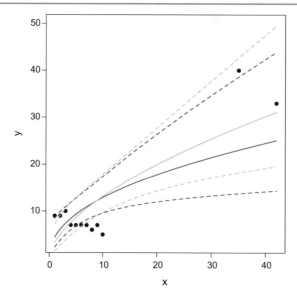

Fig. 10.4 Data and the fit of our model (negative binomial, `fm5`, dark grey; and quasipoisson, `fm6`, light grey) with 95% confidence intervals

---

Side note 8: Residual diagnostics for non-normally distributed data

The residual plots for Bernoulli and some other distributions are no fun. Even in the best cases, they never look like the "stars on a clear night" of normally distributed residuals. In fact, it is possible to achieve a similar visualisation, but only with a bit of (theoretical) effort.

The approach goes like this: A model is good if we can use it to invent new data similar to our observations. (If they were completely different, this means that our model could not describe the observed data.) So if we could simulate ("invent") data from a model, we could compare it with our observations (Dunn and Smyth 1996).

Imagine a typical regression, $\mathbf{y} \sim V(\theta = g(a\mathbf{x} + b), \varepsilon)$. After adapting this model to the data, we can use the parameters a, b and ε to simulate data. For V = normal distribution and link function $g(.)$ = identity function, the R-Code would look like `rnorm(length(y)`, `mean=a*x + b, sd=epsilon)`. For each observed y value, we now have a simulated value. We can repeat this process many times, so that for each observed y value we have *lots* of simulated values. We can plot these as an empirical cumulative density function (ECDF, see Fig. 3.6) and determine on which empirical quantile our observed y value lies. Ideally, the quantile values will be uniformly distributed and show no pattern in the scatter plot with the expected values.

The technical implementation of this idea is eased by the **DHARMa** package (Hartig 2017), here for a Bernoulli regression of the Bolger data set:

```
> bolger <- read.delim("bolger.txt")
> fm <- glm(RODENTSP ~ PERSHRUB, family=binomial, data=bolger)
> library(DHARMa)
> sim <- simulateResiduals(fm)
> plot(sim)
```

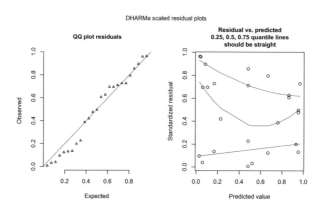

The figure on the left is a quantile plot (the little triangles should be on the line), whereas the figure on the right shows a scatter plot that should be as patternless as a starry night sky. The lines here should be horizontally parallel, but with so few data points, this will hardly ever be the case. All in all, both figures actually look quite reasonable—and much more *interpretable* than the typical residual plots (Fig. 10.2).

With the large confidence intervals starting at x values of 20, we can confidently describe the two models as very similar, even if the negative binomial model is consistently lower than the quasi-Poisson model.

Well, we haven't really done anything too useful yet. We now only know the exact shape of the models we have fitted. In this section, however, we want to formulate alternative models that take the residuals into account as well. There are two basic strategies for this:

1. Allow non-linear relationships on the link scale by using a polynomial.
2. Use a threshold value to fit two different correlations for different subranges (piecewise regression).

The effect of the second option seems very predictable: Values <12 will get their own linear and possibly non-significant regression, and the values >30 will get theirs, but the regression here will definitely not be significant (because only two data points are available for the two parameters intercept and slope). Since piecewise regression[7] is a relatively seldom used method, we'll not use it in this case. Instead, we will see if we can accomplish our goals using a polynomial.

Things are starting to get interesting!

We expand the negative binomial model ($fm5$) to include a quadratic term.[8] This can be done properly in R with the function $poly$[9] as follows:

```
> fm5.2 <- glm.nb(y ~ poly(log(x), 2), data=exa)
> summary(fm5.2)


Call:
glm.nb(formula = y ~ poly(log(x), 2), data = exa, init.theta = 196.8712563,
    link = log)


Deviance Residuals:
    Min       1Q    Median        3Q       Max
-1.33281  -0.63102  -0.02742   0.30764   1.51271


Coefficients:
                 Estimate Std. Error z value Pr(>|z|)
(Intercept)        2.2788     0.1003  22.717  < 2e-16 ***
poly(log(x), 2)1   1.4427     0.2621   5.505 3.70e-08 ***
poly(log(x), 2)2   1.4462     0.2981   4.852 1.22e-06 ***
---
Signif. codes:  0 `***' 0.001 `**' 0.01 `*' 0.05 `.' 0.1 ` ' 1
```

---

[7] Any interested readers should check out Toms and Lesperance [2003] and the **segmented** packaged.

[8] We could also use a cubic term, but here it won't make things better.

[9] Another sentence, another footnote: $poly(x, n)$ produces orthogonal polynomials of the nth order for the variable x. Unfortunately, orthogonal polynomials are not so simple as $a + bx + cx^2$, since b and c would be highly correlated. Instead, new variables, x' and x'', are defined in such a way that they are perfectly uncorrelated with (orthogonal to) each other. We do not need to understand the various ways in which this can be achieved, but we need to know that we cannot simply set the estimated parameters into the equation $a + bx + cx^2$! To make a plot, we **have to** use the $predict$ function. Many R users avoid this problem by explicitly formulating the polynomial: y ~ x +I(x^2). Then, you can directly use the coefficients and interpret them as b and c from $a + bx + cx^2$. But because both terms, x and $I(x^2)$, are not orthogonal, their standard errors are incorrectly distorted. Compare the relevant parts of the following model with the one in the main text (pay attention to the $Std. Error$ of the quadratic term):

```
> fm5.2b <- glm.nb(y ~ log(x) + I(log(x)^2), data=exa)
> summary(fm5.2b)
Coefficients:
              Estimate Std. Error z value Pr(>|z|)
(Intercept)    2.41323    0.27685   8.717  < 2e-16 ***
log(x)        -0.83080    0.28787  -2.886   0.0039 **
I(log(x)^2)    0.31320    0.06455   4.852 1.22e-06 ***
```

Since hardly anyone adheres to this knowledge, such important (in my opinion) information usually only shows up in a lengthy footnote (sic!).

```
(Dispersion parameter for Negative Binomial(196.8713) family taken to be 1)

    Null deviance: 81.4593  on 11  degrees of freedom
Residual deviance:  8.0258  on  9  degrees of freedom
AIC: 66.046

Number of Fisher Scoring iterations: 1

           Theta:   197
        Std. Err.:  873


  2 x log-likelihood:  -58.046
```

To our surprise we see that not only is this model better than fm5 (the *residual deviance* is only 8.03 instead of 11.83), but the estimator for θ jumped from 6.22 to 197! This means that it is now clearly on its way to infinity, and the negative binomial distribution is practically a Poisson distribution.

So now let's go back to our original Poisson distribution with a quadratic polynomial:

```
> fm2.2 <- glm(y ~ poly(log(x), 2), data=exa, family=poisson)
> summary(fm2.2)

Call:
glm(formula = y ~ poly(log(x), 2), family = poisson, data = exa)

Deviance Residuals:
     Min        1Q     Median        3Q       Max
-1.36588  -0.64422  -0.03574   0.31085   1.64201

Coefficients:
                   Estimate Std. Error z value Pr(>|z|)
(Intercept)          2.2795     0.0981  23.236  < 2e-16 ***
poly(log(x), 2)1     1.4457     0.2523   5.730 1.00e-08 ***
poly(log(x), 2)2     1.4362     0.2893   4.965 6.86e-07 ***
---
Signif. codes:  0 `***' 0.001 `**' 0.01 `*' 0.05 `.' 0.1 ` ' 1

(Dispersion parameter for poisson family taken to be 1)

    Null deviance: 88.2162  on 11  degrees of freedom
Residual deviance:  8.7953  on  9  degrees of freedom
AIC: 64.106

Number of Fisher Scoring iterations: 4
```

In fact, we no longer find any overdispersion in the Poisson model ($8.80 : 9 \approx 1$). And the residual deviance is practically identical to that from the negative binomial model.

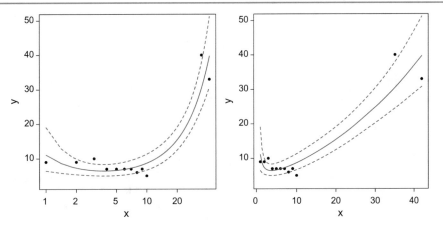

Fig. 10.5 Data and fit of our Poisson model with the quadratic term (fm2.2). Note that the *x*-axis is shown as linear on the *left* and logged on the *right*

We can also formally test these two models to see if the quadratic term in the model actually provides a significant improvement.

```
> anova(fm2, fm2.2, test="Chisq")

Analysis of Deviance Table

Model 1: y ~ log(x)
Model 2: y ~ poly(log(x), 2)
  Resid. Df Resid. Dev Df Deviance P(>|Chi|)
1        10    30.8947
2         9     8.7953  1   22.099 2.589e-06 ***
---
Signif. codes:  0 `***' 0.001 `**' 0.01 `*' 0.05 `.' 0.1 ` ' 1
```

The improvement from the linear model to a quadratic one is highly significant.

Let's now look at the functional relationship with the new model (Fig. 10.5). With the option log="x" we can easily take the log of the *x*-axis:

```
> ppreds <- predict(fm2.2, newdata=list(x=newx), se.fit=T)
> plot(exa$x, exa$y, pch=16, cex=1.5, las=1, ylab="y", xlab="x", cex.lab=1.5,
+    ylim=c(3,50), log="x")
> matlines(newx, exp(cbind(ppreds$fit, ppreds$fit-2*ppreds$se.fit,
+    ppreds$fit+2*ppreds$se.fit)), lwd=2, col="grey30", lty=c(1,2,2))
```

Or on a linear *x*-axis:

```
> ppreds <- predict(fm2.2, newdata=list(x=newx), se.fit=T)
> plot(exa$x, exa$y, pch=16, cex=1.5, las=1, ylab="y", xlab="x", cex.lab=1.5,
+    ylim=c(3,50))
> matlines(newx, exp(cbind(ppreds$fit, ppreds$fit-2*ppreds$se.fit,
+    ppreds$fit+2*ppreds$se.fit)), lwd=2, col="grey30", lty=c(1,2,2))
```

This relationship makes it possible to have a slightly negative relationship in the range of $x < 10$, and a positive relationship over the complete value range. This only looks logical in the logged plot (left). In the linear plot, the two values in the upper right still look like outliers that don't belong.

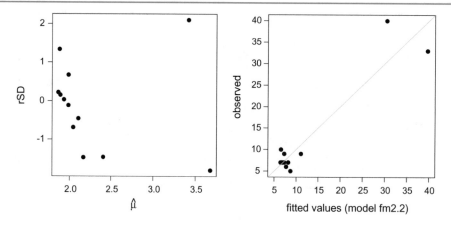

Fig. 10.6 Studentised residuals from the model fm2.2 (left) and 1:1-plot of the observed data against the fitted data (right). Compare the left plot with Fig. 10.2 to see the clear improvement (= less discernible pattern)

The main point here is that we have found a function that lies closer to the data *throughout* the range of data and is therefore much more plausible!

A quick look at the residuals (studentised on the link scale: Fig. 10.6, left) confirms our choice of model.

```
> rSD <- rstudent(fm2.2)
> muhat <- predict(fm2.2, type="link")
> plot(rSD ~ muhat, pch=16, las=1, cex.lab=1.5, xlab=expression(hat(mu)), cex=1.5)
> abline(h=0, col="grey")
```

The 1:1-plot between the observed data and model fit is also satisfactory (Fig. 10.6, right):

```
> yhat <- predict(fm2.2, type="response")
> plot(yhat, exa$y, las=1, xlab="fittet values (model fm2.2)", ylab="observed",
+     pch=16, cex=1.5, cex.lab=1.5,xlim=c(5,40), ylim=c(5,40))
> abline(0,1, col="grey")
```

What have we learned from this section? Lots of things!

1. An incomplete or falsely specified model distorts many model diagnostics, especially residual distribution and dispersion.
2. A different distribution should be tried out for comparison if the model fit leaves something to be desired.
3. A plot of the functional relationship, ideally with confidence intervals, is a great way to visualise whether different models are in fact different.
4. Tightening a "screw" on one part of the model (changing the assumed distribution, transforming the predictor, varying the functional relationship) has implications for other parts of the model as well!

10.2   Regression Diagnostics for a Linear Model (LM)

This whole chapter is constructed in such a way that the normal distribution is only one of many possible distributions. Unfortunately, this generalisation has to be dropped now, because the linear model has a few more features than the generalised linear model (not only in R). The GLM is basically correct, but there are a few substantial simplifications for the LM (see Side Note 9 for the analytical solution of the regression equation).

Side Note 9: Linear regression under the normal distribution—analytical

For the linear model, analytical solutions exist where the `glm` needs to be optimised (see Sect. 7.1). The parameters of a linear regression for normally distributed data ($\mathbf{y} \sim N(\mu = \beta_0 + \beta_1\mathbf{x}, \sigma = s)$) can be calculated as follows:

$$\beta_0 = \frac{\sum_{i=1}^n x_i^2 \sum_{i=1}^n y_i - \sum_{i=1}^n x_i \sum_{i=1}^n x_i y_i}{n \sum_{i=1}^n x_i^2 - \left(\sum_{i=1}^n x_i\right)^2} \tag{10.1}$$

and

$$\beta_1 = \frac{n \sum_{i=1}^n x_i y_i - \sum_{i=1}^n x_i \sum_{i=1}^n y_i}{n \sum_{i=1}^n x_i^2 - \left(\sum_{i=1}^n x_i\right)^2} \tag{10.2}$$

This means that with the sums of the x values, the sums of the y values, the sums of the squared x values, and the sums of the product of the x and y values you can calculate slope and intercept of the linear regression.

Each line in the linear model passes through the mean of $\mathbf{x}$ and $\mathbf{y}$, denoted as $(\bar{x}, \bar{y})$. If you calculate these values beforehand, the above formulae simplify:

$$\beta_1 = \frac{\sum_{i=1}^n (x_i - \bar{x})(y_i - \bar{y})}{\sum_{i=1}^n (x_i - \bar{x})^2} \quad \text{mit} \quad se_{\beta_1} = \sqrt{\frac{\sum_{i=1}^n (y_i - \bar{y})^2}{(n-2)\sum_{i=1}^n (x_i - \bar{x})^2}} \tag{10.3}$$

$$\beta_0 = \bar{y} - \beta_1 x_i \quad \text{mit} \quad se_{\beta_0} = \sqrt{\frac{\sum_{i=1}^n (y_i - \bar{y})^2}{n-2}\left(\frac{1}{n} + \frac{\bar{x}^2}{\sum_{i=1}^n (x_i - \bar{x})^2}\right)} \tag{10.4}$$

The formulae for calculating the standard error of the coefficient are also given.

The proximity to the correlation is shown by an equation that combines the correlation coefficients (Pearson's r) of x and y with β_1:

$$\beta_1 = r \frac{s_y}{s_x} \tag{10.5}$$

where s_y and s_x are simply the standard deviations of y and x as presented in Sect. 1.1.2.

In R, a regression for random variables from a normal distribution is fitted with the function `lm`. The syntax is identical to the `glm`, except that no `family` argument is required. The output is slightly different.

Let's go through this with an example, using the data from Fig. 10.7.

```
> x <- c(1, 2, 3, 4, 5)
> y <- c(4, 14, 25, 21, 33)
> fm <- lm(y ~ x)
> summary(fm)

Call:
lm(formula = y ~ x)

Residuals:
```

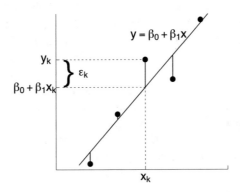

Fig. 10.7 A linear regression through five data points. The lines between the regression lines and the data points show the deviation of the model from the data. For a value of x_k, a variance value of ε_k can be calculated as the difference between y_k and the regression value $\beta_0 + \beta_1 x_k$

```
   1    2    3    4    5
-2.4  1.1  5.6 -4.9  0.6


Coefficients:
            Estimate Std. Error t value Pr(>|t|)
(Intercept)   -0.100       4.795  -0.021   0.9847
x              6.500       1.446   4.496   0.0205 *
---
Signif. codes:  0 '***' 0.001 '**' 0.01 '*' 0.05 '.' 0.1 ' ' 1

Residual standard error: 4.572 on 3 degrees of freedom
Multiple R-Squared: 0.8708, Adjusted R-squared: 0.8277
F-statistic: 20.22 on 1 and 3 DF,  p-value: 0.02054



> anova(fm)


Analysis of Variance Table


Response: y
          Df Sum Sq Mean Sq F value  Pr(>F)
x          1  422.5   422.5  20.215 0.02054 *
Residuals  3   62.7    20.9
---
Signif. codes:  0 '***' 0.001 '**' 0.01 '*' 0.05 '.' 0.1 ' ' 1
```

First, R gives us the deviations between the observed and calculated y values (the residuals). Then we get the coefficient calculations. Just as weith glm, R calls β_0 the y-intercept and β_1 is coded as the name of the explanatory variable, x. The parameter calculation is an estimation of the "true" parameters of the underlying model, which is why R calls this parameter *estimate*. In addition to the parameter value, we also get the standard error of this value and the results of a t-test, which tells us if this value is different than 0. In this case, the y-intercept is not significantly different from 0, but the slope is. The asterisks after the p-values are explained in the last line. Up to this point, the lm looks the same as a glm.

The last three lines provide a summary of the regression: Of the original total variance of y, 4.57 is left over, calculated with 3 degrees of freedom (n − number of estimated parameters). The regression reduces the variance in y by 0.87 or (with a calculation involving the number of calculated parameters: adjusted) 0.83. While the GLM specifies the deviance, here we directly get information regarding the explained deviance (which in the case of normally distributed data is equal to the sum of squares: see footnote 14 on Chap. 7). The F-test in the last line tells us whether the regression is significant or not. Since we only have one factor (x), this is identical to the t-test for the slope.[10]

Now for the output of the anova function.[11] This type of table is called an ANOVA-table (see Sect. 11.2), because it compares the sum of squares (= SS) for the individual effects of the model with the unexplained sum of squares. Accordingly, the first row contains the name of the variable, x, followed by the degrees of freedom (Df) of the effect, the sum of squares (Sum Sq) for the effect, the mean sum of squares (MSS = Mean Sq = Sum Sq/Df), the F value (F-value = Mean Sq of the effect / Mean Sq of the residuals), and their significance (Pr(>F)). In the next row, we see the corresponding results for the residuals, but here, of course, without the significance test.

The p-value of the summary and the anova commands are only the same for models with a single explanatory variable. As a rule, the anova command is the correct one for extracting a significance value, while the summary command is especially useful for figuring out the values of the coefficients and their error.

For regressions made using lm, R offers an easy solution for model diagnostics. Using the command plot(fm), we can output different diagnostic plots of the regression analysis (Fig. 10.8).

---

[10]In this case (only one predictor) it is in fact $F = t^2$.

[11]With the option test="Chisq" or test="F" (see ?anova.lm) in the anova function, we can decide which test we want to use for the significance value. For linear models, the ratio of factor and residual MSS corresponds to an F distribution. For non-normally distributed data (GLM), a χ^2-distribution is generally assumed. When applied to normally distributed data, this is somewhat more conservative, but usually very similar.

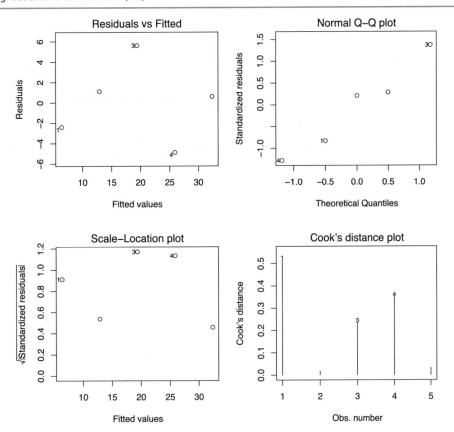

Fig. 10.8 Standard diagnostics plot for a linear model in R. See the text for explanation

```
> par(mfrow = c(2, 2))
> plot(fm)
```

This command results in the output of four plots that all tell us different things. In the first one, we only see the (Pearson = standardised) residuals plotted against the predicted values. Now we look for a systematically larger spread on the right side, which of course would not be obvious with only a few data points.

The second plot (upper right) is a variant of the qqnorm-plot. In this plot, the Pearson residuals are sorted by size and summed (y-axis). These values are then plotted against the expected values (x-axis) based on a normal distribution. The Qs stand for "Quantile": empirical quantiles on the y-axis, theoretical quantiles on the x-axis. Ideally, the points lie exactly on the 1:1 line.

The third plot is similar to the first and shows the square-root of the absolute Pearson residuals plotted against the predicted values (also called scale-location or S-L plot), which diminishes the skewness of the residuals.

And finally, the last plot shows the influence of each observation for the regression results (Cook's distance). The larger the value, the more influential the observation. Critical limits are at 1 or $4/n$ (see Sect. 9.1.2).

The model diagnostics for lm are thus much easier than for the glm in R .

10.3 Exercises

- Repeat the exercises for Regression Part I. After each model, perform model diagnostics!
- In the data set eggs.txt (Thanks to Martin Schaefer, Uni Freiburg!) we find data regarding a species of bird (the El Oro parakeet), and whether the presence of helpers (number_helpers—mostly their young from earlier years - has an effect on the number of eggs they lay (eggs_total) (Klauke et al. 2013). Analyse this relationship. Pay attention to the type of distribution (try different ones) and to overdispersion. For example, you can also use a GLM with a normal distribution and look at how the conclusions for the role of the helper depend on the model structure.

References

1. Cook, R. D., & Weisberg, S. (1982). *Residuals and influence in regression*. New York: Chapman & Hall.
2. Dunn, P. K., & Smyth, G. K. (1996). Randomized quantile residuals. *Journal of Computational and Graphical Statistics, 5*(3), 236.
3. Hartig, F. (2017). DHARMa: Residual diagnostics for hierarchical (Multi-Level/Mixed) regression models. *R Package Version,* 0.1.5.
4. Klauke, N., Segelbacher, G., & Schaefer, H. M. (2013). Reproductive success depends on the quality of helpers in the endangered, cooperative El Oro parakeet (Pyrrhura orcesi). *Molecular Ecology, 22*(7), 2011–2027.
5. McCullough, P. & Nelder, J. A. (1989). *Generalized linear models*, 2nd Ed. London: Chapman & Hall.
6. Toms, J. D., & Lesperance, M. L. (2003). Piecewise regression: A tool for identifying ecological thresholds. *Ecology, 84*, 2034–2041.
7. Ver Hoef, J. M., & Boveng, P. L. (2007). Quasi-poisson vs. negative binomial regression: How should we model overdispersed count data? *Ecology, 88*, 2766–2772.

If you give people a linear model function you give them something dangerous.

—John Fox (`fortunes(49)`)

> At the end of this (long) chapter ...
> ... you will know the t-test in its many different variations.
> ... You will understand that the idea of variance analysis is to divide the total variance into explainable and unexplained variance.
> ... you will know the F-test for calculating the significance of a variance analysis.
> ... you will understand the close relationship between ANOVA and regression.

In this chapter we will (unfortunately) leave the approach that is available for all distributions by the means of maximum likelihood, and will rather focus on **exclusively normally distributed response variables y** in a functional dependence of one (or multiple) predictor(s) $\mathbf{X}$: $\mathbf{X}$: $\mathbf{y} \sim N(\mu = f(\mathbf{X}), \sigma)$.[1] So we for the next two chapters only consider the LM (linear model), not the GLM (generalised linear model, which generalises the LM to other distributions).

For the analysis of normally distributed response variables, three approaches are important. These have historically different origins, but in principle all of them are a form of (G)LM: regression, the t-test and the analysis of variance (ANOVA). We already know the regression from the previous chapters, so we will skip over it here.

In the following sections, we will first look at the t-test and then the ANOVA. After these concepts have been introduced, we will bring them together with the linear model.

11.1   The t-test

The t-test (also called the Student's t-test[2]) tests for difference in means between (one or) two samples. The core concept of the t-test is genius in its simplicity: We compare an estimated value with its standard error. If the value is high in relation to the standard error, then it is significant. An *estimated value* can be the average of one sample, the difference between two samples or another statistic (e.g. a slope estimate in a regression) as long as it is normally distributed.[3] The t-distribution describes the significance of this difference depending on the number of data points. It looks like a somewhat flattened normal distribution and also becomes a normal distribution when the sample size goes towards infinity (Fig. 11.1).

---

[1]We will also shortly consider the case that σ depends on $\mathbf{X}$, but this is generally not possible in the linear model. The notation $f(.)$ indicates that non-linear functions could also be considered. But we're not going to do that here.

[2]At this point it is obligatory to note that "Student" was the pseudonym of W. S. Gosset when he published the t-test while working for the Guinness Brewery. His employer considered quality assurance statistics as a trade secret. However, Gosset's colleagues in mathematics knew the person behind pseudonym.

[3]The central limit theorem states that (most) parameter estimators are (asymptotically) normally distributed, even if the considered variables are *not* normally distributed. For example, if we estimate the median of a sample, this estimate has a certain error because we consider only one sample. The error of this median is normally distributed, although our sample might be crooked and skewed!

© Springer Nature Switzerland AG 2020
C. Dormann, *Environmental Data Analysis*,
https://doi.org/10.1007/978-3-030-55020-2_11

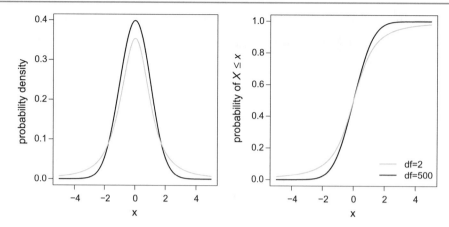

Fig. 11.1 The density and cumulative distribution function of the *t*-distribution, also called Student distribution. It only has one parameter, df, which provides the number of degrees of freedom. For high df values (black line), it is hardly discernible from a normal distribution. (As a rule of thumb, *t*-values > 2 are significant if we have more than 20 data points per group.)

11.1.1   One Sample *t*-test

Assume that we have only one sample, let's say the shoe sizes of 100 people, and we want to know if they deviate from an expected value (e.g. from the "true" mean shoe size of Germans at 40.2[4]). Let's further assume that the data are normally distributed. We can then use the *t*-test to test this comparison. The equation looks like this:

$$t = \frac{\bar{x} - A}{SE_{\bar{x}}} = \frac{\bar{x} - A}{sd_x / \sqrt{n}} \tag{11.1}$$

with $\bar{x}$ equal to the mean of the sample, A is a constant against which the value is to be tested (for us it is 40.2) and $SE_{\bar{x}}$ is the standard error of the mean of the sample (in the right side of the equation this is broken down as the standard deviation divided by the square root of the sample size).

Critical *t*-values used to be found in a table in the appendix of statistics textbooks (e.g. Zar 2013). Today, the *t*-distribution is integrated as a probability density function in all statistics programs and even in most spreadsheet programs.[5]

The fewer data points we have, the flatter the distribution, which takes into account the fact that we look at the statistics of a *sample* rather than the population as a whole. If we drew a standard normal distribution into this figure, it would be practically identical to the df=500 curve.[6] It is important to note whether the *t*-value of our data is very far left or right, i.e. whether only a very small part of the distribution is larger or smaller. So we are not interested in the $+/-$ sign (which only tells us whether the largest part of the *t*-distribution is larger or smaller), but rather the amount.

For 14 measured shoe sizes (41, 40, 37, 45, 41, 39, 38, 20, 43, 44, 44, 36, 41, 42) we get $\bar{x} = 39.36$, *sem* $= sd_x \sqrt{n} = 6.18/3.74 = 1.65$ and therefore $t = 39.36/1.65 = 23.81$. This very high value tells us that the measured values are very clearly ($p < 0.001$) different from 0 (this is the value against which the *t*-test tests by default).

However, we are much more interested in a comparison with the expected value of 40.2. To do this we subtract this value from 39.36 and divide it by *sem*: $t = -0.843/1.65 = -0.510$. That does not look significant:

```
    One Sample t-test
t = -0.5099, df = 13, p-value = 0.6186 alternative hypothesis:
true mean is not equal to 40.2 95 percent confidence interval:
 35.78636 42.92793
```

---

[4]That's a mens size 6.5 for those of you in the UK and a men's size 7 for you Yankees out there.

[5]In Libre/OpenOffice Calc with the function TDIST or in Microsoft Excel with the function TINV. The reasons why we should never rely on MS Excel for statistical calculations are given by McCullough and Heiser [2008].

[6]In fact, the normal distribution is also shown in the right figure, but is indistinguishable from the *t*-distribution with df = 500.

The result could be written in a report as follows: "The measured show sizes ($\bar{x} = 39.4$, $s = 6.18$) are not significantly different from the expected value of 40.2 ($t_{13} = -0.510$, $p = 0.619$)." The negative *t*-value shows that the sample is below the expected value. The subscript 13 for the *t*-value tells us the number of degrees of freedom (in the output above, this is `df`). We will deal more with degrees of freedom in Sect. 11.3.2 on page 160.

11.1.2 Paired Sample *t*-test

If we have two **paired samples**, i.e. each *i*-th value x_{1i} of one sample (x_1) is uniquely assigned to the *i*-th value x_{2i} of a second sample (x_2), the analysis differs little from the comparison of a sample with a constant value (Eq. 11.1). Instead of the constant *A*, the paired x_{2i} value is now subtracted. The value that is subjected to the *t*-test is thus a vector of differences between the data pairs.

The *t*-test equation now looks like this:

$$t = \frac{\frac{1}{n}\sum_{i=1}^{n}(x_{1i} - x_{2i})}{SE_{x_{1i}-x_{2i}}} \tag{11.2}$$

Here we assume that both samples stem from a normal distribution with the same standard deviation. Otherwise the denominator would have to be adjusted (see below).

Such paired samples occur both in simple experimental arrangements (e.g. treatment and control in one block) and in descriptive studies, such as the number of moss species on the southern and northern sides of several trees: (5, 8, 7, 9, 9, 9) and (12, 23, 15, 18, 20), respectively.

What we do here is to convert the two data sets into *a single* set and then perform the *one-sample t-test* as above. We first calculate the difference North−South and get (7 15 8 9 9 11). The mean is exactly 10, the standard deviation is 3.16. The sample size is now 5 instead of 10. Accordingly, $t = 10/\frac{3.16}{\sqrt{5}} = 7.07$. This value is significant ($p = 0.002$). In a report we would write "Although inference on such a small sample is unlikely to be robust, the northern sides had significantly more moss species than southern sides of the same trees (paired *t*-test, $t_4 = 7.07$, $p < 0.01$)."

A typical output from a stats software might look like this:

```
Paired t-test

data:  arten.nord and arten.sued
t = 7.0711, df = 4, p-value = 0.002111
alternative hypothesis: true difference in means is not equal to 0
95 percent confidence interval:
 6.073514 13.926486
```

11.1.3 Two Sample Test

If we are dealing with **unpaired samples**, i.e. the first value of $\mathbf{x}_1$ is unrelated to the first value of $\mathbf{x}_2$, the *t*-test can be calculated as follows:

$$t = \frac{\text{Difference of the means}}{\text{Standard error of the differences}} = \frac{\bar{x}_1 - \bar{x}_2}{SE_{\text{Differences}}}, \tag{11.3}$$

$$\text{where } SE_{\text{Differences}} = \sqrt{\frac{s_1^2}{n_1} + \frac{s_2^2}{n_2}}.$$

Here, too, the value to be tested is the difference between the samples, only the calculation of the standard error is a bit more complicated due to the different variances (and possibly different sample sizes).

This is the *t*-test in its most general form. As you can see, both samples do not have to have the same variance: they can simply be entered into the formula as s_1^2 or s_2^2.

So, for example, if we had a data set with the number of moss species on different trees instead of on the same ones, we would get the following test: $t = 10/\sqrt{(18.3/5 + 2.8/5)} = 4.87$. This value is also statistically significant ($p < 0.01$).

We see that the difference is "less significant" compared to the paired test, since in this case, less information is used.[7] This variation of the *t*-test, in which the samples can have different variances is called the Welch two sample test.

```
    Welch Two Sample t-test

data:  arten.nord and arten.sued
t = 4.8679, df = 5.196, p-value = 0.00415
alternative hypothesis: true difference in means is not equal to 0
95 percent confidence interval:
  4.778713 15.221287
```

11.2   Analysis of Variance (ANOVA): Analysing for Significant Differences

The analysis of variance, often called ANOVA, looks primarily for differences between groups and tests whether dividing the data into different groups reduces unexplained variability. It was formalised by Fisher [1918, 1925], a first comprehensive reference for ANOVA is Mann [1949] and a more current volume is Underwood [1997], which is also more accessible to ecologists.

Let's work through an example. Figure 11.2 shows the ANOVA concept for the data set we just analysed by means of *t*-test. When we start, we just have 10 values, that are mixed up in no particular order. We can calculate the variance of the data set, as described in Chap. 1: $\sigma^2 = \dfrac{\sum_{i=1}^{n}(y_i - \bar{y})^2}{n-1}$. The variance in this case is $\sigma^2 = 37.15$. This is the **overall variance** of our data set. To make things easier for us, we now focus on the numerator of this fraction for all calculations, and divide by the appropriate sample size later. The numerator is called the sum of squares, *SS*. The *SS* of the overall variance is called SS_{total} and in this case: $SS_{\text{total}} = \sigma^2(n-1) = 37.13 \cdot 9 = 334.4$.

Now we come to the core idea of the analysis of variance: can we divide this overall variance into two parts, namely the part explained by our predictor and the remaining unexplained variance? Or, in the notation with the sum of squares: Can we break down SS_{total} into one explainable part, called SS_{Factor} or SS_{Effect}, and one inexplicable part, called SS_{error} or SS_{residual}?

Important note: Even though the ANOVA was invented to be used for categorical predictors, it also works for continuous variables! More on that later, once we have really understood the principle behind ANOVA (see Sect. 11.3).

In our example, we basically ask the ANOVA if separating the data into two groups, North side and South side, will reduce the unexplained variance (or the SS_{residual}; see Fig. 11.2, right side). The idea is that the overall variance (or SS_{total}) can be split

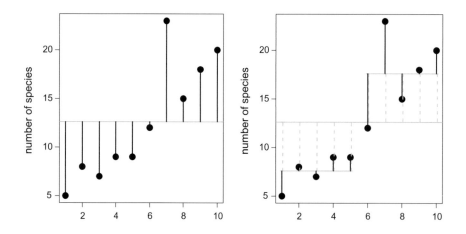

Fig. 11.2 Visualisation of an analysis of variance (ANOVA). On the left: Without explanatory variables, the sum of squares are the squared distances to the mean value. For reasons of clarity, only the distances, not their squares, are shown. On the right: In addition, the mean values per group are now displayed. The residuals (black) are only calculated within the groups, while the variance explained by the differences in the groups is shown with dashed grey lines

---

[7]The phrase "less significant" is rightly frowned upon by statisticians. If we have a threshold for significance, conventionally 0.05, then any value below it is significant, fullstop. "Less significant" is akin to "more pregnant".

up into the variance *within* the groups and the variance *between* the groups. The *SS* within the groups remains unexplained as does our SS_{residual}, while the *SS* between groups is explained by our factor levels "North" and "South": SS_{Effect}.

So we split the data into two groups, the first five data points from the South side, the second five data points from the North side and calculate a group mean for each group (this comes out to $\bar{y}_{\text{South}} = 7.6$ and $\bar{y}_{\text{North}} = 17.6$, respectively, shown as grey horizontal lines). As shown in the right panel of Fig. 11.2, we can now draw a line from any point to the group mean (black). These are the unexplained residuals, which we use to calculate the unexplained sum of squares SS_{residual} by squaring and summing:

$$SS_{\text{residual}} = (y_1 - \bar{y}_{\text{North}})^2 + (y_2 - \bar{y}_{\text{North}})^2 + \cdots + (y_6 - \bar{y}_{\text{South}})^2 + (y_7 - \bar{y}_{\text{South}})^2 + \cdots$$

In our case, $SS_{\text{residual}} = 84.4$.

SS_{Effect} can be calculated from the difference between the group mean and the overall mean (the dashed grey lines in Fig. 11.2, right side):

$$SS_{\text{Effect}} = n_{\text{North}}(\bar{y}_{\text{North}} - \bar{y})^2 + n_{\text{South}}(\bar{y}_{\text{South}} - \bar{y})^2,$$

where $\bar{y}$ is the overall mean and n_{North} is the number of data points in the North group (same for the South group). The value of $SS_{\text{Effect}} = 250$.

And in fact,

$$SS_{\text{total}} = SS_{\text{Effect}} + SS_{\text{resid}} = 250 + 84.4 = 334.4$$

Good work. We have now shown that the variance or the sum of squares can be divided into the part "explained by the predictor" and "still unexplained". What do we get from this? Similar to the *t*-test, which compares a value with its standard error, we compare the proportion of explainable variance with the proportion of inexplicable variance in an ANOVA. The greater the proportion of explained variance, the more likely it is that the grouping is a non-random predictor of our response variable.

In short: the more that we can explain through an effect, the more significant that effect is.

This is quantified by the *F*-test. We calculate the ratio of mean predictor variance to mean residual variance:

$$F = \frac{MS_{\text{Effekt}}}{MS_{\text{resid}}} = \frac{\frac{SS_{\text{Effekt}}}{df_{\text{Effekt}}}}{\frac{SS_{\text{resid}}}{df_{\text{resid}}}} \tag{11.4}$$

The degrees of freedom[8] (df) are calculated for the effect as the number of factor levels -1 and for the residuals as number of data points $-$ degrees of freedom of the effect -1.

For our example, we have

$$F = \frac{250/(2-1)}{84.4/(10-1-1)} = \frac{250}{10.6} = 23.7.$$

We can calculate the significance level of this value using the cumulative *F*-distribution (Fisher distribution), or look it up in an appropriate table. We are interested in which portion of the *F*-distribution is *larger* than our measured value. Figure 11.3 shows the decumulative *F*-distribution for 1 and 8 degrees of freedom as well as the *F*-value and the *P*-value.

We have now learned a new and alternative method in variance analysis to test for a significant difference between two groups. The result is the same as in the *t*-test. Furthermore, the results of a one-way ANOVA and a *t*-test are identical, *if the two samples have the same variance*. With the *t*-test, however, we can even examine different variances in the two samples, which is not possible with the ANOVA. So why use ANOVA in the first place?

Two points distinguish ANOVA from the *t*-test: First, ANOVA is also generalised to a factor with *several* levels. And second, ANOVA is also generalised for *several* predictors, and even combinations of continuous and categorical predictors! While the *t*-test continues to play an important role both as one- and two-sample test, the ANOVA is **the** statistical tool of the 80s and 90s; and today, the more modern GLM is still often presented as a type of ANOVA (more on this later).

---

[8]See Sect. 11.3.2 on page 160 for a more detailed derivation. For the moment, we can think of it as a measure of how much effort we put into the calculation of means: the more classes, the more degrees of freedom we use. A more reasonable explanation unfortunately has to wait until we look at ANOVA and regression as two sides of the same coin.

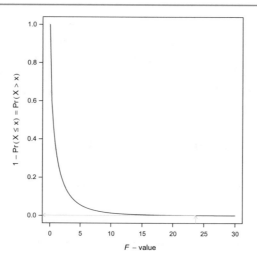

Fig. 11.3 Decumulative (beginning with 1 and decreasing to 0) F-distribution for the parameter values df1 $= 1$ and df2 $= 8$. The F-value from the example (23.7) is drawn in grey and translated to a P-value (0.00124, grey arrow). Since we are interested in the part of the distribution that is *larger* than our F-value, I computed $1-$ the cumulative distribution

In the end, we can calculate the "explained variance" using ANOVA. The explained variance is simply the portion of the overall variance that we can attribute to a predictor. The explained variance is denoted as R^2, due to its close relationship to Pearson's correlation coefficient, r.[9] It is $R^2 = \dfrac{SS_{\text{Effekt}}}{SS_{\text{total}}} \cdot 100\%$.

It is common to show the results of an analysis of variance in an ANOVA table. Such a table has the following format, for a factor with k different levels and an overall sample size of n, whereby the last row here is often left out:

Source	d.f.	SS	MS	F	P
Effect	$k-1$	SS_{Effect}	$\dfrac{SS_{\text{Effect}}}{k-1}$	$\dfrac{MS_{\text{Effect}}}{MS_{\text{Residuals}}}$	n.s., *, **, ***
Residuals	$n-k$	$SS_{\text{Residuals}}$	$\dfrac{SS_{\text{Residuals}}}{n-k}$		
Total	$n-1$	SS_{overall}			

This table contains a lot of information, some of which is redundant. The first column indicates where the variance actually is: in the effect and in the residuals, adding up to the overall sum of squares. The second column shows the respective degrees of freedom, the third column shows the numerical values for the sum of squares. The column MS shows the mean sum of squares, used to calculate the F-value as ratio $MS_{\text{Effect}}/MS_{\text{Residuals}}$. Finally, the significance level of the F value is given, possibly with asterisks to indicate a certain level of significance.

Here is the ANOVA table for our analysis, as it is output by R:

```
          Df Sum Sq Mean Sq F value   Pr(>F)
Effect     1  250.0  250.00  23.697 0.001243 **
Residuals  8   84.4   10.55
---
Signif. codes:  0 '***' 0.001 '**' 0.01 '*' 0.05 '.' 0.1 ' ' 1
```

First, we notice that R does not include a row with the overall *SS*. We can simply sum up the Sum Sq column if we really wanted to. We are looking at 10 data points in our example, so the df$_{\text{overall}} = 10 - 1 = 9$. Our explanatory variable, here

---

[9]The squared correlation coefficient between observed data $\mathbf{y}$ and the model fit $\hat{\mathbf{y}}$ is exactly R^2. You will find both r^2 and R^2 in the literature. In the simple regression model, $R^2 = r^2$ is used, but this is no longer the case with non-linear models. There, the R^2 loses its clear interpretability, since the null model is no longer necessarily a sub-model, and thus the comparison of the sum of squares is meaningless.

designated as "Effect", has 2 levels so that $k = 2$ and $df_{Effect} = 2 - 1 = 1$.[10] We have already calculated the SS-values, and they are reported here as well. The MS-values come from dividing the SS-valued by the df-values. This is then $84.4 : 8 = 10.55$. The F-value is the ratio of MS_{Effect} to $MS_{Residuals}$ and in this case is 23.7. This has a P-value (significance level) of 0.0012, which gets two asterisks according to the table of significance codes at the bottom of the output.

It is easy to tell what makes an effect significant by looking at an ANOVA table:

1. a large differences between overall mean and group mean values (which leads to a high SS_{Effect});
2. low noise within the groups (which leads to low $SS_{Residuals}$);
3. many replicates (in which a high value for degrees of freedom leads to low $MS_{Residuals}$);
4. or, more subtly, through as few levels of the predictor as possible (because many levels result in a lower MS_{Effect}).

In short: strong effects, homogeneous groups, many measurements and as few levels as possible for the explanatory variable give us some clear results.

11.2.1 ANOVA with a Continuous Predictor: An Example

Just for the sake of completeness, let's take a quick look at an ANOVA where the predictor is continuous. In such a case, we cannot talk about group differences, and the calculations we did above are not possible. But when we perform a regression, we can compare the scattering around the regression line (= $SS_{Residuals}$) with the variance explained by the regression line (SS_{Effect}), just like an ANOVA with a categorical predictor.

Think back to our example of the calibration curve of cherry tree diameter and wood volume. If we feed these data into an ANOVA, it returns:

```
          Df Sum Sq Mean Sq F value    Pr(>F)
Girth      1 7581.8  7581.8  419.36 < 2.2e-16 ***
Residuals 29  524.3    18.1
---
Signif. codes:  0 `***' 0.001 `**' 0.01 `*' 0.05 `.' 0.1 ` ' 1
```

The output looks the exact same as for with categorical predictors, and we can also calculate the R^2 with this data: $7581.8/(7581.8 + 524.3) = 0.935$. This example basically shows us that it makes no difference whether the predictors are continuous or categorical.[11]

11.2.2 Assumptions of an ANOVA

As with regression, ANOVA also requires model diagnostics. For this, we plot the residuals against the fit values. (Which residuals? This does not make a difference with normally distributed data, and there is no link scale either.)

We are primarily interested in two things:

1. Are the variances the same across all levels (variance homogeneity or homoscedasticity)?
2. Are the residuals normally distributed (assumption of normality)?

In scientific articles, we often read that "the data were normally distributed". However, this information is much less relevant than information on whether the variances are constant. A formal test for variance homogeneity is (drumroll please ...)— Surprise!—the F-test![12] This is simply the quotient of the variances of both groups, such as: $F = \frac{s^2_{North}}{s^2_{South}}$. (For our example,

---

[10]By the way, there are different ways to calculate the degrees of freedom. It is typically assumed, for example, that all groups contain the same number of data points (so-called *balanced design*), which unfortunately is rarely the case in the real world. We therefore use a generally valid equation here, which will also be useful to us later on, if we want to combine ANOVA and regression.

[11]This is especially important when considering comparisons with other literature. In other textbooks, ANOVA is often only suggested for use with categorical predictors. We see here that this line of thought is a bit narrow-minded.

[12]Well this actually shouldn't come as a surprise. We have been using the F-value throughout this chapter to test whether the predictor has a significant influence on the variances.

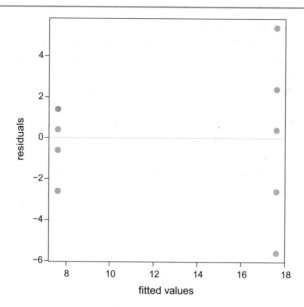

Fig. 11.4 Diagnostic plot for our ANOVA of moss diversity on the North and South side of 5 trees. The slightly darker point in the upper left contains two data points (lying on top of each other). We can clearly see that the spread is different for these two groups (South side is on the left)

$F = \dfrac{2.8}{18.3} = 0.153$, which gives us a P-value of 0.048. Note here that depending on whether North or South is in the denominator, either the right or left end of the F distribution should be considered. The variances are significantly different, and the assumption of homoscedasticity is therefore violated!)

In addition to the F-test, other model diagnostics for ANOVA include the Cochran's, Bartlett's, or Levene's test. The reason that most textbooks avoid these formal tests (see Crawley 2002; Dalgaard 2002; Quinn and Keough 2002) is that these tests are very sensitive to the number of data points. For large sample sizes, a difference in variance can always be found, although it may be practically irrelevant. Conversely, even serious differences in variance can not be detected as such with a small sample size. Therefore, the authors above argue for a graphical representation and purely visual diagnostic, such as box plots within each group, or, better yet, the residual-fitted values-plot (Fig. 11.4).

We have already met the different tests for normality (Sect. 3.5.1), so they need not be covered again here. For these tests as well, graphical diagnostics is the way to go. Here is an example for a residuals against fitted values plot (Fig. 11.4). What we can't really tell with five data points is whether the data are normally distributed or not. For this we would need some more data points!

11.2.3 Making Non-normally Distributed Data Work with ANOVA

As explained in detail in the previously mentioned volume from Underwood [1997], one can try to use transformation on the response variables to get them to a nearly normal form. This option should only be used if the many available distributions in the GLM cannot represent the data distribution. Before about 1980, when ANOVAs were still calculated by hand and GLMs were hard to calculate due to the necessary computational capacity, data were often transformed. We still see advice from this time, which is gradually proving to be unsuitable (O'Hara and Kotze 2010; Warton and Hui 2011).

Transforming the response variables is generally unproblematic and legitimate. If we measure temperature in Fahrenheit instead of Celsius (a linear transformation), the acidity of a lake as pH or as proton concentration (log-transformation), or the volume from a speaker in watts or in decibels (also a $-$log-transformation): the selection of a scale is completely arbitrary. Natural laws should be valid regardless of the measuring scale, i.e. they should be convertible from one scale to the other. Accordingly, we can also transform our response variable if we hope to gain an advantage. But there things we should consider:

- When we transform data, also the function of the relationship with x changes. If, for example, the true relationship between y and x_1 and x_2 is linear and additive (e. g.: $y = mx_1 + nx_2 + b$, then a log transformation implicitly leads to a *multiplicative* relationship:

$$log(y) = mx_1 + nx_2 + b$$
$$y = e^{mx_1 + nx_2 + b} = e^{mx_1} e^{nx_2} e^b.$$

- The aim of the transformation is to transform the response variable into the distribution that we assume for the model. This means that we still need to perform our standard model diagnostics to determine whether the transformation was appropriate. In other words, simply transforming the response variable according to any rule of thumb is not sufficient; it must also correspond to the assumed distribution afterwards.
- Binary data cannot be transformed into any other distribution than the Bernoulli distribution.
- Common transformations:
  1. For count data (that mostly follow a Poisson or negative binomial distribution), we can often get closer to a normal distribution by using a square root-transformation. Taking the log of $y + 1$ that is often seen in literature is usually too strong and does not lead to an independence of mean and variance (as with the normal distribution), but to a decrease of the variance with the mean value. For the Poisson distribution, the root transformation is analytically correct (Withers and Nadarajah 2014).
  2. Very skewed data (data that may come from a log-normal distribution) can be transformed with a log. This applies only to strictly positive data ($y > 0$). If we have zeros or negative values, then we must add a "small" amount to all y values so that $f(y) = log(y + c)$. We choose these c values so that all values are positive, and c corresponds to about half of the smallest non-zero value.[13] The logic behind c is that we don't want to set the 0-values too far off from the next value, so it depends on the absolute values. Using a value of $c = 1$ is generally wrong, and should only be used in the case that the smallest non-zero value is about 2.
  3. The Box-Cox-transformation and its generalisation, the Yeo-Johnson-transformation(Yeo and Johnson 2000), are an attempt to automate the transformation to normal distribution. A parameter λ is calculated, which determines the transformation. For specific values, λ is about the same as a root transformation ($\lambda = 0.5$), the log transformation ($\lambda = 0$) or no transformation ($\lambda = 1$). The formula is a bit confusing and consists of four conditions:[14]

$$y' = \begin{cases} \{(y + 1^\lambda - 1\}/\lambda, & \text{if } y \geq 0 \text{ and } \lambda \neq 0, \\ log(y + 1), & \text{if } y \geq 0 \text{ and } \lambda = 0, \\ -\{(-y + 1^{2-\lambda} - 1\}/(2 - \lambda), & \text{if } y < 0 \text{ and } \lambda \neq 2, \\ -log(-y + 1), & \text{if } y < 0 \text{ and } \lambda = 2. \end{cases} \tag{11.5}$$

- When we transform the response variable, we get (logically) other log-likelihoods and AIC values. Thus, models before and after transformation of response variables are *not* comparable with each other by means of AIC.

A reasonable and often used alternative is the **Kruskal-Wallis-test**. In this test, the data are rank transformed and then analysed with an ANOVA (Underwood 1997). If the data contain too many of the same value (so-called *ties*) this test cannot be used, as there is also a risk that the variances are heterogeneous. As you may be thinking, the result of a rank transformation is a uniform distribution, not a normal distribution! ANOVA, and also regressions with an assumption of normal distribution, are robust against deviations from normal distribution *as long as the residuals are symmetric*. Both methods react more sensitively to variance heterogeneity, which is prevented by the rank transformation.

The Kruskal-Wallis-test is a non-parametric variant of the ANOVA, just as the Spearman-correlation (also for rank transformed data) is a non-parametric variant of the Pearson-correlation. Let's have a look at the result in the form of an ANOVA table for our North/South moss data set:

---

[13] So we first add to all values so that the smallest value is 0. Then we look at the amount of the next-biggest value one and add half of that to all values. If possible it is better to *not* use the ANOVA but to stay with a GLM. More on this later (Sect. 11.4).

[14] For only positive y-values (so $y > 0$), the Yeo-Johnson-transformation presented here is identical to the Box-Cox-transformation. However, the Box-Cox-transformation only works for positive y-values (and shifts the values if necessary), while the Yeo-Johnson-transformation can also appropriately transform negative values without shifting. In their original work, the authors also show that their transformation often gets closer to the normal distribution, and is never worse than the Box-Cox (Yeo and Johnson 2000). Here is the original (two-parameter) Box-Cox transformation (Box and Cox 1964):

$$y' = \begin{cases} ((y + c)^\lambda - 1)/\lambda, & \text{if } \lambda \neq 0, \\ log(y + c), & \text{if } \lambda = 0. \end{cases}$$

The parameters λ and c (only if y also contains non-positive values) are calculated using the log-likelihood (i.e. adapted to a normal distribution). Since we will not deal with this transformation further (it's too old school), here are the relevant R-packages: bcPower and yjPower in car; yeo.johnson in VGAM; boxcox in MASS.

```
          Df Sum Sq Mean Sq F value    Pr(>F)
Effect     1   62.5  62.500  25.641 0.0009726 ***
Residuals  8   19.5   2.437
---
Signif. codes:  0 '***' 0.001 '**' 0.01 '*' 0.05 '.' 0.1 ' ' 1
```

Due to the rank transformation, the absolute values are lower, and so are the *SS* values. In this (not typical) case, the significance is even higher than in the standard ANOVA.

We often use the Kruskal-Wallis-test when the normal distribution or variance homogeneity assumption for ANOVA is violated and we want to check whether the result has been falsified by this violation. In this case we had seen that the variance in "North" is much higher than in "South", so the ANOVA should not have been carried out. In this particular case, however, this is qualitatively inconsequential.

11.2.4   ANOVA for More Than 2 Levels

It should be briefly pointed out that an effect in ANOVA can have not only two levels, but several. In such cases, the ANOVA table is hardly changed, only the number of degrees of freedom is adjusted. The regression output, on the other hand, will be much longer because we get the deviation (slope) from the reference level for $k-1$ levels.

Let's examine the typical weight of four dog breeds: Afghan, Boxer, Collie and Doberman (Fig. 11.5). The result of an ANOVA for these data looks like this:

```
          Df Sum Sq Mean Sq F value  Pr(>F)
Breed      3   2171   723.8   22.58 2.13e-08 ***
Residuals 36   1154    32.1
```

And for a regression (with GLM) like this:

```
Coefficients:
Estimate Std. Error t value Pr(>|t|)
(Intercept)    24.040     1.790  13.428 1.38e-15 ***
BreedBoxer      4.860     2.532   1.919   0.0629 .
BreedCollie    -5.570     2.532  -2.200   0.0343 *
BreedDobermann 14.490     2.532   5.723 1.63e-06 ***
---
Signif. codes:  0 '***' 0.001 '**' 0.01 '*' 0.05 '.' 0.1 ' ' 1

(Dispersion parameter for gaussian family taken to be 32.0535)

Null deviance: 3325.3  on 39  degrees of freedom
Residual deviance: 1153.9  on 36  degrees of freedom
AIC: 258
```

The ANOVA table indicates that the division of the data into 4 levels (with $k-1=3$ degrees of freedom) is significant. From the regression output we note that the Afghan on average weighs 24 kg (alphabetically first level as reference), the boxer weighs 4.8 kg more, the collie 5.6 kg less and the doberman 14.5 kg more than the Afghan.

The statistical tests for *difference to the reference value* are given as a *t*-test after the estimator values. The first difference (Afghan − Boxer) does not look significant, since the estimator (the "slope" from Afghan to Boxer) is not significantly different from 0.

With the ANOVA, we can not tell exactly where the significant effect comes from, and with the regression, we can not tell if the estimated slopes show a *significant* effect of dog breed. Only when considered together does the analysis become informative.

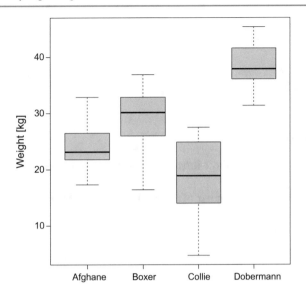

Fig. 11.5 Box plot for 4 dog breeds, with 10 weight measurements, respectively

11.2.5   *Post-Hoc* **Comparisons**

After executing an ANOVA with a predictor that has several levels, we don't know which levels are significantly different from each other. We only know that the division into the k levels significantly reduces the residual variance. To test differences between levels we use *post-hoc* tests. *Post hoc* means that we carry out them only after we execute the ANOVA. If a predictor is not significant, then we usually don't carry out *post-hoc* tests for it.

The following problem now arises. Let's say our predictor has 100 different levels (there are at least as many dog breeds). If we now do a pairwise test for all $100 \cdot (100 - 1)/2 = 4950$ comparisons, then some of them will likely be coincidentally significant. *Per definition* 5% of the 4950 comparisons (i.e. 247) will be erroneously significant when we have an error threshold of 5%. This problem is called the "multiple comparison problem".

Post-hoc comparisons and multiple tests should be carried out with a lot of cerebral fluid! Before you even do such a test, you should ask yourself whether you can bypass it somehow. Does this information really help us? That dogs are of different weight is perhaps enough; do we really need to know the significance between the breeds to make a statement?

If we do want to carry out a *post-hoc* test, then we have to correct at least for the number of comparisons, in order to actually stay at a 5% probability of error (Day and Quinn [1989], provide an excellent overview). There are two common approaches: the Bonferroni correction and Tukey's *honest significant difference* test. Furthermore, it should be noted that there are many other *post-hoc* procedures, which are not covered in this book.[15]

---

[15] For example: Duncan's new multiple range test, Dunnett test, Friedman test (non-parametric, therefore usable for the Kruskal-Wallis-test), the Scheffé method, Holm-correction, *false discovery rate*-correction.

In some of these tests (such as the Newman-Keuls test), the comparisons are first sorted by the difference in mean values and then tested one after another. As soon as a difference is no longer significant, we can abort the test, since the differences thereafter are even smaller (and the variance is the same everywhere, an assumption of ANOVA). Thus, we manage to get by with fewer comparisons, which leads to a less conservative statement than the Bonferroni correction.

With the frequently used Holm correction, all comparisons are made, but then the P values are sorted and the first comparison is corrected just like the Bonferroni, but the second one is only multiplied by $k - 1$, the third one is multiplied by $k - 2$, and so on. This makes the Holm less conservative than Bonferroni correction.

11.2.5.1 Bonferroni Correction
As a first step, we calculate all pairwise *t*-tests:

```
        Pairwise comparisons using t tests with pooled SD
              Afghan   Boxer   Collie
Boxer        0.06337  -        -
Collie       0.03375  0.00021  -
Doberman     1.6e-06  0.00053  2.1e-09
```

The values shown here are the *P*-values from the *t*-tests. In this case, all comparisons are significant, except for the one between the Afghan and the Boxer.

In step 2 we multiply these values with the number of comparisons (6) to obtain the corrected *P*-value:

```
           Afghan   Boxer    Collie
Boxer      0.38022  -        -
Collie     0.20251  0.00126  -
Doberman   0.00001  0.00317  1.2e-08
```

Now we realise that the difference between Afghan and Collie is too small to be significant in the *post-hoc* test with a 5% probability of error.

In a report, we would write something like, "we found significant differences between the four dog breeds ($F_{3,36} = 22.6$, $P < 0.001$). However, it should be noted that the Boxer and Collie were not significantly different from the Afghan (*post-hoc t*-test with Bonferroni correction, $P > 0.05$)".

11.2.5.2 Tukey's Honest Significant Difference Test
Tukey's HSD is even more strict in its correction of pairwise comparisons. It is, so to say, the "maximum standard". If *post-hoc* comparisons in Tukey's HSD are still significant, there is no reason to think otherwise.

Technically, Tukey's HSD is a variant of the pairwise *t*-test, except that instead of a *t*-distribution, a *studentised range distribution* is used, which is why this test is sometimes referred to as Tukey's range test.

The results look something like this:

```
  Tukey multiple comparisons of means
    95% family-wise confidence level

                       diff         lwr        upr       p adj
Boxer-Afghan       4.843094   -1.965632  11.651820   0.2395867
Collie-Afghan     -5.580291  -12.389018   1.228435   0.1405829
Doberman-Afghan   14.465957    7.657231  21.274684   0.0000095
Collie-Boxer     -10.423385  -17.232112  -3.614659   0.0011562
Doberman-Boxer     9.622863    2.814137  16.431590   0.0028456
Doberman-Collie   20.046249   13.237522  26.854975   0.0000000
```

In this case, we get pretty much the same result as from the Bonferroni correction.

11.2.5.3 A Priori Specified Comparisons
If we are interested in specific comparisons from the outset, then we don't need ANOVA at all. If we only make the comparisons specified in advance—a priori—, then our error rate is of course lower. However, the Bonferroni correction is still suitable for this.

To learn more about the fascinating possibilities with *post-hoc* tests, such as combining levels and testing against another level and for *pooling* levels, see Crawley [2007] and Zuur et al. [2009].

11.3 From Regression to ANOVA

In this section, we will switch gears from the ANOVA back to regression. In Chap. 7 we calculated a regression using a categorical predictor, which leads us to the question: How do the results of a GLM regression and ANOVA differ?

The short answer is: in no way at all. ANOVA and regression are two versions that operate on the same principle. With regression, the focus lies on the estimation of the model parameters, whereas with ANOVA, we are more interested in finding out if a predictor significantly reduces the variance.

Both approaches are based on a common calculation, namely the estimation of parameters by means of maximum likelihood, which is analytically solvable for normally distributed data.

We want to approach the congruence of ANOVA and regression in two ways, first by comparing the results, then by calculating the degrees of freedom and its basis in regression.

11.3.1 ANOVA and Regression: Comparing Results

Let's use the example of moss species on the North and South sides of trees again. Here is the ANOVA output:

```
          Df Sum Sq Mean Sq F value   Pr(>F)
x          1  250.0  250.00  23.697 0.001243 **
Residuals  8   84.4   10.55
```

The results from a regression (GLM) looks like this:

```
Coefficients:
             Estimate Std. Error t value Pr(>|t|)
(Intercept)    17.600      1.453  12.116 1.99e-06 ***
EffectSouth   -10.000      2.054  -4.868  0.00124 **
---
Signif. codes:  0 '***' 0.001 '**' 0.01 '*' 0.05 '.' 0.1 ' ' 1


(Dispersion parameter for gaussian family taken to be 10.55)

    Null deviance: 334.4  on 9  degrees of freedom
Residual deviance:  84.4  on 8  degrees of freedom
AIC: 55.709
```

At first glance, the two outputs look like they have nothing to do with each other. But if we look closer, we notice:

- The P-value for x in the ANOVA output is identical to the P-value for South in the regression (0.001243).
- The dispersion parameter of the regression (10.55) is identical to the Mean Sq of the ANOVA.
- The number of degrees of freedom of the residual standard errors in the regression is identical to that of the Residuals in the ANOVA (8).
- The last rows from the regression output contain much of the information from the ANOVA (specifically $SS_{overall}$ at 334.4 and the $SS_{Residuals}$ at 84.4, as well as their degrees of freedom at 9 and 8, respectively).

This can't just be a coincidence. In fact, we see here that the ANOVA and regression focus on different aspects of the model. The ANOVA describes the distribution of variances, 250 explained, 84.4 unexplained. In the regression, on the other hand, the differences between the groups are described, 17.6 for "North" and $17.6 - 10 = 7.6$ for "South".

The larger the difference between "North" and "South", the larger the explained variance in the ANOVA and the greater the slope in the regression. And indeed, the t-test of the regression and the F-test of the ANOVA describe the same thing, namely the difference between the groups. Figure 11.6 shows this visually: the box plots describe the ANOVA perspective and the regression lines the regression perspective.

Bottom line: depending on which information we are interested in, we can choose a variant to display the same analysis either as an ANOVA table or as a regression. With the ANOVA we can easily see if a factor is significant or not. With a

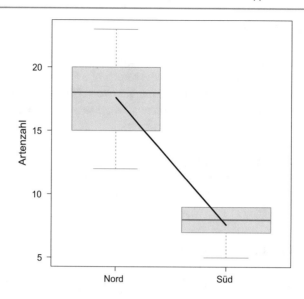

Fig. 11.6 Box plot and regression line for the example data. While the box plot shows the differences between the two levels, the regression line shows the slope. The information is practically identical, only with a different emphasis. (Note that the line connects the means, which in both cases are slightly lower than the medians of the box plots)

regression, we quantify the difference between the means of the levels. So if we want to know, for example, if a change in the pH-value has an effect on the survival of fish in experimental ponds, then we would choose an ANOVA. However, if we want to know by how much the survival rate drops in ponds with a low pH, then we would choose regression. The information from both of these variants can be converted to the other, since the same model is behind the results of both.

11.3.2   Degrees of Freedom from ANOVA and Explaining Them Through Regression

Degrees of freedom are always a bit difficult to comprehend and to calculate. For us, however, they represent something like a control calculation, to see whether our model also calculates what we want. If the degrees of freedom are not right, then the model is wrong. It is a bit like dimensional analysis[16] in physics: if the units of a calculation are not correct, then the something went wrong!

In the example above, we have two groups, the North and South sides of trees. Using the language of regression— if we fit a model here,we use the alphabetically first level as the reference (the intercept) and the second as a dummy variable with a value of 1 when from the south side, and 0 otherwise. We then have a regression model that looks like this: $\mathbf{y} = N(\mu = a + b \cdot \mathbf{SouthDummy}, \sigma)$. So we fit three parameters: a, b and σ using maximum likelihood.

What does this mean for the degrees of freedom? Well, first of all we can ignore σ, because the estimators for a and b are not affected by the value of σ: The expected value for North or South is determined solely by a and b, while σ quantifies the dispersion of the measured values around this expected value. Correspondingly, we have fitted two parameters to explain our values. Since we start with 10 data points and calculate 2 parameters, only eight freely selectable data values are left. This number, $10 - 2 = 8$, is the degrees of freedom for the model residuals.

With our ANOVA we have a test of the difference between two levels. This seems like it should only have a single parameter. In reality, we have to calculate the mean value for each group, so that their differences can be calculated. So we need two degrees of freedom here too.

For an effect with more than two levels, we use one degree of freedom for each level. This applies to both ANOVA and regression. With an ANOVA we calculate k means for k levels; in regression we calculate a reference mean and then $k - 1$ differences (= slopes). Either way, we only have $n - k$ degrees of freedom left.

On the basis of the degrees of freedom we see that regression and ANOVA are kindred spirits. For details on conversion, please refer to the next chapter.

---

[16]Analysis of whether the units on the left side of the equation are identical to those on the right.

11.4   ANOVAs for GLMs

In this last section of the chapter, we return to non-normally distributed data. Here, too, we are interested in the same questions as in ANOVA. Can we use anything from it even though our data are binomially or Poisson distributed?

Yes, in fact we can! A central piece of the ANOVA are the sum of squares, SS. Those of you who have been paying close attention to the footnotes (specifically the one) will know that these sum of squares are also known as the deviance to friends of the GLM. And there are also deviances for other distributions (also defined in the same footnote).

If we want to create an "ANOVA" table for non-normally distributed data, then we need to do two things:

1. Fit a GLM.
2. Convert the deviances from the GLM into a sort of ANOVA.

No sooner said than done. As an example, let's use the collared flycatcher data set. (e.g. Fig. 7.3) and consider the attractiveness not as a continuous variable, but as a categorical variable (only to achieve a similarity to the dog data set). The GLM output gives us:

```
Coefficients:
            Estimate Std. Error z value Pr(>|z|)
(Intercept)   1.5261     0.2085   7.319  2.5e-13 ***
attract2      0.3909     0.2700   1.448   0.1477
attract3      0.2657     0.2771   0.959   0.3377
attract4      0.6257     0.2583   2.422   0.0154 *
attract5      0.6487     0.2573   2.521   0.0117 *
---
Signif. codes:  0 '***' 0.001 '**' 0.01 '*' 0.05 '.' 0.1 ' ' 1

(Dispersion parameter for poisson family taken to be 1)

    Null deviance: 25.829  on 24  degrees of freedom
Residual deviance: 16.527  on 20  degrees of freedom
AIC: 119.62
```

The ANOVA table (in R) would look like this:

```
        Df Deviance Resid. Df Resid. Dev P(>|Chi|)
NULL                    24       25.829
attract  4   9.3021      20       16.527   0.05398 .
---
Signif. codes:  0 '***' 0.001 '**' 0.01 '*' 0.05 '.' 0.1 ' ' 1
```

Not as nice as the ANOVA tables we are used to, but it still contains all of the information for us to make a mini ANOVA table:

Source	d.f.	deviance	$P(> X^2)$
Attractiveness	4	9.302	0.05398
Residuals	20	16.527	

In place of the SS we now have the deviance. Null deviance corresponds to SS_{total}, the residual deviance corresponds to $SS_{Residuals}$ and the number below deviance corresponds to SS_{Effect}.

Instead of the F-test, we have the χ^2-test. Additionally, it is not the ratio of average `deviances` that is calculated, but rather the difference in deviance (here 9.3021) on the basis of the predictor.[17] This value is tested with the χ^2-distribution at $k - 1 = 4$ degrees of freedom—and is not very significant.

So the ANOVA table for non-normally distributed data is possible, but contains fewer numbers and uses a different test.

But why is the result not significant, since we have seen a significant effect here this whole time? This is because we have chopped the predictor (for didactic reasons) into individual categories instead of (rationally) allowing it remain a continuous predictor. With continuous attractiveness values, the ANOVA table of the GLM looks like this:

Source	d.f	deviance	$P(> X^2)$
Attractiveness	1	7.509	0.00614
Residuals	23	18.320	

So far we have only looked at the two possibilities (categorical or continuous), but what is right? Regarding the treatment of predictors there is one basic rule: *do not categorise continuous predictors!*[18] The main reason is that we would otherwise use up degrees of freedom unnecessarily, as we estimate many averages instead of a common gradient. The converse of this, however, is not permissible: if the predictor is categorical (colours, brands, nationalities, animal groups), then it cannot be converted into a continuous variable (but perhaps represented by a continuous alternative: brightness, price, population size, phylogenetic distance).

In the harsh reality, of course, borderline cases often occur. There are many predictors that are measured in three to five levels (e.g. the quality of a tree or the abdominal profile of a wild goose). Should these few levels then be represented by a continuous or a categorical variable? In general, a visualisation showing how the data points are distributed along the x-axis can help us determine if we can justify using a continuous variable (as in our example, the attractiveness of the collared flycatcher). In case of doubt, we should simply try both and become suspicious if the results are clearly different.[19] If there are only a few groups (up to about five), they should be treated as categories. From seven levels upwards, we can almost certainly do away with categorisation.

References

1. Box, G. E. P., & Cox, D. R. (1964). The analysis of transformations. *Journal of the Royal Statistical Society B, 26*, 211–252.
2. Crawley, M. J. (2002). *Statistical computing. An introduction to data analysis using S-Plus*. Chichester: Wiley.
3. Crawley, M. J. (2007). *The R Book*. Chichester, UK: Wiley.
4. Dalgaard, P. (2002). *Introductory statistics with R*. Berlin: Spinger.
5. Day, R. W., & Quinn, G. P. (1989). Comparisons of treatments after an analysis of variance in ecology. *Ecological Monographs, 59*, 433–4636.
6. Fisher, R. A. (1918). The correlation between relatives on the supposition of Mendelian inheritance. *Philosophical Transactions of the Royal Society of Edinburgh, 52*, 399–433.
7. Fisher, R. A. (1925). *Statistical methods for research workers*. Edinburgh: Oliver & Boyd.
8. Mann, H. B. (1949). *Analysis and design of experiments: Analysis of variance and analysis of variance designs*. New York: Dover Publications.
9. McCullough, B. D., & Heiser, D. A. (2008). On the accuracy of statistical procedures in Microsoft Excel 2007. *Computational Statistics & Data Analysis, 52*, 4570–4578.
10. O'Hara, R. B., & Kotze, D. J. (2010). Do not log-transform count data. *Methods in Ecology and Evolution, 1*(2), 118–122.
11. Quinn, G. P., & Keough, M. J. (2002). *Experimental design and data analysis for biologists*. Cambridge, UK: Cambridge University Press.
12. Underwood, A. J. (1997). *Experiments in ecology: Their logical design and interpretation using analysis of variance*. Cambridge, UK: Cambridge University Press.
13. Warton, D. I., & Hui, F. K. C. (2011). The arcsine is asinine: The analysis of proportions in ecology. *Ecology, 92*, 3–10.
14. Withers, C. S., & Nadarajah, S. (2014). Simple alternatives for Box-Cox transformations. *Metrika, 77*(2), 297–315.
15. Yeo, I.-K., & Johnson, R. A. (2000). A new family of power transformations to improve normality or symmetry. *Biometrika, 87*(4), 954–959.
16. Zar, J. H. (2013). *Biostatistical analysis* (5th ed.). Pearson.
17. Zuur, A. F., Ieno, E. N., Walker, N. J., Saveliev, A. A., & Smith, G. M. (2009). *Mixed effects models and extensions in ecology with R*. Berlin: Springer.

---

[17]The deviance is computed (for all practical purposes) as -2ℓ. (Actually there is another term, the log-likelihood of the so-called saturated model, that comes in here, but in the context of the GLM this is largely immaterial.) Thus, the ratio of likelihoods is equivalent to the difference of *log*-likelihoods, which is why the χ^2-test is computed on their differences.

[18]On the homepage of statistics professor Frank Harrell (Vanderbilt University, Nashville, Tennessee) this tip is listed under *Philosophy of Biostatistics* as the third point. The first two are also noteworthy: http://biostat.mc.vanderbilt.edu/wiki/Main/FrankHarrell.

[19]A possible reason is that the context is not linear and we should insert a square term: point 4 on Harrell's list.

You know my methods. Apply them.

—Arthur C. Doyle: Sherlock Holmes in the Sign of Four

> At the end of this chapter . . .
> . . . you will be able to calculate different *t*-test variants with R.
> . . . you will be able to calculate a simple ANOVA in R.
> . . . you will be able to switch back and forth between `glm` and `aov` and retrieve the information that is most relevant for you.
> . . . you will be able to calculate the corresponding *P*-value for *F*-values.
> . . . you will know some important *post-hoc* test functions for ANOVA and GLM in R.

12.1 *t*-test and Its Variants in R

The *t*-test is implemented in R with the surprisingly named function `t.test`. This function has two important arguments in addition to the variables we are comparing: `paired` and `var.equal`. With `paired=T`, paired tests are run, i.e. corresponding variables are paired together and compared. If this is the case, the data sets must be structured so that the corresponding variables are next to each other (for example in the same row of two different columns). With `var.equal=F` we allow there to be different variances for each data set.

Before we devote ourselves to these two important arguments, let's briefly look at the *one-sample t-test* as well as comparing a data set with a fixed value. Our example datetaset is once again the height of 50 men and 50 women.

```
> SiSe <- read.delim("SizeSex.txt")
> t.test(SiSe$size)


    One Sample t-test

data:  SiSe$size
t = 155.1644, df = 99, p-value < 2.2e-16
alternative hypothesis: true mean is not equal to 0
95 percent confidence interval:
 177.2934 181.8866
sample estimates:
mean of x
   179.59
```

That was trivial. Now just for fun, let's test whether all of the test subjects are on average smaller than 182 cm:

```
> t.test(SiSe$size, mu=182)

    One Sample t-test

data:  SiSe$size
t = -2.0822, df = 99, p-value = 0.0399
alternative hypothesis: true mean is not equal to 182
95 percent confidence interval:
 177.2934 181.8866
sample estimates:
mean of x
   179.59
```

The sample mean is indeed significantly lower than 182 cm. This is all that we can really do with a *t*-test for a data set.

Let's now focus on comparing two datasets. In this case, the data takes a format that is a bit unusual in my opinion: the data must be in two columns. This is not the case with our data set, GG, so we have to transform this data set first. In this case it is very easy with the function unstack. Then we compare the two sets of data, assuming the same variance:

```
> head(SiSe)

  size gender
1  184    man
2  212    man
3  170    man
4  187    man
5  176    man
6  199    man

> newSiSe <- unstack(SiSe)
> head(newSiSe)

  man woman
1 184   171
2 212   175
3 170   180
4 187   168
5 176   169
6 199   173

> t.test(newSiSe$woman, newSiSe$man, var.equal=T)

Two Sample t-test

data:  newSiSe$woman and newSiSe$man
t = -6.6928, df = 98, p-value = 1.372e-09
alternative hypothesis: true difference in means is not equal to 0
95 percent confidence interval:
 -16.724969  -9.075031
sample estimates:
mean of x mean of y
   173.14    186.04
```

The difference in height between men and women is clearly highly significant. Note that the 95% confidence interval refers to the *difference* between the two data sets.

If we now want to allow for different variances (the default of `t.test`), we get:

```
> t.test(newSiSe$woman, newSiSe$man, var.equal=F)


    Welch Two Sample t-test

data:  newSiSe$woman and newSiSe$man
t = -6.6928, df = 96.175, p-value = 1.458e-09
alternative hypothesis: true difference in means is not equal to 0
95 percent confidence interval:
-16.725877  -9.074123
sample estimates:
mean of x mean of y
173.14    186.04
```

In this case, it makes little difference. The variances are somewhat but not drastically different:

```
> apply(newSiSe, 2, var)


      man      woman
105.67184   80.08204
```

Now let's consider the example from the previous chapter: number of moss species on the North and South sides of five trees. Again, we will compare the *t*-test with and without the same variances:

```
> north <- c(12, 23, 15, 18, 20)
> south <- c(5, 8, 7, 9, 9)
> t.test(north, south, var.equal=T)


    Two Sample t-test

data:  north and south
t = 4.8679, df = 8, p-value = 0.001243
alternative hypothesis: true difference in means is not equal to 0
95 percent confidence interval:
  5.262859 14.737141
sample estimates:
mean of x mean of y
     17.6       7.6


> t.test(north, south, var.equal=F)


    Welch Two Sample t-test

data:  north and south
t = 4.8679, df = 5.196, p-value = 0.00415
alternative hypothesis: true difference in means is not equal to 0
95 percent confidence interval:
  4.778713 15.221287
sample estimates:
mean of x mean of y
     17.6       7.6
```

Although the number of effective degrees of freedom is down from 8 to just over 5 and the *p*-value has almost quadrupled, the difference remains significant. For this data set the assumption of variance homogeneity (= homoscedasticity) is certainly not met (see calculations in Sect. 11.1). The variance for the South side is only $s^2_{South} = \text{var(south)} = 2.8$, whereas $s^2_{North} = \text{var(north)} = 18.3$. Homogeneous variances look a bit different! In this case the difference between the groups was strong enough that we were able to arrive at the same conclusion with both approaches; this need not be the case. The mistake of assuming that variances are equal even though they are not, is easily avoidable with the *t*-test (and is also the default with t.test). The publication of analyses in which assumptions were violated without demonstrating the robustness of the method with regard to this violation, can be justifiably called "lies".

This specific data set was collected on northern and southern sides of the *same* tree. Accordingly, we can carry out a *t*-test for paired samples, as the first value of north refers to the same tree as the first value of south, and so forth. This is achieved by setting the paired-argument of t.test to TRUE:

```
> t.test(north, south, paired=T)

Paired t-test

data:  nord and sued
t = 7.0711, df = 4, p-value = 0.002111
alternative hypothesis: true difference in means is not equal to 0
95 percent confidence interval:
6.073514 13.926486
sample estimates:
mean of the differences
10
```

The paired *t*-test is more sensitive than its unpaired counterpart, as by computing tree-specific differences it removes some of the variation between trees (e.g. bark roughness). If oak has a higher mean number of moss species due to its bark than, say, beech, the *difference* between northern and southern sides need not be affected at all. Hence, the differences may be very similar among tree species, even if their absolute values differ substantially. We will encounter this reasoning again when designing manipulative experiments, where this approach is called "blocking" (Sect. 14.2).

Remember that the paired *t*-test is in fact a one-sample *t*-test of the difference against 0. Thus, the argument var.equal is void when using paired samples.

12.2   ANOVA in R

The R command for ANOVA is aov for *analysis of variance*. It is used in the same way as the glm function:

```
> y <- c(5, 8, 7, 9, 9, 12, 23, 15, 18, 20)
> x <- factor(rep(c("North", "South"), each=5))
> fm <- aov(y ~ x)
> summary(fm)

            Df Sum Sq Mean Sq F value   Pr(>F)
x            1  250.0  250.00  23.697 0.001243 **
Residuals    8   84.4   10.55
---
Signif. codes:  0 `***' 0.001 `**' 0.01 `*' 0.05 `.' 0.1 ` ' 1
```

Side note 10: Intuitive explanation of the *t*-distribution by simulation
The *t*-distribution is the small-sample-cousin of the normal distribution: if we draw small samples from a normal distribution, their scores are not normally distributed, although large samples are. To illustrate this point, we use simulation.
Imagine drawing samples of size *n* from a standard normal distribution:

```
> set.seed(12)
> n <- 4
> X <- rnorm(n)
```

Next, compute the mean and its standard error:

```
> mean(X); sd(X)/sqrt(n)
[1] -0.445037
[1] 0.6861172
```

This sample has a mean of -0.45 (sem $= 0.686$), which means it is $-0.445/0.686 = -0.65$ units from the true value of 0. This ratio, the signal-to-noise-ratio if you like, is called the T-score. (If we take a very large sample, this value approximates 1. This may not be intuitive, but you can think of it as the accuracy (s.e.m.) of $\bar{x}$ increasing as $\bar{x}$ getting closer to the true 0; thus, the closer $\bar{x}$ is to the truth, the higher is also our sensitivity to detect differences.) We can compute T often and turn it into a histogram:

```
> Ts <- replicate(n=1000, expr={X <- rnorm(n); mean(X) / (sd(X) / sqrt(n))} )
> hist(Ts, prob=TRUE, breaks=20, ylim=c(0, 0.4), col="grey90", las=1, main="",
+      cex.lab=1.5, border="white", xlim=c(-6, 6))
```

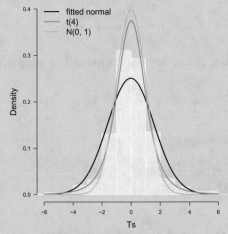

While this may look normally distributed, it is not: the distribution is more pointed. We can add both a normal distribution of the Ts as well as the original normal distribution and the *t*-distribution with *n* degrees of freedom:

```
> curve(dnorm(x, mean=mean(Ts), sd=sd(Ts)), add=TRUE, n=501, lwd=3, lty=2)
> curve(dt(x, n), add=TRUE, n=501, lwd=3, col="grey50")
> curve(dnorm(x), add=TRUE, n=501, lwd=3, col="grey70")
> legend("topleft", col=c("black", "grey50", "grey70"), lwd=3, lty=c(2,1,1),
+        legend=c("fitted normal", "t(4)", "N(0, 1)"), bty="n", cex=1.5)
```

As *n* increases, the *t*-distribution approximates the normal (with $n > 30$ being indistinguishable).
So: the *t*-distribution describes the expected T-score when drawing a small sample from a normal distribution. It is flatter than the normal, as the standard deviation of the mean (i.e. the s.e.m.) is estimated, rather than known.

For the example with continuous variables we can use the following:

```
> data(trees)
> summary(aov(Volume ~ Girth, data=trees))
```

```
            Df Sum Sq Mean Sq F value    Pr(>F)
Girth        1 7581.8  7581.8  419.36 < 2.2e-16 ***
Residuals   29  524.3    18.1
---
Signif. codes:  0 `***' 0.001 `**' 0.01 `*' 0.05 `.' 0.1 ` ' 1
```

And an example with multiple levels:

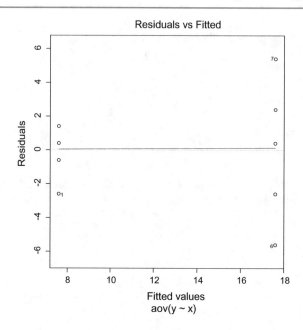

Fig. 12.1 Residual fitted values plot as a diagnostic plot for ANOVA. The spread in the group on the right ("South") is much larger. Influential points are numbered

```
> dogs <- read.csv("dogs.csv")
> summary(aov(Weight ~ Breed, data=dogs))

            Df Sum Sq Mean Sq F value    Pr(>F)
Breed        3 2171.3  723.77   22.58 2.132e-08 ***
Residuals   36 1153.9   32.05
---
Signif. codes:  0 `***' 0.001 `**' 0.01 `*' 0.05 `.' 0.1 ` ' 1
```

You can see that it is fairly simple to calculate an ANOVA in R.

We can just as easily perform diagnostics. If we use the `plot` command on an `aov` object, we get a series of diagnostic plots, the first of which is the most interesting for us:

```
> plot(fm, 1)
```

This gives us the diagnostic plot for the example data set on moss species on the North and South sides of trees (Fig. 12.1).

12.2.1   Test for Homogeneity of Variance

One of the most important assumptions of variance analysis is the homogeneity of variance or homoscedasticity. In order to test if our two samples, $\mathbf{y}_1$ and $\mathbf{y}_2$, have the same variance, we use the F-test. We calculate the variance (s^2) of the respective sample, s_1^2 and s_2^2, and *divide the larger value by the smaller value*: $F = s_1^2/s_2^2$.

It is important to divide the larger one by the smaller one, because otherwise the wrong end of the F-distribution will be used for the calculation. This means that if the F-value is less than 1, you must look at the left part of the F-distribution (see next section for details).

Critical F-values can be calculated using the F-distribution or the P-value for a certain F-ratio can be calculated with the probability function of the F-distribution: `pf (F-value, n_1-1, n_2-1, lower.tail=F)`. It is important to note that by deciding to always divide the larger variance by the smaller that we have really only done what *looks like* a one-sided test! In reality, we are interested in whether s_1^2 and s_2^2 are different, not if s_1^2 is greater than s_2^2. That is, we would describe the variances as heterogeneous if the deviations are large, no matter in which direction. This describes exactly what a two-sided test does. Our 5% significance level must therefore be divided between the right and left side of the distribution. For our F-test this means that we would not accept a $P \leq 0.05$ as significant, but rather $P \leq 0.025$ as a significant level of heterogeneity of variance.

In our North/South example the variance of the North data is 2.8 and that of South is 18.3. The larger-to-small ratio is thus:

```
> var(ySouth)/var(yNorth)

[1] 6.535714

> pf(6.536, 4, 4, lower.tail=F)

[1] 0.04815187
```

This value is higher than 0.025 and we must accept (though it is a bitter pill to swallow) that this large difference in variance is not significant.

If we wanted to know how big the F-value should have been in order for the difference to be significant, we would use the quantile function of the F-distribution:

```
> qf(0.025, 4, 4, lower.tail=F)

[1] 9.60453
```

So only if the variance ratio is at least 9.6 would our five data points be heteroscedastic. This is a function of the sample size. At 20 data points each (with 19 degrees of freedom) we get:

```
> qf(0.025, 19, 19, lower.tail=F)

[1] 2.526451
```

There is also a special R-function that runs the homogeneity of variance test for us, var.test:

```
> var.test(yNorth, ySouth)

    F test to compare two variances

data:  yNorth and ySouth
F = 0.153, num df = 4, denom df = 4, p-value = 0.09631
alternative hypothesis: true ratio of variances is not equal to 1
95 percent confidence interval:
 0.01593055 1.46954556
sample estimates:
ratio of variances
          0.1530055
```

Here, the ratio is calculated in the order of the entered values and the corresponding F-distribution side is used to calculate the P value. At 0.096 this is exactly twice our "one-sided" way of computing the P-value of 0.048. As you can see at the 95% confidence interval, our few data points are not suitable to make a clear statement about the variance ratio.

In addition to the F-test, Bartlett's and Levene's test are also often used. Bartlett's test is sensitive to deviations from a normal distribution, and therefore the more robust Levene test is usually used. Levene's test uses absolute deviations from the group mean instead of variances. In one variant of the Levene's test, (the Brown-Forsythe-test), the group medians are used instead of the group means.

Implementation in R is simple. Bartlett's test is readily available as part of the automatically loaded **stats**-package and for Levene's test and the Brown-Forsythe variant we only need the **car** package.[1]

```
> bartlett.test(y, x)

    Bartlett test of homogeneity of variances

data:  y and x
Bartlett's K-squared = 2.7582, df = 1, p-value = 0.09676
```

Here is the original Levene's test with the mean of the groups as the reference:

```
> library(car)
> leveneTest(y, x, center=mean)

Levene's Test for Homogeneity of Variance (center = mean)
      Df F value  Pr(>F)
group  1  3.5689 0.09555 .
       8
---
Signif. codes:  0 `***' 0.001 `**' 0.01 `*' 0.05 `.' 0.1 ` ' 1
```

And the Brown-Forsythe variant of the Levene-test (this is the default in R if you leave out the argument `center=...`):

```
> leveneTest(y, x, center=median)

Levene's Test for Homogeneity of Variance (center = median)
      Df F value Pr(>F)
group  1  2.8986 0.1271
       8
```

With a continuous predictor (aka: a regression) we cannot use `var.test`; in such cases, we would use the diagnostic plots we learned in Sects. 10.2 and 10.8.

12.2.2 Calculate Significance from *F*-values

F-, χ^2- and t-distributions are known as test distributions, since these values always appear in statistical test (of course, there are many other than just these three). This means that we can calculate an F-value, either with ANOVA or by hand and then look in the F-distribution to see which P-value the F-value corresponds to.

This is useful, but can be confusing to execute in R. As a starting point, we can look at the cumulative F distribution (Fig. 12.2)[2]:

```
> par(mar=c(5,5,1,1), mfrow=c(1,2))
> curve(df(x, 1, 20), from=0, to=10, ylab="Probability density",
+     xlab=expression(italic(F)-value), las=1, lwd=2, cex.lab=1.5, col="grey",
+     ylim=c(0,1))
> legend("topright", bty="n", legend="PDF", cex=2)
> curve(pf(x, 1, 20), from=0, to=10, ylab=expression(Pr(X > x)),
+     xlab=expression(italic(F)-value), las=1, lwd=2, cex.lab=1.5)
> legend("bottomright", bty="n", legend="CDF", cex=2)
```

---

[1]Levene's test is also implemented in the **lawstat** package in the function `levene.test`.
[2]Since the F-distribution at 0 goes toward infinity, we can manually limit the value range with the argument `ylim`.

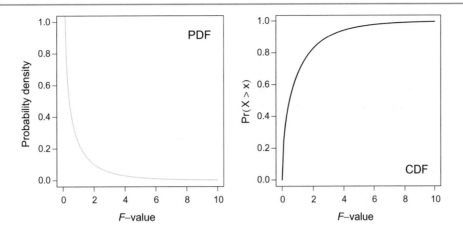

Fig. 12.2 Density function (left, grey) and cumulative density (right, black) of the F-distribution with the parameters df1 $= 1$ and df2 $= 20$

If we want to use this distribution to calculate the significance level of an F-value, then we want to look at the part of the distribution that is as large as our F-value or larger. That is to say, we are interested in the part to the *right* of the F-value, in the flat part of the PDF.

Suppose we calculated an F-value of 5. Then we want to know what proportion of the area under the PDF lies to the *right* of 5. As shown in Fig. 12.2 (right), this is only a small fraction of the distribution. However, if we use the CDF to read this value, we find out what proportion is to the *left* of 5. The CDF accumulates the area up to the respective F-value.

Mathematically, we can calculate the CDF as a definite integral from 0 (i.e. the left boundary of the F-distribution) to some value z:

$$P(F \leq z) = \int_0^z F(x)dx. \tag{12.1}$$

What we are interested in, however, is the definite integral for values $\geq z$:

$$P(F \geq z) = \int_z^\infty F(x)dx. \tag{12.2}$$

Figure 12.2 on the right visualises Eq. 12.1. For Eq. 12.2 we would need the exact opposite, a sort of "decumulative" probability that would get smaller and smaller the larger the value for z. This is shown in Fig. 11.3 on page 152.

In R, the p... functions have an option called lower.tail (by default, lower.tail=TRUE). In the default setting, the defined integral is output from the "left to z" (Eq. 12.1). With the setting lower.tail=FALSE we get the definite integral of "z to the right" (Eq. 12.2), which is exactly what we are looking for here.

```
> pf(5, 1, 20, lower.tail=T)

[1] 0.9630952

> pf(5, 1, 20, lower.tail=F)

[1] 0.03690484
```

The P-value we get for our F value of 5 is 0.0369!

If we want to determine the level of significance of a particular F value, then we need to choose the option lower.tail=FALSE*!*

The "decumulative F distribution" in Fig. 11.3 on page 152 was created with the following R code:

```
> curve(pf(x, 1, 8, lower.tail=F), from=0, to=30, las=1, cex.lab=1.5, xlab=
+    expression(italic(F)-value), ylab=expression(1 - Pr(X <= x) == Pr(X > x)))
```

12.2.3 *Post-Hoc* Comparisons with R

The *post-hoc* tests introduced in the previous chapter (Bonferroni and Tukey's HSD) can be easily implemented in R. Furthermore, the package **multicomp** makes it possible to calculate different *post-hoc* tests for GLMs and even for more complex models with multiple predictors. In particular, the function `ghlt` should be noted.

The simple uncorrected pairwise *t*-test can be implemented in R as follows:

```
> pairwise.t.test(dogs$Weight, dogs$Breed, p.adjust.method="none")

    Pairwise comparisons using t tests with pooled SD

data:  dogs$Weight and dogs$Breed

          Afghan Boxer   Collie
Boxer     0.06288 -        -
Collie    0.03432 0.00021 -
Doberman  1.6e-06 0.00053 2.1e-09

P value adjustment method: none
```

The last line here tells us that we did not do a correction for multiple comparisons. We can either do this the long way[3] or we can use the option `p.adjust`:

```
> pairwise.t.test(dogs$Weight, dogs$Breed, p.adjust.method="bonferroni")

    Pairwise comparisons using t tests with pooled SD

data:  dogs$Weight and dogs$Breed

          Afghan Boxer  Collie
Boxer     0.3773 -       -
Collie    0.2059 0.0013 -
Doberman  9.8e-06 0.0032 1.3e-08

P value adjustment method: bonferroni
```

Tukey's *Honest Significant Difference*-test can be called with the function `TukeyHSD`. It can be used on an `aov` object:

```
> TukeyHSD(fm.aov)

  Tukey multiple comparisons of means
    95

Fit: aov(formula = Weight ~ Breed, data = dogs)

$Breed
                   diff        lwr       upr    p adj
Boxer-Afghan       4.86  -1.959079 11.679079 0.2380529
Collie-Afghan     -5.57 -12.389079  1.249079 0.1426264
Doberman-Afghan   14.49   7.670921 21.309079 0.0000095
```

---

[3] `6*(pairwise.t.test(dogs$Weights, dogs$Breed, p.adjust.method="none")$p.value)`.

```
Collie-Boxer        -10.43 -17.249079  -3.610921 0.0011685
Doberman-Boxer        9.63   2.810921  16.449079 0.0028693
Doberman-Collie      20.06  13.240921  26.879079 0.0000000
```

12.3   ANOVA to Regression and Back

If instead of using `aov` we are using the `glm` (or `lm`) function to fit a linear regression, we can still use `aov` to make it into an ANOVA object:

```
> fm.glm <- glm(Weight ~ Breed, data=dogs)
> summary(aov(fm.glm))

            Df Sum Sq Mean Sq F value   Pr(>F)
Breed        3 2171.3  723.77   22.58 2.132e-08 ***
Residuals   36 1153.9   32.05
---
Signif. codes:  0 `***' 0.001 `**' 0.01 `*' 0.05 `.' 0.1 ` ' 1
```

The opposite works exactly the same way. If we first calculate an ANOVA and are then interested in the coefficients or other (g)lm information, we can make an `aov` object into a (g)lm object:

```
> fm.aov <- aov(Weight ~ Breed, data=dogs)
> summary(lm(fm.aov))

Call:
lm(formula = fm.aov)

Residuals:
    Min     1Q  Median     3Q    Max
-13.770 -2.547  -0.105  3.625  9.030

Coefficients:
              Estimate Std. Error t value Pr(>|t|)
(Intercept)     24.040      1.790  13.428 1.38e-15 ***
BreedBoxer       4.860      2.532   1.919   0.0629 .
BreedCollie     -5.570      2.532  -2.200   0.0343 *
BreedDoberman   14.490      2.532   5.723 1.63e-06 ***
---
Signif. codes:  0 `***' 0.001 `**' 0.01 `*' 0.05 `.' 0.1 ` ' 1

Residual standard error: 5.662 on 36 degrees of freedom
Multiple R-squared: 0.653,   Adjusted R-squared: 0.6241
F-statistic: 22.58 on 3 and 36 DF,  p-value: 2.132e-08
```

We can even query coefficients directly from the `aov` object, just like we would with a `glm` or `lm`:

```
> coef(fm.aov)

  (Intercept)     BreedBoxer   BreedCollie  BreedDoberman
        24.04           4.86         -5.57          14.49
```

The output differs depending on whether we are dealing with an `lm` (linear model, only for normally distributed data) or a `glm` (other distributions can be specified):

```
> summary(glm(fm.aov))

Call:
glm(formula = fm.aov)

Deviance Residuals:
     Min        1Q     Median        3Q        Max
  -13.770    -2.547    -0.105     3.625      9.030

Coefficients:
               Estimate Std. Error t value Pr(>|t|)
(Intercept)      24.040      1.790  13.428 1.38e-15 ***
BreedBoxer        4.860      2.532   1.919   0.0629 .
BreedCollie      -5.570      2.532  -2.200   0.0343 *
BreedDoberman    14.490      2.532   5.723 1.63e-06 ***
---
Signif. codes:  0 `***' 0.001 `**' 0.01 `*' 0.05 `.' 0.1 ` ' 1

(Dispersion parameter for gaussian family taken to be 32.0535)

    Null deviance: 3325.3  on 39  degrees of freedom
Residual deviance: 1153.9  on 36  degrees of freedom
AIC: 258
```

The output of the `glm` is just as valid in its form for the other distributions, while the `lm` output only makes sense for normally distributed data. For example, R^2, F-test and SS are only appropriate here, but not for Poisson or γ distributed data.

12.4   ANOVAs for GLM

As shown above, we can also create an ANOVA table from a (G)LM for normally distributed data using `summary(aov(.))`. Now we want to look at how to extract an ANOVA-like table for GLMs with non-normally distributed data. The function here is `anova`, which can be applied to an object produced using the `glm` function.

Let's fit a model using GLM looking at the effect of the passenger class on the survival of passengers on the Titanic and extract an ANOVA-like table from this model:

```
> library(effects)
> data(TitanicSurvival)
> fm.tita <- glm(survived ~ passengerClass, data=TitanicSurvival, family=binomial)
> anova(fm.tita, test="Chisq")

Analysis of Deviance Table

Model: binomial, link: logit

Response: survived

Terms added sequentially (first to last)
```

```
            Df Deviance Resid. Df Resid. Dev P(>|Chi|)
NULL                         1308     1741.0
passengerClass  2  127.77    1306     1613.3 < 2.2e-16 ***
---
Signif. codes:  0 `***' 0.001 `**' 0.01 `*' 0,05 `.' 0.1 ` ' 1
```

You need to specify the type of test, otherwise the last column will not be included in the output. For all distributions other than the normal distribution you should choose `test="Chisq"`, otherwise you will get a warning. `test="F"` is only a valid option when working with normally distributed data.

For the sake of completeness, here is a brief example of a *post-hoc* test for GLMs. Unfortunately, I only know three options for *post-hoc* after a GLM: Dunnett, Sequen(tial) and Tukey. All are available in the **multcomp** package and can be implemented as follows (we have to adjust the argument accordingly, shown here for `"Tukey"`.):

```
> library(multcomp)
> summary(glht(fm.tita, mcp(passengerClass="Tukey")))


     Simultaneous Tests for General Linear Hypotheses


Multiple Comparisons of Means: Tukey Contrasts


Fit: glm(formula = survived ~ passengerClass, family = binomial, data = TitanicSurvival)


Linear Hypotheses:
              Estimate Std. Error z value Pr(>|z|)
2nd - 1st == 0  -0,7696    0,1669  -4,611   <1e-04 ***
3rd - 1st == 0  -1,5567    0,1433 -10,860   <1e-04 ***
3rd - 2nd == 0  -0,7871    0,1488  -5,289   <1e-04 ***
---
Signif. codes:  0 `***' 0,001 `**' 0,01 `*' 0,05 `.' 0,1 ` ' 1
(Adjusted p values reported -- single-step method)
```

12.5   Exercises

- The data set `cormorant.txt` contains diving times (`divetime`) for two subspecies of the cormorant *Phalocrocorax carbo* (subspecies *carbo* and *sinensis*), `subspecies` coded as C and S.
    1. Perform a *t*-test for this data set.
    2. Perform an ANOVA for this data set.
    3. Calculate a GLM for this data set.
    The same data set contains diving times for the four seasons (`season` with levels "spring", "summer", "autumn", "winter").
    1. Perform a pairwise *t*-test for these data: with and without the Bonferroni correction.
    2. Perform an ANOVA for these data. Then do a `post-hoc` test using Tukey's HSD.
- The data set `ancova.data.txt` contains data for an experiment in which bacteria cells are cultivated with or without an addition of glucose (`glucose` no or yes). The variable `diameter` describes the diameter of the cell in μm.
    1. Perform a *t*-test for this data set.
    2. Perform an ANOVA for this data set.
    3. Calculate a GLM for this data set.
- Implement a *t*-test for unequal variances as an optimisation task. This is a relatively simple extension of the hyena example from Sect. 8.4 on page 112. Use the data for height and sex as an example data set.

Hypotheses and Tests

13

Errors using inadequate data are much less than those using no data at all.

—Charles Babbage

At the end of this chapter ...
... you will have an idea of the role that hypotheses play in scientific research.
... you will know that you can not verify a hypothesis, but rather can only falsify, and that for this reason, the falsified null hypothesis has an important role in statistical hypothesis testing!
... you will know what type 1 and type 2 errors are and will know the meaning of the term "power calculations".
... you will know that a non-significant result is not the same as "no correlation".

Research in the natural sciences is supported by a philosophical structure that provides a basis for such work. This is not a philosophy book and the author is incompetent in this field, but a few words may be helpful before we get to grips with hypotheses and testing them.

When we conduct scientific research, we do so on the assumption that by measuring, thinking and calculating we can understand the processes of the world. We assume that there is a set of rules followed by all observable processes.[1] This set of rules may be so complex that we will never be able to truly understand it. However, the things that we measure are a result of these rules. This outlook ("philosophy") might be referred to as "realism".

To put it bluntly: If you are not convinced that we can approach the truth by measuring, thinking and calculating, I do not see any basis for scientific research. Those who want to feel the truth or seek to find it through meditation and faith will not benefit from the statistical tools discussed here. I do not want to attempt to rise above this approach, or say that it is somehow inferior, but I just want to make it clear that it is fundamentally incompatible with a statistical approach.

13.1 The Scientific Method: Observations, Ideas, Conceptional Models, Hypotheses, Experiments, Tests and Repeat

For most environmental researchers, it is the fascination with nature and natural processes that led them to this field of work. Our daily interactions with an extremely diverse and complex world provides the impetus for environmental research, and the task of understanding the processes and patterns in the environment provides the challenge. For many decades we have looked

---

[1] I am aware that this is a philosophical minefield that I am stumbling into. I want to use these sentences to describe the (in my opinion) prevailing "realistic" world view. The scientific realism is opposed by various post-modernistic points of view, some more extreme than others, but all allowing for multiple points of view to be "true". Personally, I consider post-modernism to be illogical, disproved by the extreme reproducibility of results and unsuitable as a basis for scientific research (Sokal 2008).

© Springer Nature Switzerland AG 2020
C. Dormann, *Environmental Data Analysis*,
https://doi.org/10.1007/978-3-030-55020-2_13

177

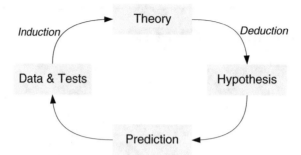

Fig. 13.1 The hypothesis based deductive approach of the scientific method. The goal is to generate a better understanding of natural processes, here designated as "Theory". Such a theory could be developed based on an observed pattern ("induction"). From an existing theory, we can derive further concrete hypotheses, which is a process referred to as "deduction". These concrete hypotheses make a prediction that we can test with the collection of experimental or observational data. If the hypothesis makes a prediction that is contradicted by the data, then we have to adjust the theory accordingly

at nature, reconstructed it, measured it, manipulated it or mathematically simulated and described it. The typical, iterative course of a scientific process is briefly traced here to help you better understand the role that statistics plays in this process (Fig. 13.1).

It usually starts with observation; we notice a pattern or regularity amongst the confusion. After some thought, we may develop an idea as to what might have given rise to this pattern. So we develop a **theory**, a model idea of how this small part of the world is composed. From here it is only a short step to consider a **hypothesis** (or several hypotheses), from which **predictions** could be tested by suitable experiments. We collect the **data** in an experiment or by observation, and can now check this data for consistency with our expectations using **tests**. Now finally comes the time for statistics (in this case formally allowing to draw inference from data)! If the test is positive, we can exclaim "Eureka!" and then relax in our favourite armchair, or press on and address new topics. If the test turns out negative, then we assume that the mechanism suspected by us was not responsible for the pattern we observe *in this case*. Then we can begin to look for new options and the game starts all over again.

What I so simply and crudely explain here is basically the daily life and work of many researchers. Accordingly, this system has been developed and refined over time. There is a scientific-philosophical basis that presents contrasting research approaches (keywords "inductive" and "deductive").

Inductive research assumes that by observing phenomena in case studies, we can apply what we learn to a larger set of conditions (from the specific to the general). This was the prevailing approach in ancient Greece, for example.[2] From then to now, the natural sciences have adopted a more hypothesis-based, deductive approach. In such an approach, a general hypothesis is established, and then we try to find support for this hypothesis through observation and experimentation (from the general to the specific).

An important point to consider is *causality*. Just because there's a connection does not necessarily mean that it's causal. In fact, we can only establish causality in one way: through manipulative experiments. Only by experimentally turning off or amplifying a process and looking at the reaction of the system do we get indisputable evidence of causality.[3, 4, 5] Since

---

[2]One important philosopher of science, Karl Popper, argued that induction typically doesn't exist: we already have some form of theory that enables us to go from the case study to the population.

[3]There is a current field of statistical research called "causal modelling", which by its name suggests that they are investigating causality. They do not. In this approach, the analyst proposes plausible causal processes and links, and assess the capacity of such a system to describe the observed pattern. However, even a very good representation is no proof of causality. As an example, we can think of autonomous driving vehicles; their algorithms are trained on real-world situations, but we have no idea what the trained algorithm actually represents. The predictions are superb, but for all we know they are completely ignorant of the causal process. In technical jargon: the model is phenomenological, i.e. describing a phenomenon. While this is an active field of research for some statisticians, it is primarily a field of philosophy, and relevent works for the natural sciences are Pearl [2009] and Illari and Russo [2014].

[4]However, we do not necessarily find out something relevant. I could not find the source of the quotation that sums this idea up nicely: "*[Manipulative] experiments force nature to give answers even to ill-posed questions.*"

[5]Computer experiments are also used in climate research. In such a case, for example, the oceanic circulation is stopped (within the computer model) and the effect on the climate is observed. The information received by it is whether *the used model* causally connects circulation and climate. We do not get (unfortunately) any information about the effects of circulation in the real world! Unfortunately, this distinction is sometimes lost by computer scientists and some climate researchers. They think their model is so good that they consider it to be a true reflection of the world. However, we also have to see the value of such simulation experiments, because since we cannot manipulate the earth in a reasonable design (lack

manipulative experiments can also contain so-called experimental artefacts, the design of such experiments is by no means trivial (see next chapter).

13.2 Testing Hypotheses

Statistics is a method used to support our decisions. One element of statistics is hypothesis testing. Once we have collected data, we still do not know if these data were generated from a random process or not. We do not know this even after we perform statistical analysis! However, what statistics does for us is provides an estimate for the *probability* that the data have been generated randomly. In order to understand the role of statistics, we first need to look at the recipe for testing hypotheses.

13.2.1 Recipe for Testing Hypotheses

To start, we have a simple question: "Are all swans white?", "Can polar bears swim longer than penguins?" or "Does the enzyme kinetics of esterase X follow the Michaelis-Menton-equation?" We then reformulate these questions to generate a working hypothesis: "All swans are white.", "Polar bears can swim longer than penguins." and "Esterase X follows the MM-kinetics". Now we realise that it will not be easy to verify these working hypotheses. We would have to look at all swans, the endurance of all polar bears and penguins, and measure the reaction of esterase X under all environmental conditions and substrate concentrations. Since we can realistically only measure a small portion of what is required, the conclusions we draw are only valid until further data may disprove them.[6]

Since the 1950s, a philosophy of science that is essentially based on the ideas of the philosopher Karl Popper has established itself as the standard (see Popper 1972). Popper argued that we cannot prove a hypothesis, but can only refute it. Therefore, he suggested that each working hypothesis (H_1) should be accompanied by a complementary null hypothesis (H_0), and that this should be examined instead. If we can refute the null hypothesis, the working hypothesis is accepted until further notice. Some examples would be: "There are non-white swans.", "Penguins swim at least as long as polar bears do." and "Esterase X does not exhibit MM-kinetics."

Now we are carrying out our measurements and experiments: we look for 100 swans and record their colour; we go by dinghy next to polar bears and penguins and measure their swimming time; we pipette different substrate concentrations to our esterase and measure their production rate.

Finally the hour of statistics is upon us! If all swans are white, we reject the null hypothesis that there are swans of a different colour. A single green, blue or black swan, on the other hand, confirms the null and disproves the working hypothesis. That was easy. With penguins and polar bears it's getting harder. Some individuals have only been swimming for a short time, sometimes the penguins have been in the water for longer, sometimes the polar bears. How should we proceed? Well, let's imagine that penguins and polar bears would have the same endurance. This means that their data *distributions* should not differ, and that they originate from the same statistical population. The exact subsequent statistical tests do not matter here. The probability of this can be calculated using statistical methods (e.g. with the GLM). And finally to esterase. We can fit a regression line to measure the reaction rate per substrate concentration. The better the fit, the more likely it is that esterase X actually follows MM-kinetics. If other enzyme kinetics models result in a better fit, then we may have to accept our null hypothesis.

---

of replicas and control), hypotheses from computer simulations are very important building blocks for interpreting descriptive measurements of the earth system (Edwards 2010).

[6]That does not mean that all hypotheses will someday be disproved! In many areas of the natural sciences, knowledge is consolidated over decades and a theory that has been tested a thousand times over is not, in a sensible way, overturned by a single (and possibly faulty) measurement. One example is Einstein's Theory of Relativity, which sets an upper limit for the speed of light. For almost 100 years it has been confirmed hundreds of times—exclusively confirmed! But then researchers supposedly detected neutrinos that travel faster than the speed of light (Adam et al. 2011). Is this the end of the Theory of Relativity? Such a well-tested system is not so easy to write off. In my opinion it was probably a measurement error—and the refutation of this claim did not take long (Antonello et al. 2012).

13.2.2 Error Types

We can distinguish four situations when considering our decision for or against the working hypothesis (Table 13.1): Depending on the true state (columns 2–3) we can either accept the null hypothesis or the working hypothesis based on our statistical analysis (rows 2 or 3). If we accept the *working hypothesis* even though it is false, we commit a type 1 error (lower left). However, if we accept the *null hypothesis* even though it is false, we commit a type 2 error (upper right). The probabilities (P) calculated in a statistical analysis indicate the type 1 error (α). The type 2 error ($1 - \beta$) is rarely calculated. On the other hand, the power of a test ($1 - \beta$) is occasionally reported for non-significant effects.

If we decide to accept our working hypothesis for our swan example (falsely, because there are in fact black swans), we would be committing a type 1 error. Additionally, if our enzyme kinetics data do not fall in line with the MM-kinetics (even though the enzyme indeed follows the MM-kinetics), we would commit a type 2 error.

The critical value for the probability of type I error is $P = 0.05$. This means that we reject our null hypothesis if the probability of a type 1 error (often referred to as the "significance level") is under 5%.[7] Within the scientific literature, this P-value provides an important orientation. During statistical analysis, if a test gives us a P-value < 0.05, then we identify the result as "statistically significant".

Of course, even better than 5% are error rates of 1% or 0.1%. In publications (and with R), these three significance levels are usually designated with asterisks: $P < 0.05$ get one star (*), $P < 0.01$ gets two (**), $P < 0.001$ gets three (***) and values greater than 0.05 are designated as "not significant" (*n.s.*)[8]

For β, the type 2 error, there are no such hard conventions. A common value is $\beta = 0.2$ (Cohen 1969), which is clearly less strict than the type 1 error. Occasionally it is argued that a type 2 error is not so bad. But it's not that simple; type 2 errors can have considerable consequences. Let's imagine that we should check the environmental compatibility of a drilling permit for gas in the North Sea (e.g. for the destruction of habitat by lowering the sea floor in shallow areas). We do not find any significant effect of previous drilling in our investigations ($P > 0.05$). Can we therefore conclude that the gas wells will not lead to subsidence in the sea floor? To accept the null hypothesis ("doesn't matter") *can* lead to the destruction of whole ecosystems! The often-quoted notion is: "Absence of evidence is not evidence for absence."[9]

What we can learn from this are primarily two things:

1. If a working hypothesis could not be confirmed ($P > 0.05$) this does not mean that the null hypothesis is correct! It may just as well be that the noise in the data is so high that we could not have recognised a true working hypothesis. This type 2 error is very human, very frequent and very wrong! For example, if we can not find a significant correlation between x and y in a regression, it does **not** mean that there is no correlation. Unfortunately, however, such a conclusion is often inferred. A somewhat brutal example: If we find that in a group of adolescents who either smoke or do not smoke there are no deaths due to lung cancer within the next 10 years ($P > 0.05$), we must not conclude that smoking does not lead to lung cancer.

Table 13.1 Type 1 and type 2 errors, depending on whether we falsely accept a working hypothesis (type 1) or falsely reject it (type 2)

	Null hypothesis true	Working hypothesis true
Null hypothesis accepted	True; $P = 1 - \alpha$	False; Type 2 error $P = \beta$
Working hypothesis accepted	False; Type 2 error $P = \alpha$	True; $P = 1 - \beta$

---

[7]Here you can go into historical and statistical details for as long as you like. The 5% rule is a convention, and not a derived value. Ronald A. Fisher [1925] suggested it, and since then it has become dogma. There are many good scientific articles dealing with the sense and nonsense of hypothesis tests based on a $P < 0.05$ (Cohen 1994; Gill 1999; Gliner et al. 2002; Hobbs and Hilborn 2006; Johnson 1999; Stephens et al. 2007). I tend to side with Simberloff [1983], who advises to break the conventions only when one has mastered them.

[8]Often you will see the phrases "almost significant" or "marginally significant". Such phrases describe P-values between 0.5 and 0.1 and are often designated with a dagger (†) or in R with a ".".

[9]The problem of the lack of scientific verifiability leads us to the ethics of the "precautionary principle". This states that an intervention should not be carried out until it has been clearly shown that it is free of consequences. In statistical terms, this means that a test for the effect of an intervention should not only be $P > 0.05$, but also that $\beta < 0.2$. From an economic point of view, the precautionary principle is problematic, since under its application, a new technology could only be introduced if its safety had been proven. This would dramatically slow down innovation processes. In fact, the precautionary principle, and with it the type 2 error, is usually not applied in politics, whatever politicians may claim.

2. There is also the possibility of drawing conclusions from non-significant results that the null hypothesis is correct. We cannot draw any conclusions from the α value, but we can draw some conclusions from the β. We have rejected the working hypothesis by testing it (top line of Table 13.1) and now want to state that the null hypothesis is true.[10] The test strength is $= 1 - \beta$, where β is the tolerable level of type 2 error. (According to Cohen 1969, p. 56, β should not exceed four times the α, typically 0.2, so that we can claim that the null hypothesis is true.[11]

To make this clearer, consider the following example. The following data set is examined for a difference between **A** and **B**: **A** is always smaller than **B** in all but two cases (value pairs 4 and 7). We would find that to be a fairly consistent difference. Right?

```
   [,1] [,2] [,3] [,4] [,5] [,6] [,7] [,8] [,9] [,10]
A  8.7 10.4  8.3 13.2 10.7  8.4   11 11.5 11.2   9.4
B 13.0 10.8  8.8  5.6 12.2  9.9   10 11.9 11.6  11.2
```

A t-test shows us that the two groups are not significantly different from each other:

```
    Welch Two Sample t-test

data:  A and B
t = -0.2636, df = 16.615, p-value = 0.7953
alternative hypothesis: true difference in means is not equal to 0
95 percent confidence interval:
 -1.984051  1.544051
sample estimates:
mean of x mean of y
    10.28     10.50
```

So, given the results of this test, can we conclude that the two groups are from the same population? To do so we calculate the power of the test.[12] We get:

```
Two-sample t test power calculation

              n = 10
          delta = 0.22
             sd = 1.866309
      sig.level = 0.05
          power = 0.04362753
    alternative = two.sided


NOTE: n is number in *each* group
```

Our accepted significance level, β, for this power test would be 0.2, with the *power* being $= 1 - \beta = 0.8$. As we see in the output, the *power* value is 0.04, far from our desired power of 0.8! On the basis of this analysis we can *not* say that both groups are equal. To achieve the desired power level of 0.8, we would need a group size of more than 1000 (with a difference between the group mean of 0.22 and a standard deviation of 1.87).[13]

---

[10]The probability that the null hypothesis is true, has nothing to do with the probability of rejecting the working hypothesis. It must be evaluated separately. This is done by calculating the strength (*power*) of a test. Even a *P*-value of 0.99 can have a low test power of 0.90!

[11]This 0.2 is a similar convention as the 0.05 for *P*. But because it has not been in the literature for 60 years, and the type 2 error receives so little attention, the 0.2 value rarely appears in statistical literature. Cohen's book is quoted a lot, but rarely for this value.

[12]Since we do not want to go into the derivation of these calculations, here is the simple R code for this calculation:
`power.t.test(n=10, delta=0.22, sd=sqrt((sd(A)^2+sd(B)^2)/2), sig.level=0.05)`.

[13]`power. t.test (power=0.8, delta=0.22, sd=1.866, sig.level=0.05)`.

```
Two-sample t test power calculation

            n = 1130.652
        delta = 0.22
           sd = 1.866309
    sig.level = 0.05
        power = 0.8
  alternative = two.sided
```

NOTE: n is number in *each* group

We can see: *A non-significant difference may still be far from an indication of equality!* So the t-test may not support our first impression of differences between the groups, but we can not claim that the two data sets are the same.

13.3   Tests

The examples show that the sample size, i.e. the number of measurements, will have an influence on our conclusion. At some point we will be walking though a park and a black swan (*Cygnus atratus*) from Australia will cross our path, and our H_1 is toast.

In addition to the sample size, there are other factors that have an influence on the power of a test (and thus our ability to recognise a true working hypothesis as true). The greater the difference between the data of the null hypothesis and the working hypothesis, the more likely we will find this difference. Similarly, highly variable measurements (= large standard deviation) also lead to a reduction in power.

We find that a possible difference between the working hypothesis and the null hypothesis is more easily detectable the greater the true difference, the smaller the scattering of the data, and the larger the sample size.[14]

13.3.1   Further Test Related Terminology

A test is considered a *one-tailed test* if a directional hypothesis is tested: male ptarmigans are larger than female ptarmigans. If we want to test whether there is any difference at all in size (regardless of in which direction), we need to use a *two-tailed test*. With the test distributions, this is done by simply doubling the P-value of a one-sided test. Most statistics programs are automatically set to the more frequently used two-tailed test.

All tests are based on **assumptions**. Which assumptions in particular can be determined by referring to the documentation of a specific test. For all tests it holds true that the data must be **independent** from each other. If we measure the weight of one boy and one girl 17 times, this does not give us information about a difference between boys and girls, but only between *this* boy and *this* girl. The simplest way to generate independence is to randomly select the sample. Randomness itself is not a statistical necessity, however, as even regular ("every fifth") selections can also generate independent data. It is important that each object has the same chance of being selected. More on this in the next chapter.

A problem arises when we take many measurements on objects (approximately 50 people, each with shoe size, nose length, arm span, step length, ear length, etc.) and subsequently test for differences between objects (such as between the sexes). Simply from pure coincidence we will find a significant difference; the more comparisons we make, the more differences will be significant. Then we accept this hypothesis, even though it is wrong. This phenomenon during **multiple testing** is called type 1 error inflation. There are several ways to correct it (for example, divide the P-value received by the number of comparisons made: Bonferroni correction (Sect. 11.2.5 on page 156); simultaneous testing of all measured variables in a MANOVA). The important thing is to be aware of the problem.

Finally, the problem of **data ties** occasionally arises during parametric and non-parametric tests. In this case, there are one or more measurement values that occur several times in the data. Especially in non-parametric tests, this can lead to distortion of the results. While alternatives exist (permutation tests are less susceptible to data ties), knowledge of this problem is the most important.

---

[14]It is exactly these data that are needed in calculating the power in the two previous footnotes.

13.3.2 Final Comments on Tests

The magic P-value of 0.05 (or 5% significance level) is simply a convention and is therefore arbitrary (as a guideline rather than a rigid value proposed by Fisher 1925). Some people find this significance level to be too high, others too low. The fixation on significance is often (rightfully) criticised (see Hobbs and Hilborn 2006; Johnson 1999; Shaver 1993; Stephens et al. 2007), but is still the prevailing paradigm (Ford 2000; Gibbons et al. 2007; Oakes 1986). Above all: only those who master this classical form of the significance test should start to modify the critical P-values (Simberloff 1983).

A common mistake is to confuse the P-value with the strength of an effect. A smaller P-value does not mean that the tested effect is also ecologically strong. This can be easily illustrated by increasing the number of data points from 10 to 1000 for a constant effect (for example, a slope of around 0.5). At first the slope is not significant, then just barely and finally highly significant—without having changed! Especially with large amounts of data we will be able to recognise almost every influencing factor as statistically significant, even if it explains only 0.1% of the variance.[15] Remember: the P-value is not a measure of the strength of an effect.

A temptation to which many people succumb is **extrapolation**. While it seems quite understandable that we interpolate values between the values measured by us, it is problematic to infer values outside of the measured range. When we do a regression of the basal metabolic rate against body weight, we cannot extrapolate to the basal metabolic rate of a 0.1 g size animal, or that of a mammal the size of our planet. In some cases, other factors may play a limiting role here than for the body weights that are in the data set. If we have to make such predictions, we must do so with many explanatory words and an appropriate error estimate.

13.4 Exercises

Since multiple comparisons are covered in Sect. 11.2.5 (under "*post-hoc*-Tests") and errors on predictions are covered in Sect. 8.1, here we will just deal with exercises regarding power.

1. The `power.t.test` function calculates the fifth value if the other four arguments are given. Calculate the necessary n-values for different power values. Make a figure showing power on the y- and n on the x-axis (maybe 10 different values). Note: For `sig.level` take 0.05, for `delta` maybe 1 and for `sd` maybe 2. As long as these values are not totally absurd, all of them should result in a similar figure (of course with different absolute values).
2. What happens if we select a larger group difference (e.g. `delta = 2`)? Is the curve higher or lower than it was before? Why?
3. A well-known Canadian forestry company carries out a major management comparison: after a clear-cut, half of the 200 plots are planted with new trees, the other half the plots are allowed to regenerate naturally. After 20 years the basal area (area covered by trees at breast height) in each plot is measured and compared between treatments. A t-test says that the difference of 7.3 m^2/ha is not significant ($P > 0.05$). The two treatments have an average variance (!) of 321.44. With which level of conviction (= power) can we say that the treatments are the same?
4. In a re-analysis it turns out that the basal area had been calculated incorrectly: a factor 2 had been forgotten somewhere. The difference is now 14.6 m^2/ha and the averaged variance is 1286. How does this change the power?

References

1. Adam, T., Agafonova, N., Aleksandrov, A., Altinok, O., Sanchez, P. A., Anokhina, A., et al. (2011). Measurement of the neutrino velocity with the OPERA detector in the CNGS beam. arXiv1109.4897.
2. Antonello, M., Aprili, P., Baibussinov, B., Ceolin, M. B., Benetti, P., Calligarich, E., et al. (2012). Measurement of the neutrino velocity with the ICARUS detector at the CNGS beam. arXiv1203.3433v.
3. Cohen, J. (1969). *Statistical power analysis for the behavioral sciences*. New York: Academic Press.
4. Cohen, J. (1994). The earth is round (p < 0.05). *American Psychologist, 49*, 997–1003.
5. Edwards, P. (2010). *A vast machine: Computer models, climate data, and the politics of global warming*. Cambridge, MA: MIT Press.

---

[15] A colleague once analysed ringing data from geese, a record with over 50, 000 entries. In this analysis, even the initial letter of the first name of the ringer had a significant influence on geese growth.

6. Fisher, R. A. (1925). *Statistical methods for research workers*. Edinburgh: Oliver & Boyd.
7. Ford, E. D. (2000). *Scientific method for ecological research*. Cambridge, UK: Cambridge University Press.
8. Gibbons, J. M., Crout, N. M. J., & Healey, J. R. (2007). What role should null-hypothesis significance tests have in statistical education and hypothesis falsification? *Trends in Ecology and Evolution, 22*, 445–446.
9. Gill, J. (1999). The insignificance of null hypothesis significance testing. *Political Research Quarterly, 52*, 647–674.
10. Gliner, J., Leech, N. L., & Morgan, G. A. (2002). Problems with null hypothesis significance testing (NHST): What do the textbooks say? *The Journal of Experimental Education, 71*, 83–92.
11. Hobbs, N. T., & Hilborn, R. (2006). Alternatives to statistical hypothesis testing in ecology: A guide to self teaching. *Ecological Applications, 16*, 5–19.
12. Illari, P., & Russo, F. (2014). *Causality: Philosophical theory meets scientific practice*. Oxford, UK: Oxford University Press.
13. Johnson, D. H. (1999). The insignificance of statistical significance testing. *Journal of Wildlife Management, 63*, 763–772.
14. Oakes, M. (1986). *Statistical inference: A commentary for the social and behavioral sciences*. New York: Wiley.
15. Pearl, J. (2009). *Causality: Models, reasoning and inference* (2nd edn.). New York, NY, USA: Cambridge University Press.
16. Popper, K. (1972). *Objective knowledge: An evolutionary approach*. Oxford University Press
17. Shaver, J. (1993). What statistical significance testing is, and what it is not. *Journal of Experimental Education, 61*, 293–316.
18. Simberloff, D. (1983). Competition theory, hypothesis-testing, and other community ecological buzzwords. *The American Naturalist, 122*, 626–635.
19. Sokal, A. (2008). *Beyond the hoax: Science, philosophy and culture*. Oxford, UK: Oxford University Press.
20. Stephens, P. A., Buskirk, S. W., & del Rio, C. M. (2007). Inference in ecology and evolution. *Trends in Ecology and Evolution, 22*, 192–197.

Experimental Design

14

To consult the statistician after an experiment is finished is often merely to ask him to conduct a post mortem examination. He can perhaps say what the experiment died of.

Ronald A. Fisher

> At the end of this chapter ...
> ... you will know the basics of designing an experiment.
> ... you will have gotten to know the most important experimental designs.
> ... The terms "pseudoreplication", "control" and "random effect" will mean something to you.
> ... you will know that Stuart Hurlbert's publication from 1984 is required reading.

There are a few basic things to consider for all kinds of sampling procedures. While the following points seem to follow common sense, you should be aware of the reasons that they are so important. In fact, statisticians around the world spend a considerable part of their time "rescuing" poorly planned data collection. All too often they give in to the wishes of those seeking help, instead of simply describing a data set as lost beyond repair. A side effect is that statisticians become more important—although actual scientific progress is made by well thought-out questions and well executed data collection, not by the most complicated analysis possible. In fact, it would be my ideal that the analysis of a large data set from a multi-year study is carried out in just a few minutes because the *sampling* has been thought through and *the analysis was considered from the beginning*.

14.1 Design Principles

14.1.1 Principle 1: Representativity

As described in Chap. 1, we are usually interested in the whole population (such as all beech trees), but can only measure a small part of the population (the sample). In order to draw conclusions from this small sample that relate to the whole population, the sample has to be *representative*, or a typical representation of the total population.

To make sure that we are including representativity in our sample, we must make sure that:

1. the total range of values of the population is also included in the sample;
2. the sample is not biased.

The first point can be achieved by simply collecting data points for our sample (**replication**). The second point can be covered if we either randomly or systematically (and not haphazardly or arbitrarily) collect data. What this means exactly is best described with some examples. In the field, these considerations sometimes abandon us. If you want to dig up 200 random plants, there is a slim chance that you will tag 10000 plants with small numbers and then generate a random number table and choose which plants to dig up. Instead, you would probably throw a pocket knife over your shoulder or throw out a net every 5 min during a walk and choose your plants to sample in this manner. Of course, these are not random samples. They

© Springer Nature Switzerland AG 2020
C. Dormann, *Environmental Data Analysis*,
https://doi.org/10.1007/978-3-030-55020-2_14

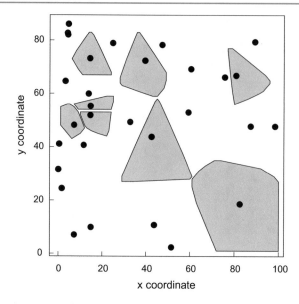

Fig. 14.1 If we randomly select *x* and *y* coordinates, and then choose the next closest tree to be a part of our sample, this is *not* random! For trees that stand further apart from each other, more (*x*, *y*) pairs will be selected (grey areas around points) than for trees that are closer together (Adapted from Crawley [1993])

assume that our arbitrariness is impartial and therefore has the same randomness as a truly random sample. Often this will be fine, but we must always be aware of the problem!

A classic example, courtesy to Mick Crawley, showing us how this arbitrariness can go awry is depicted in Fig. 14.1. If the decrease in tree density occurs for an ecological reason and we use the described method for selecting a tree to be part of our sample, we will likely be taking our samples along this ecological gradient and our data will be masked by this effect, resulting in the disappearance of the effect that we are trying to investigate!

A clear violation of the principle of representativity can be seen in the design of the North American *breeding bird survey*,[1] in which volunteers collect bird data on over 4000 transects—along roads! Such data is worthless for describing bird populations away from roads. As much as this will pain all of those volunteers to hear, their data is extremely biased and not representative. Of the over 900 North American bird species, trends are only available for around 400 species, according to the BBS.[2]

Other factors may lead to similar bias: insufficient knowledge of species (leads to systematically more false identifications in groups with many similar species), a preference for "attractive" species (leads to an uneven coverage of the total biodiversity), collecting data from species that live near settlements (leads to geographic clumping of data), accessibility of the study sites (also leads to geographic bias), the number of observers (leads to many observations in wealthier areas), level of organisation and professionalism of the observer, etc. Much more subtle and partly even unknown are the effects of education, research traditions, perceptual limitations (we don't hear bats) and similar issues.

We have to separate such bias from measurement errors that make the result noisy, but do not systematically distort it if they deviate equally upwards or downwards. For example, a GPS device indicates the point of a measurement with a measurement error in meters, but this inaccuracy is the same in all directions, i.e. it is not systematically distorted.

---

[1] https://www.pwrc.usgs.gov/bbs/index.cfm.

[2] Much time has been invested in "rescuing" these valuable data. And for many interesting questions these data can be justifiably analysed. But they still do not represent the birds of North America.

14.1.2 Principle 2: Independence

Independence simply means that the likelihood of an event does not change because another event occurs (or fails to occur).[3]

Independence ensures that the calculation of the likelihood is correct. To check this, we multiply the probability density of the individual values in order to calculate the probability of a common occurrence. However, only for **independent** events A and B is $P(A, B) = P(A) \cdot P(B)$ valid![4]

Randomisation is the most common way to achieve independence. Nevertheless, we still need to check the results of the randomisation. If by pure coincidence all samples in a treatment condition are on the right and all of the controls are on the left, this is **not acceptable**: here, the location is mixed with treatment (*confounded*) and afterwards we do not know what causes the difference in these groups: is it the location or the treatment? The solution to this problem is another (and possibly repeated) randomization.

Whether the experimental subjects are systematically or randomly arranged depends on the case. Neither system is better than the other *per se*, even if Sokal and Rohlf [1995] suggests otherwise (see Side Note 11). The difficult thing about randomisation is to be sure that there is no systematic gradient in the setup.

Side note 11: Random or systematic sampling? In the majority of statistics textbooks you will find lines such as "The methods of inductive statistics can only be applied to purely random samples". This suggests that we need to select the objects of our sampling randomly. This is not the case. *A random selection ensures that each object can be drawn with the same probability*. Every approach that ensures this is acceptable. The selection does not have to be random—it can also be done systematically. (Every tenth person that enters a rock concert is searched for drugs; which person you start with, however, *must* be randomly chosen).

Two big names in statistics, William S. Gosset (inventor of the t-distribution and the t-test, both published under his pseudonym, "Student") and Ronald A. Fisher (probably the most important statistician of the 20th century and the founder of many standard statistical approaches such as ANOVA), argued about this topic throughout their lives. The older Gosset was defeated mainly because he died first (1937, Fisher died in 1962). These 25 years after Gosset's death led to the cementation of randomisation in experimental design, although Gosset's arguments for systematic sampling could not be disproven by Fisher (for details on this historically fascinating dispute, see Hurlbert 1984, p. 196 et seq.).

To make it perfectly clear, the goal of sampling is to ensure that each experimental unit has the same probability of being sampled! If this can be ensured using a systematic approach, then by all means, this is also fine.

14.1.3 Principle 3: Control Group

In an experiment, we actively change something; we manipulate some process that is of interest to us. We might fertilise a plant, put a mouse onto a nuts-only diet, or freeze a mushroom. We raise the ambient temperature around a plant community using infrared lamps, turn off a biochemical pathway in a knock-out-mouse, or thin a forest stand. In order to see if the *treatment* has an effect, we need a comparison group in which there is *no* treatment: the control.

Without a control there is no effect!

How do we know that fertilising makes a plant grow faster if we don't know how it grows without fertiliser? How can a mouse become obese if we don't know whether this is simply the fate of every mouse? If the mushroom tastes bad after thawing, how do we know how this is due to freezing? Maybe every mushroom tastes different after a certain of time has passed? I think it's clear, why a control is essential.

The control only serves as a control for the treatment effect, not for possible side effects of the treatment. For example, if you want to investigate the effect of stones in the soil on plant growth and sift the stones out of the soil while the control keeps the stones, you have to sieve the control exactly the same way, except that the stones are put back. Otherwise, loosening and aeration of the soil may yield your 'treatment' effects, but these would have nothing to do with the stones.

A perfect isolation of treatment from side effects is not always possible in the control, and yet we must quantify the side effects of the treatments themselves. Therefore, in some situations we may need to perform an additional 'treatment artefact control'. For example, if you keep moths away from leaves by using small cloth bags around the leaves, you reduce the wind

---

[3]This concept is sometimes difficult for people to grasp: if we roll a die and roll a 6 five times in a row, then the probability that the next roll is a 6 must be smaller, right? But indeed, it is not. The die does not recall what was rolled before, and therefore the likelihood that a 6 is rolled (or any other number) is independent of the previous rolls: 1/6.

[4]In fact, there are also statistical procedures for taking systematic dependencies into account (such as Generalised Least Squares and Generalised Linear Mixed Models). These are sometimes inevitable, but are a more advanced topic.

Table 14.1 Potential sources of confusion in an experiment and how to minimise their effect (Adapted from Hurlbert 1984). Note that replication features strongly as strategy to overcome various sources of variability

	Source of confusion	Design features that can reduce this confusion
1.	Temporal changes	Control treatments
2.	Procedure effects	Control treatments
3.	Experimenter bias	Randomised assignment of experimental units to treatments Randomisation in conduct of other procedures "Blind" procedures (for large subjective elements)
4.	Experimenter generated variability (Random error)	Replication
5.	Initial or inherent variability among experimental units	Replication; interspersion of treatments; Concomitant observations
6.	Nondemonic intrusion (chance)	Replication; interspersion
7.	Demonic intrusion	Eternal vigilance, exorcism, human sacrifices, etc.

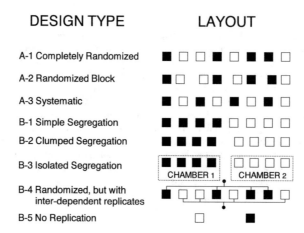

Fig. 14.2 Scheme of acceptable designs (A), how to mix replicated (boxes) of two treatments (white and black), and different designs that violate the randomisation principle (B) (Adapted from Hurlbert 1984)

speed on the surface of the leaf at the same time. A treatment artefact control would be an open cloth bag that leaves the moths on the leaf, but not the wind. This gives an estimation of the wind effect, which is conflated here with the moth-eating.[5]

14.1.3.1 Everything that Can Go Wrong

It is impossible to create a list of all possible errors that can occur in an experimental design. However, there is a long (but very readable) article from Hurlbert [1984], which provides many thoughts on the typical mistakes that occur in ecological experiments and how to counteract them (see Table 14.1).

For example, point 5, knowledge of the initial condition of an experimental unit *before* the experiment increases the statistical test strength, just as gathering of relevant environmental parameters (covariates) can also dramatically improve test strength.

This provides us with a manageable set of acceptable experimental designs (Fig. 14.2). It is easy to dismiss the B-designs as 'obviously flawed', but climate chamber experiments are largely of type B-3, and using the same fertiliser solution for your enriched mesocosms (rather than a new preparation for each single one of them) yields B-4. A comparison of forst growth (Morford et al. 2011) on two sites differing in bedrock-N (design type B-1) tells you something about differences between these sites, but we cannot attribute this difference to bedrock-N (it could be other elements, the topography, a climatic difference, management).

---

[5]In medical studies, this could be, for example, a placebo. The patients in medical studies don't receive either nothing or a medicine ("study drug"), but rather a third group receives a pill with no effect: a placebo. The "placebo effect" is actually quite strong: in 20–80% of the cases it actually helps, even though there is nothing in it! For a review of this effect, see Finegold et al. [2014].

Apart from all statistical considerations, ecological understanding should guide experimental design. For example, the reasonable minimum distance between two test areas depends on the target organisms and treatments. Clonal plants may transport nutrients over several meters, and pollinators interfere with plots dozens of meters away. As a rule of thumb, the distance between two plots should be 2–3 times the plot size (i.e. at least 2 m between plots of 1×1 m), but as the pollinator example shows, it could be hundreds of times the size of the plot to achieve full separation. If at all possible, we should visit similar experiments elsewhere to improve our intuition about plot layouts. The Park Grass experiment (Silvertown et al. 2006) had virtually no buffer around the treatments for 150 years, yet the borders between treatments are razor-sharp, suggesting that lateral nutrient transfer is practically absent.

Financial, spatial, logistical or temporal restrictions will always force us to compromise on the "ideal" design. Mention these limitations and their possible effect on the results of the experiment when writing up the results. Be aware, however, that the use of an inappropriate design is to be equated with lying (Magnusson 1997).

14.2 Important Designs for Manipulative Experiments

To understand the most important designs, we need to consider two important concepts: orthogonality and random effects. Often we are not just interested in a single influencing factor, but rather in multiple factors. The reaction of plants to nitrogen fertilisation may depend on phosphate availability. We could establish three treatments: a control (0), with nitrogen (N) and with nitrogen and phosphorus (NP). Unfortunately, we cannot now clearly interpret the NP effect because we do not know what the reaction would have been to phosphorous alone (P). Is the NP effect now simply the sum of N and P effects? Do the two fertilisers act synergistically, or even antagonistically? Only a **factorial design**, where *all* levels of one treatment are combined with all levels of the other treatment(s) allows for a complete interpretation. In our case, we would need a 2×2 factorial design with the treatments, 0, N, P, NP.[6] Now N and P are also independent of each other, or, as the statistician says, orthogonal. *Orthogonality* means that the explanatory variables in the analysis (= the manipulated effects in the design) are independent of each other. In the design, this is achieved by a factorial combination of the treatments. The strength of the factorial design lies in the fact that in a 2×2 factorial experiment with 10-fold repetitions, we examine the main effects (e.g. N$\pm$) with 40 plots (namely 10 plots P$-$N$-$, 10 plots P$-$N$+$, 10 plots P$+$N$-$ and 10 plots P$+$N$+$). If I wanted to test N and P effects in separate experiments, I would only have 20 plots (10 times N$+$ and 10 times N$-$). This means that the factorial design doubles the number of test plots without additional costs. In addition, I get the possibility to test for an interaction of N and P.

Sometimes we combine experimental plots into **blocks** (see below) or we measure several things on a randomly selected object. In such cases, blocks or objects differ from each other, but we are not really interested in that. Nevertheless, we must also include these effects in the statistical analysis. Effects that are of interest to us are called "fixed effects" (e.g. the nitrogen fertilisation in the above example or the sex in the analysis of the Titanic data). Those effects that are somehow included, but for which we are not really interested, are called "random effects". Examples are the number of the block in which the plots are located, the name of the mountain on which we have replicated our 3 treatments, etc. Whether something is a random or fixed effect depends on whether we learn something important for our hypothesis from its calculation. If so, then it is a fixed effect, otherwise it is a random effect. Statistical analyses in which both fixed and random effects occur are called mixed effects models. They are an advanced topic (see Pinheiro and Bates 2000; Zuur et al. 2009), but appear briefly in the analyses below.

14.2.1 Fully Randomised Block Design

1. Definition   The treatment plots for all treatment combinations are put together in a single block. This block is then replicated. The treatments are assigned randomly within the block.
2. Example   We want to investigate the effect of nematodes on competition. We will examine whether the competition between *Festuca* and *Artemisia* is dependent on nematode predation damage. We have the following treatments: *Festuca rubra* in competition with *Artemisia maritima* (F+); *Artemisia maritima* in monoculture (F−); untreated soil with nematodes

---

[6]It is called "factorial" because the number of levels per treatment is multiplied, i.e. they are factors that determine the number of test variants. Two treatments with 4 levels each result in $4 \times 4 = 16$ variants in a factorial design. To emphasise this special quality, such a design is often referred to as a "full factorial design", in which no variant is missing.

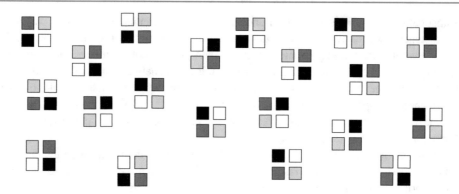

Fig. 14.3 A 20-fold replicated fully randomised block design

(N+); soil treated with nematicide and thus without nematodes (N−). With a factorial experiment we then have four treatment combinations (F+N+, F+N−, F−N+, F−N−) in each block with, let's say, 20-fold replication. In the end we have a fully randomised block design that looks something like Fig. 14.3.

3. Details   This is the most common, the most important and the most intellectually satisfying experimental design. It has two core elements: (1) Randomise whatever you can, and (2) put all treatment combinations in a single block and then replicate the block. The statistical analysis is uncomplicated, because there are no dependencies of the treatments or the blocks. The common analysis for such a set up is an ANOVA (for normally distributed data). Since we are not really interested in the difference between blocks, the block label is taken into the model as a (categorical) random variable.

4. Strengths & Weaknesses   Randomisation can take quite a lot of time (although we can do this ahead of time while sitting in our favourite armchair). If we know about the existence of a gradient in the study area that might influence our experiment, we would have to take this into account both in the spatial distribution of the blocks and the treatment areas within the blocks. This is not possible with a randomised design, of course, but influences of this kind have to be compensated for by increasing the number of samples. This means we have to use more replicates than would be necessary without this gradient.

5. References   Hurlbert [1984], Mead [1990], Underwood [1997], Potvin [2001], Crawley [2002].

6. Example analysis   In the following we will look at a typical analysis. It is typical because some of replicates have been lost and the design is thus unbalanced. If there were also a whole combination missing (e.g. all black fields in Fig. 14.3), then we would be dealing with a design with *missing cells*.

Let's stay with our example of nematodes and competition between *Festuca* and *Artemisia*. Our dataset is a little unbalanced, because we only have 11 replicates for the monoculture, whereas we have 16 replicates for competition. Additionally, the plants in two pots have died. But this just a small matter. First we read in the data, encode the factors as factors and get the mean values for the treatment combinations. Then we calculate an ANOVA in which we tell the model that we have used blocks.

```
> nema <- read.table("nematode.txt", header = T)
> nema$comp <- as.factor(nema$comp)
> nema$block <- as.factor(nema$block)
> nema$nematodes <- as.factor(nema$nematodes)
> attach(nema)
> names(nema)
"pot"       "comp"      "nematodes" "block"     "Artemisia"
> tapply(Artemisia, list(comp, nematodes), mean, na.rm = T)
          0          1
0 15.216455 14.339500
1  2.069063  2.116063
```

Now we first do a "normal" ANOVA, in which the `block` is a fixed effect. This is not entirely correct, but easier to understand:

```
> summary(aov(Artemisia ~ block + nematodes * comp))
               Df  Sum Sq Mean Sq  F value  Pr(>F)
block          15  159.64   10.64   1.6581 0.11253
nematodes       1   20.74   20.74   3.2313 0.08169 .
comp            1 1896.81 1896.81 295.5305 < 2e-16 ***
nematodes:comp  1    2.08    2.08   0.3236 0.57344
Residuals      32  205.39    6.42
---
Signif. codes:  0 `***' 0.001 `**' 0.01 `*' 0.05 `.' 0.1 ` ' 1
2 observations deleted due to missingness
```

Although we will discuss multiple regressions in the next chapter, the output here is relatively straightforward: block with 16 levels, explains 160 SS-units and is not significant, while comp explains 1900 SS-units and is highly significant. Neither nematodes nor the interaction between nematodes and comp is significant.

If we were to leave out the block effect, then the 160 SS-units explained by the block would fall out and the F-value would decrease. In this way, the block helps us to increase the sensitivity of the analysis.

What follows is the R syntax that does justice to the design by indicating the block as a random effect:

```
> fm <- aov(Artemisia ~ nematodes * comp + Error(block))
> summary(fm)

Error: block
          Df  Sum Sq Mean Sq F value Pr(>F)
nematodes  1   0.018   0.018  0.0017 0.9678
comp       1  25.197  25.197  2.4368 0.1425
Residuals 13 134.422  10.340


Error: Within
               Df  Sum Sq Mean Sq  F value  Pr(>F)
nematodes       1   20.74   20.74   3.2313 0.08169 .
comp            1 1896.81 1896.81 295.5305 < 2e-16 ***
nematodes:comp  1    2.08    2.08   0.3236 0.57344
Residuals      32  205.39    6.42
---
Signif. codes:  0 '***' 0.001 '**' 0.01 '*' 0.05 '.' 0.1 ' ' 1
```

Before we take a closer look at the issue in detail, let's have a quick comparative look at this and the previous model: The SS-values for nematodes, comp and their interaction are exactly identical. The SS for block is missing here (because we are not really interested in block). So does it matter if we declare a random effect as such? Short answer: If we only have *single* random effect, then no. However, as soon as the random effect is not orthogonal to the treatments (e.g. in the *split-plot* and in the *nested* design discussed below) the results differ! Now for a detailed consideration of the output.

By calling the model (fm), we get some important additional information: In stratum 1, one of the three effects cannot be calculated. This is the interaction of the two treatments. It is missing in the above output. The reason for its absence is that the design is unbalanced and therefore there are not enough data for all blocks to calculate the interaction effect (only blocks 2, 3, 4, 6 and 8: table(block)).

The upper part of the output refers to the upper stratum: the blocks. We are not really interested whether blocks differ, and usually we would ignore this stratum. The point is that *if* a treatment is missing in a block, the difference between this block's mean and those of complete blocks indicates the effect of the missing treatment! Thus, we can use block means to test for treatments *if* our design is incomplete. This is unintended, but logical. Within each block, only one value is calculated for the respective treatment, i.e. by averaging over the other treatment. This means that a total of 16 values are available for each treatment (one per block). The statement of the non-significant treatments here is basically saying: "If we only compare the mean values per block, we see no difference between the treatments."

In the second stratum, on the other hand, all data is used, so we have more degrees of freedom for the residuals. This is the actual analysis our design demands. Note that the degrees of freedom of the residuals from the 1st stratum are deducted from those in the 2nd stratum. The statement here is: "If we calculate the block effect from each individual measurement, then we find a significant effect of the competition, and a very slight nematode effect."

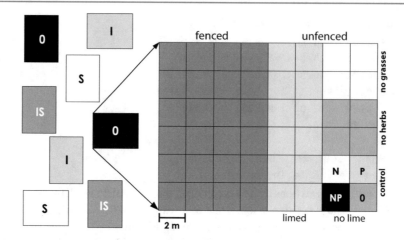

Fig. 14.4 Example of a split-split-split-split-plot design. On the level of the 12 × 16 m large plot there are two treatments factorially applied on two levels: slug poison (0/1) and Insect poison (0/1). Then each plot is repeatedly split into different levels: first, one half is fenced and one half left unfenced, then the half with no fence is split into lime/no lime treatment. This split-split-plot is then split with three levels of different species compositions (without grass, without herbs, and a control), and each of these subplots is then factorially assigned nitrogen and phosphorous fertilisation treatments. At the end, we are left with 384 plots that are 4 m² each. The mastermind behind this experiment, Mick Crawley, Imperial College London, is renowned for his efficient use of resources (Adapted from Crawley 2002)

To repeat the explanation for this analysis: The output for the first stratum only contains the treatments themselves if the design is unbalanced, i.e. treatments in one or more blocks are missing. The reason for this is that the block effect cannot be properly estimated if there are missing combinations. If, for example, a competing plant's pot is missing, the mean value is higher (competition-free *Artemisia* are "heavier"). If the design is balanced, the block effect can be calculated independently of the treatment effects and deducted from the 2nd stratum.

14.2.2   Split-Plot Design

1. Definition   The different levels of a treatment are applied within each level of a different treatment.
2. Example   We want to investigate the effect of fertilisation and grazing on vegetation. To do this, we build exclusion cages, *within* which we include the two levels of fertilisation (with/without). We can recognise this as a split-plot design because the two treatments (grazing/fertilising) are applied on two areas that differ in size: grazing exclusion on the whole area of the exclusion cages, with the fertilisation on only half of this area.
3. Details   Split-plot designs are usually used for practical reasons. In the example above, we only need 10 exclosures for a 10-fold replication instead of 20 exclosures that would be needed in a non-split design. An illustrative, but perhaps extreme example can be seen in Fig. 14.4.
4. Advantages & Disadvantages   The practical advantages come with two disadvantages: (1) Statistics become more complex as we have to include the *split-plot* structure in our analysis. In the above example, the exclosure effect is analysed first, and then the exclosure effect and fertilisation together. (2) We apply our treatments on different plot sizes. However, if now, for example, the effect of competition on a single plant is examined on a 1/2 m² plot, but fertilisation on 2 m², then the sphere of influence of the individual plant differs accordingly. With the fertilisation, many more species and individuals can participate in the use of the fertiliser, whereas with the competition between individuals, the number of individuals and species involved is lower. In practice, this will often not be of much consequence, but we should nevertheless keep this problem in mind.
5. References   Crawley [2002], Quinn and Keough [2002].
6. Example analysis   Instead of reproducing the entire split-split-... -plot design of Crawley, we will only look at the highest levels. The experiment looks like this (see Fig. 14.4): A treatment with insecticide is factorially combined with one with molluscicide and replicated twice. These are the treatments on the whole area of a plot (left half of the figure). Now one half of each plot is surrounded by a fence against rabbits. Each of these sub-plots is in turn divided into two halves, which

are limed (or not). So we have a split-split-plot-experiment. We first load the data, then average the values over the ignored, lower levels and generate a new data set, which we then analyse.[7] Pay attention to how the splits are coded in the formula!

```
> crawley <- read.table("splitplot.txt", header = T)
> crawley$Insect <- ifelse(crawley$Insect == "Unsprayed", 1, 0)
> crawley$Mollusc <- ifelse(crawley$Mollusc == "Slugs", 1, 0)
> crawley$Rabbit <- ifelse(crawley$Rabbit == "Fenced", 0, 1)
> attach(crawley)
> splitplot <- aggregate(Biomass, list(Block, Insect, Mollusc,
+     Rabbit, Lime), mean)
> colnames(splitplot) <- colnames(crawley)[c(1:5, 8)]
> fm <- aov(Biomass ~ Insect * Mollusc * Rabbit * Lime +
+     Error(Block/Rabbit/Lime), data = splitplot)
> summary(fm)

Error: Block
               Df Sum Sq Mean Sq F value   Pr(>F)
Insect          1 34.551  34.551 34.2712 0.004248 **
Mollusc         1  0.729   0.729  0.7229 0.443088
Insect:Mollusc  1  0.921   0.921  0.9139 0.393209
Residuals       4  4.033   1.008
---
Signif. codes:  0 '***' 0.001 '**' 0.01 '*' 0.05 '.' 0.1 ' ' 1


Error: Block:Rabbit
                      Df Sum Sq Mean Sq  F value    Pr(>F)
Rabbit                 1 32.399  32.399 4563.5924 2.877e-07 ***
Insect:Rabbit          1  0.033   0.033    4.6985   0.09607 .
Mollusc:Rabbit         1  0.001   0.001    0.1600   0.70963
Insect:Mollusc:Rabbit  1  0.021   0.021    2.9078   0.16335
Residuals              4  0.028   0.007
---
Signif. codes:  0 '***' 0.001 '**' 0.01 '*' 0.05 '.' 0.1 ' ' 1


Error: Block:Rabbit:Lime
                           Df Sum Sq Mean Sq   F value    Pr(>F)
Lime                        1 7.2198  7.2198 1918.2643 8.148e-11 ***
Insect:Lime                 1 0.0028  0.0028    0.7556   0.41001
Mollusc:Lime                1 0.0102  0.0102    2.7005   0.13894
Rabbit:Lime                 1 0.0122  0.0122    3.2284   0.11010
Insect:Mollusc:Lime         1 0.0043  0.0043    1.1425   0.31631
Insect:Rabbit:Lime          1 0.0003  0.0003    0.0794   0.78529
Mollusc:Rabbit:Lime         1 0.0075  0.0075    2.0043   0.19458
Insect:Mollusc:Rabbit:Lime  1 0.0389  0.0389   10.3353   0.01233 *
Residuals                   8 0.0301  0.0038
---
Signif. codes:  0 '***' 0.001 '**' 0.01 '*' 0.05 '.' 0.1 ' ' 1
```

For interpretation, we have to work our way through the individual strata. At the top level, i.e. the whole plot level, we only have molluscicide and insecticide as treatments. Their effects are displayed first. There are 8 experimental units, each treatment and interaction has a degree of freedom, leaving $8 - 1 - 1 - 1 = 5$ degrees of freedom (the residuals then have $5 - 1$ degrees of freedom). In this top-most stratum, the effect of the insecticide is significant.

---

[7]We can also simply calculate the model defined below with Crawley's full data set. Then, in addition to the listed *output*, R also gives us an indication of the variance within in the remaining strata. Our reduced model delivers the identical statements as Crawley's complete model (see Crawley 2002, p.354 et seq.). However, our SS and MS values are different, since we consider averaged data and fewer thus fewer data points.

In the next stratum the effect of rabbit exclusion is added. First, we can evaluate this effect directly (and it is significant). Then we can analyse interactions with *all* factor combinations of the previous stratum. For the analysis we now have two values for each plot (in this case, 16). For the calculation of the degrees of freedom we deduct the sum of all degrees of freedom of the previous strata ($16 - 7 = 9$). The remaining degrees of freedom are distributed to the effects (one each) and $5 - 1$ degrees remain for the residuals of this stratum. One more word regarding the interpretation: In this second stratum, the treatments of the first stratum do not appear as main effects. If we wanted to determine the effect of molluscicide, we would have to average the rabbit plots in order to avoid pseudoreplication. However, this is exactly what is done in the first stratum.

In the third stratum (as in the second) a new factor is added. This time it is the effect of calcification (`Lime`), in addition to all interactions with effects of the previous stratum. The degrees of freedom are calculated as usual: $32 - 7 - 8 = 17$ for this stratum, 8 of which are for the effects, $9 - 1 = 8$ remain for the residuals. The significance of the 4-way interaction must be interpreted in the same way as in a conventional design: The effect of each treatment depends on the level of each other treatment.

There are also at least two specific packages for analysing mixed models, **nlme** and **lme4**. The use of `lme` (linear mixed effects) can be challenging, but in this case it is rather simple. The main difference to `aov` is that we also declare random effects as such and code the splitting and nesting accordingly: .

```
> library(nlme)
> fme <- lme(Biomass ~ Insect * Mollusc * Rabbit * Lime, random = ~1 |
+     Block/Rabbit/Lime, data = splitplot)
> anova(fme)
```

	numDF	denDF	F-value	p-value
(Intercept)	1	8	922.328	<.0001
Insect	1	4	34.271	0.0042
Mollusc	1	4	0.723	0.4431
Rabbit	1	4	4563.592	<.0001
Lime	1	8	1918.264	<.0001
Insect:Mollusc	1	4	0.914	0.3932
Insect:Rabbit	1	4	4.698	0.0961
Mollusc:Rabbit	1	4	0.160	0.7096
Insect:Lime	1	8	0.756	0.4100
Mollusc:Lime	1	8	2.701	0.1389
Rabbit:Lime	1	8	3.228	0.1101
Insect:Mollusc:Rabbit	1	4	2.908	0.1634
Insect:Mollusc:Lime	1	8	1.143	0.3163
Insect:Rabbit:Lime	1	8	0.079	0.7853
Mollusc:Rabbit:Lime	1	8	2.004	0.1946
Insect:Mollusc:Rabbit:Lime	1	8	10.335	0.0123

We see that the values are identical to the previous `aov`-analysis, although they appear in a different order. In mixed models, in addition to the degrees of freedom of the effect (`numDF` = numerator degrees of freedom = counter degrees of freedom) the degrees of freedom against which are tested (`denDF` = denominator degrees of freedom = denominator degrees of freedom) are also output. If the denominator degrees of freedom are incorrect, the model is incorrectly coded!

For completeness, here is the same analysis with **lme4**. The function `lmer` does not provide us with any significance values, unless we call it from the package **lmerTest**.[8]

```
> library(lme4)
> library(lmerTest)
> fmer <- lmer(Biomass ~ Insect * Mollusc * Rabbit * Lime +
+   (1|Block/Rabbit/Lime))
> summary(fmer)

Linear mixed model fit by REML
```

---

[8] Have a look at the help page of `pvalues` (`?lmer::pvalues`) in the **lme4**-package to read more about the problem of correctly estimating confidence intervals, degrees of freedom and significance values for mixed effect models.

```
t-tests use  Satterthwaite approximations to degrees of freedom ['lmerMod']
Formula: Biomass ~ Insect * Mollusc * Rabbit * Lime + (1 | Block/Rabbit/Lime)

REML criterion at convergence: 1659.4

Scaled residuals:
Min      1Q  Median     3Q     Max
-1.8168 -0.8576 -0.1928  0.6993  1.8578

Random effects:
Groups              Name         Variance Std.Dev.
Lime:(Rabbit:Block) (Intercept)  0.0000   0.0000
Rabbit:Block        (Intercept)  0.0000   0.0000
Block               (Intercept)  0.1565   0.3956
Residual                         4.5843   2.1411
Number of obs: 384, groups:  Lime:(Rabbit:Block), 32; Rabbit:Block, 16; Block, 8

Fixed effects:
                                   Estimate Std. Error     df t value Pr(>|t|)
(Intercept)                         7.65892    0.51892  18.00  14.759 1.68e-11 ***
Insect                             -1.92213    0.73387  18.00  -2.619 0.017379 *
Mollusc                             0.50447    0.73387  18.00   0.687 0.500582
Rabbit                             -2.10984    0.61808 364.00  -3.414 0.000713 ***
LimeUnlimed                        -0.98390    0.61808 364.00  -1.592 0.112287
Insect:Mollusc                     -0.39132    1.03784  18.00  -0.377 0.710543
Insect:Rabbit                       0.35798    0.87410 364.00   0.410 0.682383
Mollusc:Rabbit                      0.15581    0.87410 364.00   0.178 0.858621
Insect:LimeUnlimed                  0.13590    0.87410 364.00   0.155 0.876536
Mollusc:LimeUnlimed                 0.19570    0.87410 364.00   0.224 0.822969
Rabbit:LimeUnlimed                 -0.01212    0.87410 364.00  -0.014 0.988944
Insect:Mollusc:Rabbit              -0.48212    1.23617 364.00  -0.390 0.696758
Insect:Mollusc:LimeUnlimed         -0.37166    1.23617 364.00  -0.301 0.763850
Insect:Rabbit:LimeUnlimed          -0.25448    1.23617 364.00  -0.206 0.837015
Mollusc:Rabbit:LimeUnlimed         -0.15609    1.23617 364.00  -0.126 0.899587
Insect:Mollusc:Rabbit:LimeUnlimed   0.55785    1.74821 364.00   0.319 0.749837
---
Signif. codes:  0 '***' 0.001 '**' 0.01 '*' 0.05 '.' 0.1 ' ' 1
...
```

Compared to the significance value found with lme, we see that both the 4-fold interaction and calcification effect are *not* significant in lmer. Is this perhaps because the two models estimate different parameter values? To check this, we compare the parameters:

```
> summary(fme)

...
Fixed effects: Biomass ~ Insect * Mollusc * Rabbit * Lime
                  Value Std.Error  DF   t-value p-value
(Intercept)    7.658921 0.5189223 352 14.759284  0.0000
Insect        -1.922129 0.7338669   4 -2.619179  0.0589
Mollusc        0.504468 0.7338669   4  0.687411  0.5296
Rabbit        -2.109841 0.6180847   4 -3.413514  0.0269
LimeUnlimed   -0.983897 0.6180847   8 -1.591848  0.1501
Insect:Mollusc -0.391318 1.0378446  4 -0.377049  0.7253
Insect:Rabbit  0.357982 0.8741037   4  0.409541  0.7031
```

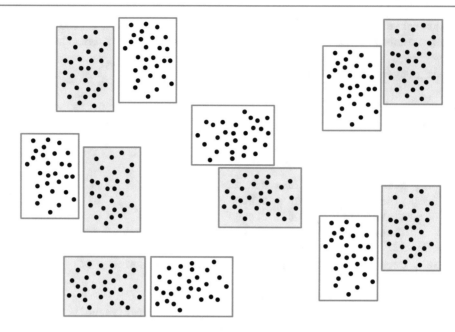

Fig. 14.5 Example of a blocked nested design. The experimental units are the 12 plots of the two treatments (grey and white). These are blocked as pairs. In each plot, we take 28 sub-samples (black dots). Since they are not independent within the treatment (they are all in the same plot), they cannot be considered replicates

```
Mollusc:Rabbit                        0.155814 0.8741037  4  0.178256 0.8672
Insect:LimeUnlimed                    0.135898 0.8741037  8  0.155471 0.8803
Mollusc:LimeUnlimed                   0.195702 0.8741037  8  0.223889 0.8285
Rabbit:LimeUnlimed                   -0.012120 0.8741037  8 -0.013866 0.9893
Insect:Mollusc:Rabbit                -0.482117 1.2361694  4 -0.390009 0.7164
Insect:Mollusc:LimeUnlimed           -0.371659 1.2361694  8 -0.300654 0.7713
Insect:Rabbit:LimeUnlimed            -0.254478 1.2361694  8 -0.205860 0.8420
Mollusc:Rabbit:LimeUnlimed           -0.156092 1.2361694  8 -0.126271 0.9026
Insect:Mollusc:Rabbit:LimeUnlimed     0.557845 1.7482075  8  0.319096 0.7578

...
```

The comparison of parameter estimators between lme and lmer shows that both deliver the same values. The difference in significance is actually due to the fact that lme and lmer follow different philosophies regarding statistical significance. In lme, the degrees of freedom are computed in the same way as explained for the aov earlier. lmer questions this degree of freedom, and rather computes something known as "Satterthwaite approximation", which also gives some degrees of freedom to the random effects, making it more conservative. The issue of inference in mixed models is somewhat in flux and hotly debated. As a consequence, model output and R-functions may change from one version of **lme4** to the next.[9]

14.2.3 Nested Design

1. Definition   Multiple measurements are taken within a single experimental unit, either in a parallel (multiple objects measured next to each other: Fig. 14.5) or sequential manner (repeated measurements).
2. Example   We want to investigate whether the exclusion of grazers has an impact on the sex ratio of a willow species (willows are dioecious, i.e. they have separate male and female individuals). To do this, we build n fences and controls, wait ten years, and then record the sex ratio on each of these $2n$ plots by randomly placing, for example, 28 squares and

---

[9]If you want to read more about this, check out the lively and extensive discussion with Douglas Bates (the statistical mind behind both of these approaches!) in multiple R-Wiki and mailing lists as well as the documentation (so called vignettes) from the **lme4** package (http://glmm.wikidot.com/faq, https://stat.ethz.ch/pipermail/r-help/2006-May/094765.html, http://rwiki.sciviews.org/doku.php?id=guides:lmer-tests).

recording the sex of the willow plants that occur in them. The nesting arises from the fact that 28 squares lie within a single treatment unit, and thus are not replicates but sub-samples. The same holds true if instead of waiting ten years, we record the sex ratio in a square (or the total area) each year. These data points are of course not independent, but are nested within the fenced areas or controls.

3. Details   The evaluation of nested designs is not simple. Especially if we implement temporal nesting, the factor of `time` will not necessarily affect the response variable in a linear way. Accordingly, we have to include the structure of this correlation into the model.

4. Advantages & Disadvantages   Nesting *has* to be taken into consideration in the analysis, otherwise our estimates have incorrect (too narrow) standard errors, leading to wrong significance levels. Accounting for nesting is statistically possible, so we do not need to average our 28 measurements within a treatment unit before the analysis. Use of these sub-samples or repeats is one of the great strengths of nesting. At the same time, the analysis is no longer as straightforward as a GLM.

5. Literature   Pinheiro and Bates [2000], Bolker et al. [2009], Zuur et al. [2009].

6. Example calculation   Let's take a look at the example above. A 10×30 m exclusion fence around a plot in the Arctic tundra prevented reindeer from grazing in the plot. An equally large control area is located next to it. This arrangement (`block`) is replicated 6 times. After six years, we determine the sex of 10 randomly selected willow plants per treatment unit (Dormann and Skarpe 2002). Thus each plant is a sub-sample or pseudo-replicate, nested within the treatment `exclosure`. We must also inform the GLM that `block` is a random effect, which is not really of interest to us, and that all measurements of subsamples are nested in the treatment effect `excl` (the structure goes behind the vertical line, "|", in the following R-code, starting with the largest unit). The syntax in the corresponding generalised linear mixed model (GLMM) in R looks like this:

```
> willowsex <- read.table("willowsex.txt", header = T)
> library(MASS)
> summary(glmmPQL(fixed = female ~ excl, random = ~1 | block/excl,
+      family = binomial, data = willowsex))
iteration 1
Linear mixed-effects model fit by maximum likelihood
 Data: willowsex
  AIC BIC logLik
   NA NA    NA


Random effects:
 Formula: ~1 | block
         (Intercept)
StdDev: 9.199434e-05


 Formula: ~1 | excl
         (Intercept)  Residual
StdDev: 3.397999e-05 0.9999998


Variance function:
 Structure: fixed weights
 Formula: ~invwt
Fixed effects: female ~ excl
               Value Std.Error  DF   t-value p-value
(Intercept) 0.2006707 0.2616895 108  0.7668274  0.4449
excl       -0.9698038 0.3831561   5 -2.5310931  0.0525
 Correlation:
      (Intr)
excl -0.683


Standardized Within-Group Residuals:
       Min        Q1       Med        Q3       Max
-1.1055418 -0.6807455 -0.6807455 0.9045342 1.4689772
```

```
Number of Observations: 120
Number of Groups:
        block excl
           6           12
```

The interpretation is a bit complicated, so let's take it step by step. The result that we are interested in, namely whether the exclosure has shifted the sex ratio of willows, can be found in the middle of the output, under the heading `Fixed effects: female ~ excl`: The effect `excl` is bordering on significance. Here you can also see whether the model structure is implemented correctly. 5 degrees of freedom are available for the effect `excl`, i.e. the number of units (12) − the number of estimated block values (6) −1. That is correct. Another thing to check is the number of groups listed below: 6 for `block`, 12 for `excl within block`, also correct.

Let's come now to the top part of the outputs. First, the model fit is quantified: `AIC` and `BIC` stand for Akaike and Bayes' Information Criterion, `logLik` stands for log-likelihood. This information would be of interest for model selection, but they are not defined for the used quasi-likelihood, so we are left with `NA`. The random effects are then fit for each stratum individually. With the standard deviations we see that there is more variability between blocks (9.2e-05) than between exclosures within the block (3.4e-5).

If we back-transform the parameters received for the proportion of female plants (they are logit-transformed by default in binomially distributed data) we get a value of $\frac{e^{0.2}}{1+e^{0.2}} \approx 0.55$ for the controls and a value of $\frac{e^{0.2-0.97}}{1+e^{0.2-0.97}} \approx 0.32$ for the exclosure. Not too bad at all for true values of 0.6 and 0.4, respectively.

It is worth mentioning that while we can force an ANOVA-like table via `anova.lme(.)`, this is incorrect, because an F-test is used instead of the correct χ^2-test.

14.3   Survey Designs for Descriptive Studies

Survey design exists on a continuum of possible sample designs used to gain knowledge about something. Especially in the area of monitoring, i.e. the permanent observation of environmental changes, there are many anecdotal observations, but also well-replicated manipulative experiments (Elzinga et al. 1998, S.4).

In the vast majority of cases, it is impossible to observe all the events of interest to us (so-called population-level observation), so we have to rely a sample. If we want to know how many ground beetles live on one hectare of wheat fields, we can't simply round up all of the beetles on a hectare sized plot! Instead, we draw several smaller samples (plots of about 1 m² each) and use this to make an estimation on the hectare population. It is a bit more difficult if we want to know, for example, the timber stock of a forest with an area of 244 ha. Then we have to estimate both the number of trees and the average wood volume per tree. These are typical questions for descriptive samples that require a survey design.

The goals of a survey design are the same as that of manipulative experiments: **independence**, **randomness**, and **replication**. Independence of the data points is a prerequisite for statistical evaluation. Randomness is necessary to prevent overlooked gradients (e.g. light in a greenhouse, soil moisture on a slope, availability of nutrients on a geologically variable subsoil) from leading to a systematic error. Replication is necessary in order to calculate not only the mean value but also a measure of the variability of the results. This is the only way to find out whether our value is relevant.[10]

The diversity of possible survey designs makes it difficult to write a systematic description of them. A first distinction can be made depending on whether all sample units have the same probability of being selected (simple random sampling as opposed to stratified sampling). Furthermore, it is possible to differentiate whether one or more values are sampled per sample unit (the latter is called cluster sampling).

To make this as confusing as possible, the evaluation of survey designs has developed in two different directions: survey statistics or mixed models. The older methodology of survey statistics, originating mainly from forestry and fishing and nowadays typically used in social sciences, has developed a detailed statistical approach for each design. In this approach, there is a formula for the population level estimator and its variance for each of the dozens of sampling methods.[11] In forestry

---

[10]This reminds me of a colleague who took water samples in two depths in the Indian Ocean and determined iron concentrations. One value was 0.01 nMol, the other value was 0.03 nMol. Is this difference significant? Without a definition of variability *per water depth* it is not possible to answer this question!

[11]This is somewhat of a caricature. Of course, the surveys also have mean estimates.

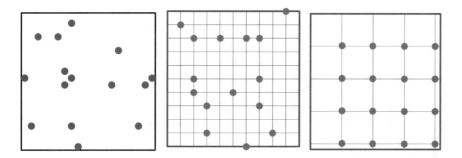

Fig. 14.6 Random (left), restricted random (middle) and random systematic (right) selection of 15 sampling points in a square sampling area (simple random sampling = SRS). Restricting the samples to the grid corners (middle) creates a minimum distance between the sampling points and ensures a level of mixing. In order for systematic sampling (right) to be considered statistically acceptable, the first point needs to be randomly selected. For this reason, the lines are not equidistant to the edge of the sampling area

science, for example, we record tree volumes in random samples to calculate the timber volume of the entire forest. For this reason, we're only interested in the estimator for the total volume and its variance.

Conducting consumer surveys about a new product in pedestrian zones is similar. Here it is of interest to estimate the potential market, i.e. extrapolate the opinions of some respondents to the entire population.

Those who are active in these fields cannot avoid these tools of the trade. For the environmental sciences (with emphasis on forestry and terrestrial sampling) the extremely solid and complete book of Gregoire and Valentine [2008] is a standard; for the social sciences, the respective tome is Lohr [2009]; for an implementation in R, see Lumley [2010].

The other branch for the analysis of sample designs does not focus on extrapolation to population-level values, but on the analysis of the correlation between the measured values and environmental factors (climate, soil, management regime, etc.), i.e. above all it focuses on the mean value. Of course, the population-level value can also be estimated by multiplying it by the size of the population.

The main tool for dealing with complex designs is the mixed model (Bolker et al. 2009; Pinheiro and Bates 2000; Zuur et al. 2009). In extreme cases, the analyst is not interested in absolute values at all, but only in the connections and correlations with the driving forces. This type of analysis logically arises from regression, but is (substantially) complicated by sample designs. As this kind of analysis is identical to that used for manipulative experiments, we will only look at the evaluation of sample designs according to the survey statistical approach in the following section.

14.3.1   Simple Random Sampling

Simple random sampling, or SRS[12] is the simplest case of sampling design. We randomly (or systematically with a random start point: see Fig. 14.6 and Side Note 11) select n units (e.g. trees in the forest). Since all N units have the *same probability* of being selected (n/N), this is referred to as equal probability sampling.

The selection probability is often denoted as π. Here we will stick to the nomenclature used by Gregoire and Valentine [2008]. In the case of SRS, $\pi = n/N$. So if we have 10000 trees in a forest and we select 20, then the probability of selecting any one tree is 20/10000.

If we calculate a mean value of $\hat{y} = 4$ m³ for the wood volume from the 20 trees we selected, then the timber stock of the forest would be $10000 \cdot 4$ m³ $= 40000$ m³. Another approach would be to say that our 20 sample trees together have a total wood volume of $\hat{\tau}_y = 80$ m³. Since $\pi = n/N = 20/10000$, the total timber stock is calculated as

$$\hat{\tau}_y/\pi = \frac{80}{20/10000} \text{ m}^3 = \frac{80 \cdot 10000}{20} \text{ m}^3 = 40000 \text{ m}^3.$$

This calculation is known as the Horvitz-Thompson estimator. In its general form, the HT-estimator is defined as follows:

---

[12]If each tree can only be sampled once, then this is called simple random sampling without replacement. This is the normal case. We will ignore the case of SRS with replacement here.

$$\hat{y}_{HT} = \frac{1}{N} \sum_{i=1}^{n} \frac{y_i}{\pi_i}, \text{ for the mean.} \tag{14.1}$$

or

$$\hat{\tau}_{HT} = N\hat{y}_{HT} = \sum_{i=1}^{n} \frac{y_i}{\pi_i}, \text{ for the total value;} \tag{14.2}$$

In the SRS, π_i is the same for all units, so the subscript is omitted here. However, if units have different probabilities of being selected in other designs ($\pi_i \neq$ constant), then these formulas are still valid. That point is rather trivial. It is now important to estimate the variance of the mean value or the total value.

$$\text{var}(\hat{\tau}) = N^2 \text{var}(\hat{y}) = N^2 \left(1 - \frac{n}{N}\right) \frac{s_y^2}{n}, \tag{14.3}$$

where the standard deviation is calculated as $s_y = 1/(n-1) \sum (y_i - \bar{y})^2$. The origin of the term $(1 - n/N)$ is explained in Side Note 12.

So if a forester goes to collect data in the forest, she does this first to determine the wood volume of a typical tree (y_i), and second to figure out how many trees are in the forest (N, or N/forest area = trees per hectare).

Side note 12: Finite population correction In the majority of cases, we can only sample a small portion of the total population that we are interested in. However, if we do sample a substantial proportion of the population, say a quarter of all trees in a forest plot, then we overestimate the standard error of the mean (and total) of our sample. In the extreme case that we measure all trees in a forest plot (total population), we still get a value for the standard error of the mean ($s/\sqrt{n}$), even though we have measured all units and therefore have an error-free (population) mean!

This problem arises because our total population is finite, not infinite! For sampling a substantial proportion of a finite population, the standard error of the mean needs a correction, the finite population correction:

$$\text{FPC} = \sqrt{\frac{N-n}{N}}$$

This correction term is relevant if we sample *more* than one twentieth (5%) of the total population (then the FPC is $= \sqrt{(N - 0.05N)/N} = \sqrt{1 - 0.05} = 0.97$). When we sample 25%, the FPS $= \sqrt{(N - 0.25N)/N} = \sqrt{0.75} = 0.87$; we overestimate the standard error by 13% if we use Eq. 1.8 from Chap. 1.

Conversely, if we sample less than 5% of the total population, the FPC term becomes negligible and we safely employ the statistical machinery of Chap. 1.

We have seen this FPC before: you can see it in Eq. 14.3, but transformed and squared: $(1 - n/N) = (N - n)/N$.

(*Nota bene*: If we use an SRS with replacement, which is rarely the case, this correction would be incorrect, as by replacement of our sampling units we essentially generate an infinite population!)

14.3.2   Multi-level Sampling Methods

In the case of multi-level sampling methods, samples are grouped together as part of the sampling design before data collection begins.[13]

14.3.2.1 Stratified Sampling

One disadvantage of the SRS is that we are ineffective in sampling. For example, if we have three forms of forest management in our forest (A, B, C) and two of them (B and C) occupy relatively small areas, then we will take a lot of samples for A, but very few for B and C. This means that we estimate our variance for A very accurately, but for B and C very imprecisely. Wouldn't it be better if we drew a minimum number *per management form*? Or if we distribute our 90 samples evenly over A, B and C?

---

[13] Or afterwards. *Post-hoc*-stratification, however, is not the focus here. For more, see Gregoire and Valentine [2008].

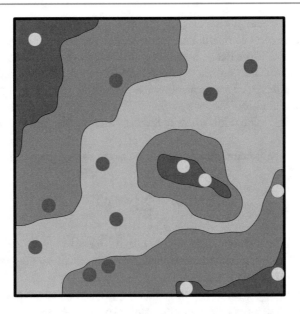

Fig. 14.7 Stratified sampling of three locations on a hillside (valley, slope, peak) . In this case, the same number of plots were placed in each area (5)

This is the idea of stratification: we break down our population into different groups on the basis of one (or more) characteristic(s) and sample each of them separately.[14] Stratification can be carried out on the basis of management forms, altitudes, administrative districts, tree species, etc. (see Fig. 14.7).

So how should we distribute our n sampling points across the different strata? There are three basic possibilities:

1. **Evenly**. Each stratum is sampled the same number of times, irrespective of size, i.e. for k strata and n samples:

$$n_i = n/k$$

2. **Proportionally** to the area A (or population size, time interval, ...). Larger strata are sampled more often:

$$n_i = n \frac{A_i}{A_{\text{total}}}$$

3. **Variance-optimally** (or "Neyman allocation"). The more variable the measurements from a stratum, the more samples we should allocate there. To do this, we need to know the variability (e.g. standard deviation s_i) in the different strata i. More precisely: the share of the total variance determines the number of sample points: (Cochran 1977):

$$n_i = n \frac{s_i}{\sum_{i=1}^{k} s_i}$$

4. **Proportional to variance *and* area**. This simply combines the two previous allocation schemes into one[15]:

$$n_i = n \frac{A_i s_i}{\sum_{i=1}^{k} A_i s_i}$$

---

[14]Each stratum can be sampled in a different way. The designs per stratum are independent of each other.

[15]If the cost of data collection (κ_i in Euros, time or personnel requirements) differs between the strata, then this can also be taken into account as a correction term:

$$n_i = n \frac{A_i s_i / \sqrt{\kappa_i}}{\sum_{j=1}^{k} A_j s_j \sqrt{\kappa_j}}$$

Although logistical and practical considerations are extremely important (easy to locate the sampling points, simple, easy to explain design), options 1 and 2 are the most common. Probably it is often difficult to specify the costs for sampling or know the variability within a stratum beforehand. More likely, however, is that people simply are not aware of the Neyman-allocation.

In the statistical analysis, an important difference compared to SRS is that we get a different π value for each stratum, since the strata are usually different in size. This means that the probability that an area or an individual is sampled is no longer the same for all experimental units, but varies by stratum. Accordingly, the Horvitz-Thompson estimator (Eqs. 14.1 and 14.2) retains the subscript i. The total value of all k Strata is simply the sum of the total values of each stratum.: $\hat{\tau}_G = \hat{\tau}_1 + \hat{\tau}_x \cdots + \hat{\tau}_k = \sum_{h=1}^{k} \hat{\tau}_h$.

The variance of the total value of a stratified sample is calculated as the sum of the variances of the total values per stratum:

$$\text{var}(\hat{\tau}_G) = \sum_{h=1}^{k} \text{var}(\hat{\tau}_h) \tag{14.4}$$

This means that our uncertainty adds up and we cannot simply take the mean!

If we wanted to think about the analog in experimental design, the stratified sample would correspond to a completely randomised design with the stratum as an additional factor.

An example We have determined the wood volume in three forest types (spruce, beech, and mixed forest). The three forest types exist in stands of different sizes, therefore our measurements look as follows:

Forest type	Trees (N)	n_i	$\bar{y}_{\text{volume}}$	s^2_{volume}
Spruce (S)	333	10	12	9.2
Beech (B)	420	12	16	8.4
Mixed (M)	121	5	14	15.2

The estimated total wood volume $\hat{\tau}_D$ is calculated as the sum of the volumes in the individual strata:

$$\hat{\tau}_D = \sum_{i=1}^{3} N_i \bar{y}_i = 333 \cdot 12 + 420 \cdot 16 + 121 \cdot 14 = 12410.$$

The variance of $\hat{\tau}_D$ is calculated according to Equation 14.4 as the sum of the variances of the individual strata. The variance for each individual strata is calculated according to Equation 14.3.[16]

$$\text{var}(\hat{\tau}_F) = N_F^2(1 - \frac{n_F}{N_F})\frac{s^2_{y_F}}{n_F} = 333^2(1 - 10/333)\frac{9.2}{10} = 98954$$

and accordingly for B and M:

$$\text{var}(\hat{\tau}_B) = 420^2(1 - 12/420)\frac{8.4}{12} = 119952 \text{ and } \text{var}(\hat{\tau}_M) = 121^2(1 - 5/121)\frac{15.2}{5} = 42669$$

The variance of the population-level estimator is therefore $\text{var}(\hat{\tau}_G) = 98954 + 119952 + 42669 = 261575$, and the standard deviation is $\sqrt{261575} = 511$. If we assume that the data are normally distributed, the distribution of estimated total volume follows the shape in Fig. 14.8.

14.3.2.2 Cluster Sampling

Stratified sampling still requires the collection of many independent data points. This can be very complex in terms of logistics. If we were now able to collect fewer samples, but could measure several values in each case, could such an approach save us time and money?

With *cluster sampling*, clustering (or grouping) of samples is used to simplify the logistics of data collection. For example instead of searching for 90 sampling areas in our forest in which we sample a single tree, we only visit 10 plots, but sample

---

[16]Since we sample less than 10% of the population, we can disregard the calculation of the finite population correction factor $(1 - \frac{n_i}{N_i})$.

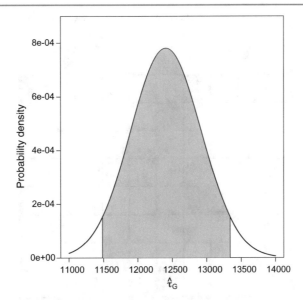

Fig. 14.8 Estimated total wood volume ($\hat{\tau}_G$) of the three forest types from the table on page 202 as a probability density. The grey area represents the 95% confidence interval. Even though the variance is enormous, the estimate is still quite good (coefficient of variance $= \sqrt{\mathrm{var}(\hat{\tau}_G)}/\hat{\tau}_G = 511/12410 = 0.041$, or 4%)

12 trees at each plot. We then have 10 clusters with 12 subplots. Now we have the issue that the subplots of a cluster are not independent from each other and therefore we can only analyse the data on the level of the cluster, not on the level of the individual data points!

If the elements of a cluster vary a lot, then cluster sampling can be efficient: with relatively few clusters, we can already calculate a reasonable estimate of the variability of the sample. If, however, the elements of a cluster are very similar, then we have a problem, since we only efficiently sample the variability between clusters and not that within the cluster itself.

For each cluster i the total value is calculated from the individual elements (i.e. the n trees of the cluster):

$$\tau_i = \sum_{i=1}^{n} y_i = N_i \bar{y}_i \tag{14.5}$$

The total value for all K clusters is their sum, weighted for the probability that cluster i is in the sample:

$$\hat{\tau}_G = \sum_{k=1}^{K} \frac{\bar{y}_k}{\pi_k} \tag{14.6}$$

We have seen this equation before as well: it is the HT-estimator for the SRS, only this time at the cluster level and not for each individual experimental unit.

The variance is separated into two parts: one part is the variance between the clusters, V_1, and one part is the variance within a cluster, V_2. The formulas for this are long and ugly, as they combine the probability that a cluster was selected, with the probabilities that a tree was selected within a cluster (consult Gregoire and Valentine 2008, for details).

For cluster sampling in the strict sense of the word, the problem is simplified *when all elements of the cluster are sampled.* Then the variance within the cluster = 0 (because we sampled the entire population of that cluster), and the variance between clusters is calculated as in SRS. While a complete sampling with trees might be feasible, this is hardly ever the case in the social sciences: within a single high-rise building it is rare that we can survey all of the residents.

In an experimental setting, cluster sampling is analogous to the nested design. Subplots are "only" used to provide a more detailed description of value of experimental units (e.g. plots).

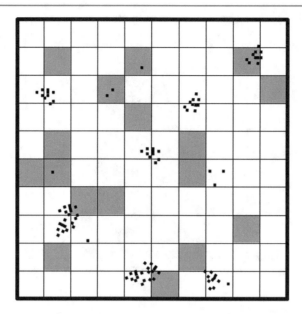

Fig. 14.9 A highly clumped distribution of the objects cannot be effectively covered by random (or systematic) sampling. For example, if we only measure those trees that fell exactly on the grid axis points, we would only have measured three values. Instead, we randomly select cells (grey) and collect data for all points (cluster sampling) or collect points in a way that it is proportional to the size of the cluster. In this case, the cells are too small, because none of them contains a complete cluster of points (10 of 17 are empty)

14.3.2.3 Multi-stage Sampling

With multi-stage sampling, you first select units to sample, but within this first selection (possibly according to other criteria) sub-units are selected again. Cluster sampling is an example of *two-stage* sampling (first the cluster, then units within the cluster in incomplete sampling).

The selection within a unit (2nd, 3rd, 4th level) may depend on environmental conditions, for example: On a map we select 20 forest plots to search for tawny owls. After arriving at a forest plot, we find out that the trees are all less than 30 cm thick and therefore are not suitable for the as a habitat tree for the tawny owl. Or, perhaps we have trees that are thick enough and then only sample the trees over 30 cm. Or we only sample the conifers over 30 cm, etc.

As with cluster sampling, the problem arises that we have to combine several different selection probabilities: that of the forest type, that of the tree species, and that of the tree size. So, in order to be able to extrapolate our tawny owl habitat trees to the total forest area, we need to know how likely it was that we had chosen this forest plot in particular. Then we need to know what percentage of the trees are over 30 cm thick.

The big advantage of multi-stage sampling is its efficiency; we sample only that which provides us with new knowledge. However, this means that we need a thicker book at our side for the analysis. (such as Gregoire and Valentine 2008, p. 373 et seq.).

A forest inventory using the "angle-count" method developed by Walter Bitterlich is an example of cluster sampling in a systematic design: Each sample plot (cluster) is located on a grid point (see Fig. 14.6 middle). At each plot, we collect data on several trees. Since the number of trees sampled depends on their diameter, plot area changes from site to site (leading to the name "variable area cluster sampling"). With fixed sample plot sizes, we would have a "fixed area sampling", but still a variable number of trees.

Multi-stage sampling is especially well suited if the objects to be measured are highly clumped. In such a case, it makes sense to first randomly select a large cell, and then sample it proportionally to the number of objects (i.e. large clusters more intensively than small ones). In fact, we use this method to test two different questions: 1. How many cells contain a cluster? 2. If there is a cluster, what are the values of its objects? As Fig. 14.9 shows, good planning is required to determine the appropriate size of the cells, so that the objects of the study question are properly included.

14.3.2.4 Transects and Line Transects

These procedures are only mentioned in passing. Since they require care both in their application and calculation, we cannot adequately cover them here.

With **transects**, samples are taken along a (not necessarily straight) line. All objects in the vicinity of this line are measured. Butterfly transects are an important example. In this case, all butterflies that occur about 2 m to the right and left of the transect are recorded. In Africa, the same thing happens with elephants and in Australia with whale censuses - both with somewhat larger distances to either side.

Since it is easier to overlook objects the further away they are, it is necessary to correct the probability of detection depending on the distance. This is the subject of *distance sampling* (Buckland et al. 2015).

With the **line intersect(ion) method**[17] only objects that are exactly on the line are recorded (Gregoire and Valentine 2003, 2008; Kaiser 1983). Let's imagine a one-dimensional line through a forest, along which we record deadwood. We would typically only measure the diameter at the exact point where the line intersects the fallen tree (if we also measure the stumps and branches to the left and right of the line, we are back at the transect method!). Line transects are difficult in practice (how can you maintain a perfectly straight line in the forest with so many obstacles?), but are theoretically well described. These approaches were in common use in remote sensing (e.g. virtual lines across a country to estimate how much area of it is forested), but are less common nowadays, where land cover is more completely classified.

References

1. Bolker, B. M., Brooks, M. E., Clark, C. J., Geange, S. W., Poulsen, J. R., Stevens, M. H. H., et al. (2009). Generalized linear mixed models: A practical guide for ecology and evolution. *Trends in Ecology and Evolution, 24*, 127–35.
2. Buckland, S. T., Rexstad, E., Marques, T., & Oedekoven, C. (2015). *Distance sampling: Methods and applications*. Berlin: Springer.
3. Cochran, W. G. (1977). *Sampling techniques* (3rd edn.). Oxford, UK: Wiley.
4. Crawley, M. J. (1993). *GLIM for ecologists*. Oxford, UK: Blackwell.
5. Crawley, M. J. (2002). *Statistical computing. An introduction to data analysis using S-Plus*. Chichester: John Wiley & Sons Ltd.
6. Dormann, C. F., & Skarpe, C. (2002). Flowering, growth and defence in the two sexes: Consequences of herbivore exclusion for Salix polaris. *Functional Ecology, 16*, 649–656.
7. Elzinga, C. L., Salzer, D. W., & Willoughby, J. W. (1998). *Measuring and monitoring plant populations*. Denver, Colorado: Bureau of Land Management.
8. Finegold, J. A., Manisty, C. H., Goldacre, B., Barron, A. J., & Francis, D. P. (2014). What proportion of symptomatic side effects in patients taking statins are genuinely caused by the drug? Systematic review of randomized placebo-controlled trials to aid individual patient choice. *European Journal of Preventive Cardiology, 21*, 464–474.
9. Gregoire, T. G., & Valentine, H. T. (2003). Line intersect sampling: Ell-shaped transects and multiple intersections. *Environmental and Ecological Statistics, 10*(2), 263–279.
10. Gregoire, T. G., & Valentine, H. T. (2008). *Sampling strategies for natural resources and the environment*. New York: Chapman and Hall/CRC.
11. Hurlbert, S. H. (1984). Pseudoreplication and the design of ecological field experiments. *Ecological Monographs, 54*, 187–211.
12. Kaiser, L. (1983). Unbiased estimation in line-interception sampling. *Biometrics, 39*, 965–976.
13. Lohr, S. L. (2009). *Sampling: Design and analysis* (2nd edn.). Pacific Grove, CA: Duxbury Press.
14. Lumley, T. (2010). *Complex surveys: A guide to analysis using R*. Hoboken, NJ: Wiley.
15. Magnusson, W. E. (1997). Teaching experimental design in ecology, or how to do statistics without a bikini. *Bulletin of the Ecological Society of America, 78*, 205–209.
16. Mead, R. (1990). *The design of experiments: Statistical principles for practical Applications*. Cambridge, UK: Cambridge University Press.
17. Morford, S. L., Houlton, B. Z., Dahlgren, R., & a., (2011). Increased forest ecosystem carbon and nitrogen storage from nitrogen rich bedrock. *Nature, 477*(7362), 78–81.
18. Pinheiro, J. C., & Bates, D. M. (2000). *Mixed-effects models in S and S-Plus*. Berlin: Springer.
19. Potvin, C. (2001). ANOVA: Experiments layout and analysis. In S. Scheiner & J. Gurevitch (Eds.), *Design and analysis of ecological experiments* (pp. 63–76). Oxford, UK: Oxford University Press.
20. Quinn, G. P., & Keough, M. J. (2002). *Experimental design and data analysis for biologists*. Cambridge, UK: Cambridge University Press.
21. Silvertown, J., Poulton, P., Johnston, A., Edwards, G., Heard, M., & Biss, P. (2006). The Park Grass experiment 1856–2006: Its contribution to ecology. *Journal of Ecology, 94*, 801–814.
22. Sokal, R. R., & Rohlf, F. J. (1995). *Biometry* (3rd edn.). New York: Freeman.
23. Underwood, A. J. (1997). *Experiments in ecology: Their logical design and interpretation using analysis of variance*. Cambridge, UK: Cambridge University Press.
24. Zuur, A. F., Ieno, E. N., Walker, N. J., Saveliev, A. A., & Smith, G. M. (2009). *Mixed effects models and extensions in ecology with R*. Berlin: Springer.

---

[17]Note that this term is easily and often confused with "line intercept", which is part of distance sampling!

Multiple Regression: Regression with Multiple Predictors 15

At the end of this chapter . . .
. . . you will understand how a factorial design leads to higher sensitivity in the analysis than two separate experiments.
. . . you will be able to calculate a *two-way* ANOVA by hand if necessary.
. . . you will know what a statistical interaction is and how it should be interpreted.
. . . you will be familiar with principal component analysis (PCA) and cluster analysis for addressing the problem of collinear predictors.
. . . you will understand how AICc and BIC can be used for model simplification.

The next step on our journey is the expansion of statistical analysis to more than a single predictor. In the context of regression, this is known as "multiple regression".[1]

In principle, with multiple regression we are "only" including another predictor.[2] This should not give us any trouble in principle. Thus, a model with a predictor for a negative binomial regression simply becomes a model with two:

$$\mathbf{y} \sim \text{NegBin}(\text{mean} = a + b\mathbf{x}_1 + c\mathbf{x}_2, \text{size} = \mathbf{d})$$

This means that we have to determine multiple parameters using maximum likelihood. The result of a multiple regression looks something like the output below (here for the Titanic survival data):

```
Coefficients:
             Estimate Std. Error z value Pr(>|z|)
(Intercept)  1.235414   0.192032   6.433 1.25e-10 ***
age         -0.004254   0.005207  -0.817    0.414
sexmale     -2.460689   0.152315 -16.155  < 2e-16 ***
---
Signif. codes:  0 `***' 0.001 `**' 0.01 `*' 0.05 `.' 0.1 ` ' 1
```

For each of the two predictors, `age` and `sex`, we get the corresponding maximum likelihood estimator. So other than an additional row for a second predictor, nothing in the output changes, right?

Wrong! With the inclusion of more than one predictor variable the following complications arise:

1. Visualising multiple predictors;
2. Interactions between predictors;
3. Collinearity of the predictors;
4. Model simplification/variable selection;

We will now explore these issues one at a time.

---

[1] Unfortunately and confusingly, this is sometimes introduced under the heading of "multivariate" statistics (e.g. Izenman 2008; Tabachnick and Fidell 1989). The term has a specific meaning (Anderson 1958; Chatfield and Collins 1980 and also in CRAN Task View "MultivariateStatistics": https://cran.r-project.org/view=Multivariate): Multivariate statistics deals with multidimensional *response* variables. In contrast, here we are considering univariate statistics, i.e. with a single response variable.

[2] Or multiple other predictors.

© Springer Nature Switzerland AG 2020
C. Dormann, *Environmental Data Analysis*,
https://doi.org/10.1007/978-3-030-55020-2_15

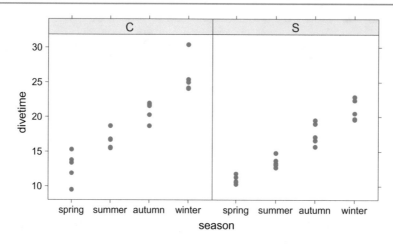

Fig. 15.1 Panel plot of the cormorant data. On the left, the subspecies "C", on the right "S". The seasons are listed in order on the x-axis for each panel

15.1   Visualising Multiple Predictors

As a general rule, we cannot represent more than three dimensions in a single figure, i.e. two explanatory and one response variable. Of course we can simply include multiple two dimensional figures, one for each additional dimension. This results in so-called *panel plots*.

15.1.1   Visualising Two Categorical Predictors

Let's start with two categorical predictors, using the cormorant data set as an example. We want to know whether the two European subspecies of cormorant (*Phalacrocorax carbo sinensis* and *P. c. carbo*) differ in dive times, depending on the season. Our response variable (divetime) should be explained by the categorical variables subspecies (S and C) and season (spring, summer, autumn, winter). Let's take a look at the data (Fig. 15.1).

We could visualise the data by including two panels—one for each subspecies (Fig. 15.1). As an alternative, we can also put all of the data points in a single plot, but distinguish the subspecies using different symbols (Fig. 15.2). Panel plots are extremely helpful in providing a quick overview of the data. All-in-one-plots require much more effort and graphics know-how, but take up less space on a page (A better "data-to-ink ratio" in the words of Tufte 1983).

What we can see in both of these visualisations is that the dive time increases throughout the course of the year for both subspecies and that the (smaller) subspecies, *sinensis*, has a slightly lower dive time.

15.1.2   Visualising a Categorical and a Continuous Predictor

With a categorical and a continuous predictor we can also make a panel for each level of the categorical predictor, which gives us a figure such as Fig. 15.1, only with the x-axis now as continuous. We can also create a figure like Fig. 15.2 just as well for a continuous variable instead of the categorical predictor "season".

Visualisation becomes more difficult if our data are binary, such as the survival data from the Titanic disaster. In this case, neither the panel plot nor the scatter plot are acceptable options (Fig. 15.3). Apparently, we must first prepare the data so that the chances of survival for every age group can be easily read. This can be done elegantly by calculating and applying a moving average to the data. Fig. 15.4 makes it clear what a moving average is: For each x value, an area ("window") is defined to the right and left, for which the mean value of all points is then calculated.[3] The size of this window determines how much

---

[3]More precisely, a weighted average is calculated where points in the middle of the window are fully weighted, and the weight becomes less and less towards the edge. The most common form shown here is the LOESS or LOWESS (*locally weighted scatterplot smoothing*). Mathematically, there is a lot going on: A quadratic function is placed through the data inside the window, whereby the weight of the data decreases with the cube

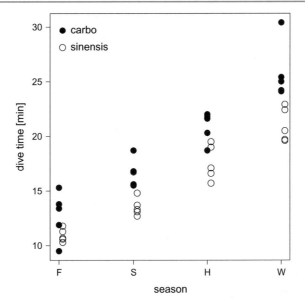

Fig. 15.2 Scatter plot of the cormorant data. For clarity, the data for "S" are shifted slightly to the right

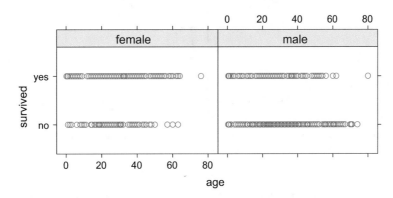

Fig. 15.3 Panel plot of binary data: there is no way we can detect a pattern in the Titanic survival data with such a visualisation!

smoothing occurs. For example, a wide window (length = 1) always contains all data points and leads to very smooth lines. A narrow window traces the data more exactly and is more wiggly (usually lengths between 0.2 and 0.5 are used, with large data sets lengths of 0.67 or 0.75 may be used).

Similarly to the previous figures, we can also unify these two panels and bring all of the information into a single figure (Fig. 15.5). A more recent alternative is a visualisation technique using supersmoother, which calculates and continuously changes the window length using the variability of the data for each window itself (Fig. 15.5, right).

15.2   Interactions Between Predictors

A statistical interaction of two predictors occurs if the effect of one variable depends on value of another variable. This can easily be explained using a simple example. What does a plant need to grow? Water and light. If we give a plant more light, it will grow better. But only if it has enough water: light has no effect without water. And vice versa: We can drown a plant in water, but it will only grow when it receives light.

---

to the edge. From this fit, however, only the center point is used, then the window is moved one unit further and all points are connected by straight lines. Exciting as it is, it serves us here only for visualisation and the calculations are not of paramount importance.

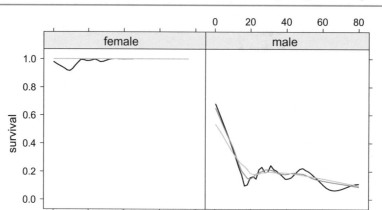

Fig. 15.4 Panel plot of binary data using a moving average (LOESS with a range of: 0.2 (black), 0.5 (grey) and 0.75 (light grey))

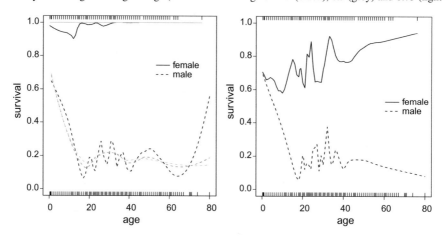

Fig. 15.5 LOESS-plot (left) and supersmoother plot (right) showing the survival data from the Titanic separated by sex (solid line female, dashed line male). As in Fig. 15.4, the different shades of grey represent different window lengths (0.2, 0.5 and 0.75). The small vertical lines (rug) show the data points for females (top) and males (bottom). The supersmoother is a new variant of the moving average which adjusts the length according to the number of data points in the window. So while the length of 0.2 on the left shows a high value for 80-year-old men, this value is lost by all other variants, as we only have one data point in this age class

In the humid tropics, trees grow so well because there is rain *and* heat in an ideal combination. When the rainfall diminishes, there might so much sunshine that the lack of water becomes the limiting factor for growth. Statistically, this would be expressed as an interaction of temperature and precipitation.

We have such an interaction between the age and sex of Titanic passengers affecting their survival. For females, the survival rate was fairly evenly high; for males, it was strongly dependent on age. This is an interaction because the effect of age depends on sex (and vice versa).

In a figure one can recognize a statistical interaction by the fact that lines connecting values of one predictor have a different slope for values of the second predictor (Fig. 15.6). But how do we recognize such a statistical interaction in our analysis?[4]

Let's stay with the Titanic example. If we only put the two predictors *age* and *sex* into the model, we end up with the following:

$$\mathbf{y} \sim \text{Bern}\left(p = \frac{e^{a+b\,\mathbf{age}+c\,\mathbf{sex}}}{1 + e^{a+b\,\mathbf{age}+c\,\mathbf{sex}}} \right)$$

---

[4]In fact, it happens that in an analysis we find a significant interaction that actually goes back to a non-linear effect of one of the two predictors (or in the words of Marsh et al. (2012) "It is well known that the presence of unmodeled quadratic effects may give the appearence of a significant interaction effect that is spurious (Ganzach 1997; Kromrey and Foster-Johnson 1999; Lubinski and Humphreys 1990; MacCullum and Mar 1995)", p. 445). Since interactions allow the model to fit non-linear functions, the interaction here leads us astray. Therefore, it is important to include both interactions *and* non-linear effects in the model.

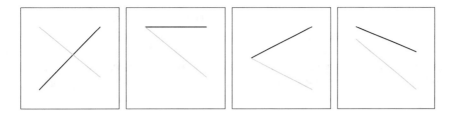

Fig. 15.6 Typical interaction plots. The effect of predictor 1 (on the *x*-axis) depends on the level of the second predictor (grey/black). All lines have different slopes. If there If there was no interaction, the grey and black lines would be parallel

(The Bernoulli distribution only has a single parameter, *P* and the link function is the *logit*). Here is the model output:

```
Coefficients:
             Estimate Std. Error z value Pr(>|z|)
(Intercept)  1.235414   0.192032   6.433 1.25e-10 ***
age         -0.004254   0.005207  -0.817    0.414
sexmale     -2.460689   0.152315 -16.155  < 2e-16 ***
---
Signif. codes:  0 `***' 0.001 `**' 0.01 `*' 0.05 `.' 0.1 ` ' 1
```

We assume that the two predictors are independent of each other, i.e. their effects in the model are purely additive. For any single case, say a 42-year-old man, we would only need to add the effects for age and sexmale to the intercept and we get our expected value after back-transformation:

$$\text{logit}^{-1}(1.235 - 0.00425 \cdot 42 - 2.46) = \frac{e^{-1.404}}{1 + e^{-1.404}} = 0.197$$

Sex and age in this model do not interact with each other here—each effect is simply additive to each other effect.

So what is a statistical interaction when when we include it in our model? What we want is that for one of the sexes there is an *additional* effect. For example, if we use the model to model the female survival probability, we want to add the male effect and also how the difference between men and women changes with age. The statistical trick is to simply generate a new variable, namely the interaction, by multiplying the two interacting terms!

Let's use this example to have a closer look: "Female" is coded as "0", "Male" as "1" (since it is coded as a dummy variable: see Sect. 7.2.2). If we now multiply the age of a person by these *dummy* values, we get "0" for all females and the respective age for all males:[5]

	sex	age	sex by age
Allen, Miss. Elisabeth Walton	female	29.0000	0.0000
Allison, Master. Hudson Trevor	male	0.9167	0.9167
Allison, Miss. Helen Loraine	female	2.0000	0.0000
Allison, Mr. Hudson Joshua Crei	male	30.0000	30.0000
Allison, Mrs. Hudson J C (Bessi	female	25.0000	0.0000
Anderson, Mr. Harry	male	48.0000	48.0000
...			

If we include the predictor sex by age in the model, this represents an effect of age that exclusively impacts *males*. This is exactly what we want! (The "by" is usually replaced by a "×" or ":" in the output.)

---

[5]The 11 month old child whose age is given exactly to 4 decimal places seems a bit unnecessary here. The data set comes from an R package (see next chapter), although this is just a weak excuse for not changing it.

If the two predictors are continuous, the procedure is the same. The interaction is then calculated as a product of the values and taken into the model.[6] As a result, synergistic (positive estimators for interaction) and antagonistic effects (negative estimators) between predictors become visible. This will hopefully become clearer in a later example (Sect. 16.1.3).

Let's continue with the Titanic survival data and include an interaction in the model. We do not actually need to calculate the interaction to include it, we just need to specify an interaction term in the model. We then get the following output:

```
Coefficients:
             Estimate Std. Error z value Pr(>|z|)
(Intercept)  0.493381   0.254188   1.941 0.052257 .
age          0.022516   0.008535   2.638 0.008342 **
sexmale     -1.154139   0.339337  -3.401 0.000671 ***
age:sexmale -0.046276   0.011216  -4.126 3.69e-05 ***
---
Signif. codes:  0 `***' 0.001 `**' 0.01 `*' 0.05 `.' 0.1 ` ' 1
```

As we see, the estimator for the interaction age:sexmale[7] is higly significant from 0. This means that the effect of age is highly dependent on the sex (as we suspected from looking at the plots), or the other way around: the effect of sex depends on age (which means the same thing).

We would calculate the value for our 42-year-old man as follows:

$$\text{logit}^{-1}(0.493 + 0.0225 \cdot 42 - 1.154 - 0.046 \cdot 42 \cdot 1) = \frac{e^{-2.09}}{1 + e^{-2.09}} = 0.110$$

Since we are dealing with a male, we multiply the interaction by 1. For a 42-year female, the sexmale value and the interaction value would be 0, which means that we can leave both out in our calculation.

$$\text{logit}^{-1}(0.493 + 0.0225 \cdot 42 - 1.154 \cdot 0 - 0.046 \cdot 42 \cdot 0) = \frac{e^{0.994}}{1 + e^{0.994}} = 0.730$$

Finally, we can have a look at the ANOVA table for the model, where we see both the interaction and the "normal" effects:

```
Terms added sequentially (first to last)
        Df Deviance Resid. Df Resid. Dev P(>|Chi|)
NULL                    1045      1414.6
age      1    3.238    1044      1411.4   0.07196 .
sex      1  310.044    1043      1101.3 < 2.2e-16 ***
age:sex  1   17.903    1042      1083.4 2.324e-05 ***
---
Signif. codes:  0 `***' 0.001 `**' 0.01 `*' 0.05 `.' 0.1 ` ' 1
```

It may be tempting to use the explained deviance to judge how important the interaction is. But beware: it is misleading to think that the interaction explains more deviance than age itself! If there is an interaction, then we always have to explain it, but not the main effects, because it *contains* the main effects. And more still: the significance of an interaction (in ANOVA) is evidence of the fact that the main effects do not sufficiently explain the pattern on their own. In our example, the two sexes differ strongly, but not for young children (Fig. 15.5, right). To claim that male travelers had lower chances of survival is therefore not the whole truth. Because of the interaction, we know that *depending on age*, the chances of survival between the sexes vary strongly.

In the next chapter, we will look at another example involving interactions (Sect. 16.1.2).

---

[6]The predictors should be standardised so that the interaction is not dominated by the numerically larger predictor. A **standardisation** is carried out by subtracting the average from all predictor values and then dividing it by the standard deviation of the predictor: $x' = \frac{x - \bar{x}}{s_x}$. This results in values with an average value of 0 and a standard deviation of 1. Such standardisation is also referred to as the "standard score" or "z-score" (https://en.wikipedia.org/wiki/Standard_score).

[7]age:sexmale is identical to sex by age above. The order is simply alphanumeric so that age comes before sexmale.

15.3   Collinearity

Collinearity (also referred to as multicollinearity) arises when predictors are highly correlated among each other. We looked at multiple correlations in Chap. 5, but now we have to do something about it!

Collinearity causes (at least) two types of problems: first, if two predictors are very similar, we cannot uniquely attribute an effect to one of them (the "interpretational problem"). Second, collinear predictors lead to unstable estimates (the "estimation problem"). This means that our optimisation algorithm working behind the GLM has difficulties finding the optimum. The reason is that we can substitute effects of predictor $\mathbf{A}$ by those of a collinear predictor $\mathbf{B}$. If, for example, $\mathbf{B} = 2\mathbf{A}$ and $\mathbf{y} = 5 + 4\mathbf{A}$, then we could alternatively write $\mathbf{y} = 5 + 2\mathbf{B}$ or $\mathbf{y} = 5 + 2\mathbf{A} + \mathbf{B}$ or many other such combinations. As the GLM attempts to estimate the coefficients for $\mathbf{A}$ and $\mathbf{B}$, it fails, because there are infinitely many possibilities.

In reality, $\mathbf{A}$ and $\mathbf{B}$ are rarely perfectly linear combinations of each other, and the optimisation does not fail. Still, the higher the correlation, the less "peaked" the optimum, i.e. the more parameter combinations with similarly high likelihood exist, leading to high standard errors for the GLM estimates. This phenomenon, the increased errors of the estimates due to collinearity, is referred to as "variance inflation". It is quantified by the variance inflation factor (VIF), whose value should not exceed 10—if it does, we have to remove one of the correlated predictors from the model or do something else about it (Dormann et al. 2012).

Strategically, we analyse collinearity before submitting the predictors to the multiple regression, e.g. by using Pearson's correlation coefficient (r). The literature is equivocal, as to what "too highly correlated" means, but the current consensus seems to be an $r^2 > 0.5$ (Dormann et al. 2012). If all correlations are $|r| < 0.7$ we are on the safe side.

15.3.1   Principal Component Analysis

A common way to analyse and visualise correlations between many predictors is the principal component analysis, or PCA. It brings order into our disarray of interrelated covariates by constructing a new coordinate system. Mathematically, the PCA is an eigenvalue decomposition of the correlation matrix $\mathbf{R}$.[8] We learned what such a matrix looks like in $\mathbf{R}$ from Sect. 5.1.3; it is a symmetric matrix, depicting the correlation of each covariate with each other. Typically Pearson's correlation coefficient is used, since he invented the PCA (Pearson 1901), but Spearman's or Kendall's correlation coefficients are also valid computations.

The idea of the PCA is relatively intuitive. Each covariate in our data set has many different values. When we plot two covariates against each other, we get a more or less undirected point cloud. If the variables are correlated, we recognise a pattern. The PCA puts a straight line (the first principal component) through the longest extension of this point cloud, like a skewer through a bun.[9] This first principal component (PC) describes the main trend in the data, it explains the most variance. For k covariates, there are k principal components, all perpendicular to each other. Figure 15.7 shows this for the case of only two variables $\mathbf{X}_1$ and $\mathbf{X}_2$, and the two principal components through their point cloud (left).

Here we are making use of the fact that the principal components are orthogonal to each other: whatever the original correlation within the predictors, we can generate uncorrelated new predictors by means of the PCA. *We can now use the principal components in lieu of the original variables in our multiple regression: they are uncorrelated, and hence we have solved our collinearity problem!* Details about the actual computation of the PCA will be the topic of the next chapter. One important caveat to mention at this point, is that PCA is restricted to continuous covariates, strictly for multivariate-normally distributed data.[10]

---

[8]We remember that the correlation is a standardised version of covariances. Some algorithms use the covariance matrix, rather than the correlation matrix, which is problematic. The resulting principal components will be dominated by those covariates that have the highest *absolute* values, although the units are completely arbitrary! To avoid such distortion, we have to ensure that the algorithm uses the correlation matrix, or that we standardise the covariates before we employ the algorithm..

[9]Mathematically this is the eigenvector with the largest eigenvalue. It actually is not the longest extension as such, but its positions in such a way that the data have minimal squared distance to this line. For typical data clouds, this will look like the longest extension of the data, which is more intuitive and easier to remember.

[10]The multivariate normal distribution (MVN) is an extension of the normal distribution to more than a single response variable. In our case, we have two variables, $\mathbf{X}_1$ and $\mathbf{X}_2$, each normally distributed but additionally correlated. The MVN has parameters for the means of each variable, plus a symmetric covariance matrix, where the diagonal contains the variances of each variable, and the off-diagonal elements are the covariances between $\mathbf{X}_1$ and $\mathbf{X}_2$. This is typically expressed like this: $\begin{pmatrix} \mathbf{X}_1 \\ \mathbf{X}_2 \end{pmatrix} \sim \text{MVN} \left(\mu = \begin{pmatrix} 5 \\ 8 \end{pmatrix}, \sigma = \begin{pmatrix} 1.2 & 0.5 \\ 0.5 & 0.9 \end{pmatrix} \right)$. 5 and 8 are the means of $\mathbf{X}_1$ and $\mathbf{X}_2$; 1.2 and 0.9 are the variances of $\mathbf{X}_1$ and $\mathbf{X}_2$, respectively, while the covariance is 0.5.

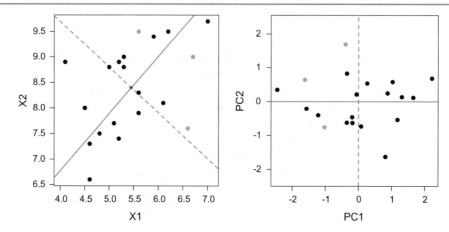

Fig. 15.7 A principal component analysis places orthogonal axes through the point cloud of covariates X_1 and X_2 (left). Principal component 1 (solid grey line) bisects the longest extent of the point cloud, PC 2 lies perpendicular to it (dashed grey line). We can now project the data points, with the PCs becoming the new coordinate system. This actually means rotating the point cloud (in this case counter-clockwise) to turn the PC 1 into the new x-axis. The point pattern remains unchanged, i.e. all relative positions are identical, as the three grey points illustrate

This should become more apparent with an example. On 20 experimental plots in a forest in the Appalachian mountain range (USA), a vegetation survey was done and the plant diversity was converted to an index (Shannon's H[11]). At the same time, soil samples were taken and examined to determine their nutrient contents and other soil properties (Bondell and Reich 2007). Figure 15.8 shows how these soil variables correlate with each other.

We see that there are some there are very high correlations, for example between Ca and Mg or density and pH. In a so-called *biplot* we can show all variables relative to each other after a PCA (it is called a biplot, because it brings together two pieces of information in one picture: the position of data points and the weight of the variables). So we first perform a PCA, map the values of the 20 points for the first two principle components, and then use arrows to show how important each variable is for these two principle components. The result (Fig. 15.9) is a bit confusing at first, but with some practice you can begin to understand it.

There are many things to discuss when it comes to principal component analysis (Fahrmeir et al. 2009; Joliffe 2002), but we can only scratch the surface here. For example, if two arrows have extremely similar positions (or exactly opposite ones), then these variables are highly correlated. We must not forget that the actual point cloud does not have only two dimensions, as shown here, but 13 (= number of variables in the PCA). The arrows are therefore projections from the 13-dimensional space into two dimensions. What looks like a high correlation here can be completely uncorrelated in two other dimensions.

These leads us to the question as to why we are only looking at two dimensions, and how much information are we overlooking because of this. The principal components record a decreasing amount of variance: the first most, the k-th the least. The so-called eigenvalues of the principal components show how much variance this is. In our Appalachian example, these are the following values: 5.147, 3.377, 1.449, 1.056, 0.648, 0.498, 0.411, 0.219, 0.112, 0.047, 0.032, 0.003 and 0.000. A visualisation using a screeplot (Fig. 15.10) shows that the first two principal components are much more important than the rest. They contain around 66% of the variance:

```
Importance of components:
                        PC1     PC2     PC3     PC4      PC5      PC6     PC7 ...
Standard deviation     2.2687  1.8377  1.2036  1.02749  0.80509  0.70591  0.6410 ...
Proportion of Variance 0.3959  0.2598  0.1114  0.08121  0.04986  0.03833  0.0316 ...
Cumulative Proportion  0.3959  0.6557  0.7671  0.84835  0.89821  0.93654  0.9681 ...
```

(The standard deviation is the square root of the variance, whose values are shown in the text above.)

The only thing left to explain is what these principal components have to do with the raw data (X). Mathematically, the PCA represents a rotation of the coordinate system. The new, rotated values, X_{PCA}, are calculated by means of a transformation

---

[11]Shannon's $H = -\sum_{i=1}^{N} p_i \ln p_i$, where p_i is the percent cover of species i, N is the total number of species, and $\sum p_i = 1$. The larger the value for H, the more diverse the plot is, or the higher the coverage of the plants in the plot is.

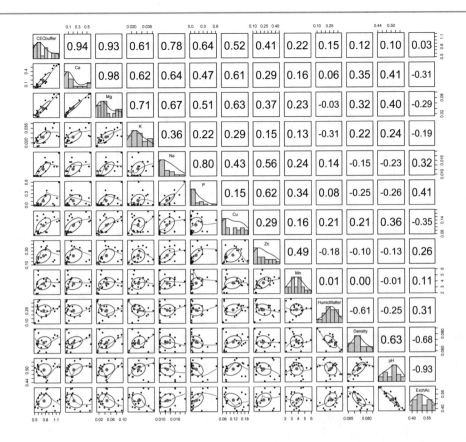

Fig. 15.8 Analysis of multiple correlations for soil variables from 20 forest plots in the Appalachian mountains (data from Bondell and Reich 2007). The numerical values in the upper right triangle show the values of the Pearson's correlation coefficient. Histograms of the variables are shown on the diagonal and pairwise scatter plots are shown below the diagonal

matrix Λ (also called "rotation matrix") from $\mathbf{X}$:

$$\mathbf{X}_{PCA} = \Lambda^T \mathbf{X}. \tag{15.1}$$

Λ is the matrix of the eigenvalues of the correlation matrix for $\mathbf{X}$. We can write this in algebraic notation as a system of equations:

$$
\begin{aligned}
\mathbf{x}'_1 &= \lambda_{11}\mathbf{x}_1 + \lambda_{21}\mathbf{x}_2 + \ldots + \lambda_{j1}\mathbf{x}_i \\
\mathbf{x}'_2 &= \lambda_{12}\mathbf{x}_1 + \lambda_{21}\mathbf{x}_2 + \ldots + \lambda_{j2}\mathbf{x}_i \\
&\ \ \vdots \\
\mathbf{x}'_i &= \lambda_{1i}\mathbf{x}_1 + \lambda_{2i}\mathbf{x}_2 + \ldots + \lambda_{ji}\mathbf{x}_i
\end{aligned}
\tag{15.2}
$$

There are i variables ($\mathbf{X}_1$ to $\mathbf{X}_i$) with j values. In this notation we recognise that the new principal components $\mathbf{x}'_i$ arise as a linear combination of the raw data. The transformation matrix Λ with its entries λ_{ij} contains the coefficients for this linear combination. So the greater the value of λ_{ij} is in the sum, the more a variable contributes to a principal component. We then say that "a variable has a certain *loading* on a component":

In our example, Λ looks as follows:

	PC1	PC2	PC3	PC4	PC5	PC6	PC7	PC8	PC9	PC10	PC11	PC12	PC13
CECbuffer	0.41	-0.08	-0.13	0.16	0.06	-0.16	0.15	-0.31	-0.04	0.01	-0.21	-0.24	0.73
Ca	0.41	0.10	-0.17	0.08	0.01	-0.20	0.06	-0.25	0.03	-0.22	-0.13	-0.46	-0.63
Mg	0.43	0.09	-0.08	0.08	0.06	-0.05	-0.04	-0.09	0.01	-0.13	-0.28	0.82	-0.12
K	0.29	0.11	0.11	0.43	0.56	0.43	-0.16	0.08	-0.08	0.12	0.40	-0.06	-0.02
Na	0.34	-0.26	-0.01	0.10	-0.25	-0.02	0.13	0.45	0.69	0.17	0.13	-0.03	0.00
P	0.28	-0.32	0.16	0.04	-0.19	-0.40	-0.31	0.39	-0.56	-0.03	0.18	-0.02	0.00
Cu	0.29	0.10	-0.30	-0.37	-0.24	0.59	0.22	0.27	-0.34	0.07	-0.14	-0.09	0.01

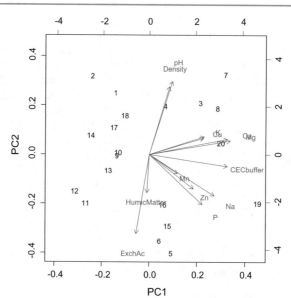

Fig. 15.9 Biplot of the principal component analysis of the soil variables for 20 forest plots in the Appalachian mountains. Shown here are the first two principal components (PC1 and PC2). The arrows shows the "loading" of the variables to each of the principal components. For example, HumicMatter only contributes to PC2, whereby P and Na contribute rather equally to PC1 and PC2. CECBuffer, on the other hand, contributes mostly to PC1

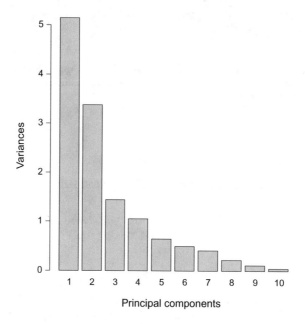

Fig. 15.10 Screeplot of the principal component analysis of soil variables for 20 forest plots in the Appalachian mountains. For each principal component, the (real part of the) eigenvalue is shown. The larger this value is, the more variance explained by the principal component in question

```
Zn            0.23 -0.22  0.42 -0.31 -0.31  0.29 -0.28 -0.53  0.11 -0.05  0.27  0.02 -0.01
Mn            0.15 -0.12  0.34 -0.61  0.60 -0.15  0.22  0.16  0.07 -0.04 -0.11 -0.04  0.00
HumicMatter  -0.01 -0.24 -0.68 -0.28  0.18 -0.16  0.00 -0.19  0.01  0.07  0.53  0.14 -0.01
Density       0.11  0.43  0.25  0.02 -0.21 -0.24  0.60 -0.08 -0.14  0.16  0.45  0.12 -0.01
pH            0.12  0.47 -0.06 -0.23  0.00 -0.20 -0.46 -0.01  0.09  0.66 -0.11 -0.07  0.01
ExchAc       -0.07 -0.51  0.07  0.18  0.05  0.06  0.28 -0.22 -0.19  0.65 -0.23  0.06 -0.25
```

For the first principal component, the loading variables are mainly Mg, CECbuffer and Ca. In Fig. 15.9, these variables have the longest arrows pointing along PC1 (either to the right or the left). For PC2, the main loading variables are ExchAc, pH and Density, which have similarly long up/downward pointing arrows with slight deviations to the right/left. Without being able to see this in the figure, PC3 is characterised by HumicMatter and Zn. This axis goes off "into the back", so to speak; it is lost in the 2-dimensional representation.

In summary, we can sketch out the following procedure for using the PCA for collinearity correction:

1. Transform the explanatory variables so that they are (relatively) normally distributed
2. Execute a PCA with these variables (including standardisation).
3. Look at the screeplot to determine how many components are needed to explain the data set. A common rule of thumb is to achieve 90% of the explained variance (for calculation see Sect. 16.2.1).
4. Use the principal components identified in the PCA as the new predictors in a multiple regression model.

A rather sloppy solution (because it does not eliminate collinearity) is to select a representative from each principal component (e.g. the most loading variable) and use them as predictors. In our example, we could select Mg (for PC1), ExchAc (interchangeable acidity; for PC2) and HumicMatter (for PC3) and hope that they represent the other variables as well.

15.3.2 Cluster Analysis

PCA is a useful and widespread method of summarising variables that provide us with similar information. The biplot is the corresponding visualisation that shows us what variables are connected to each other. But what do we do if we do not have normally distributed multivariate data, or if there are categorical predictors in the data set?

A nice alternative to the PCA is the cluster analysis. This type of analysis also has many different variations and is the topic of many books (such as Evritt et al. 2001; Fraley and Raftery 1998; Kaufman and Rousseeuw 2005). The basic idea is can be summarised as follows: As with the PCA, the similarity of the variables is calculated first. With cluster analysis, this can be any similarity matrix, in our example Spearman's ρ^2. In the second step, two variables that are similar to each other are grouped together ("clustered"). The total similarity with the other variables is then used to add another variable, and so on.[12] The result is a dichotomous tree.[13]

With an example we can see that the cluster analysis does not share the assumptions of PCA and can calculate both categorical and continuous variables together. Figure 15.11 shows the result for a data set in which different land-use variables were collected to describe the vegetation.

A cluster diagram shows the similarities of the variables as an upside-down tree. The similarity is shown by the horizontal line that connects the respective branch. In this case, ManagementNM and Manure are highly correlated ($\rho^2 > 0.6$), while this cluster has little to do with the cluster ManagementSF/UseHayPastu ($\rho^2 \approx 0.1$).

For our handling of correlated variables, this means that we select only one variable for our analysis from a highly correlated pair of variables (or variable group). The threshold value above which we consider variables to be correlated depends on the distance measure. For Spearman's ρ^2 0.5 is a good value. In this case there is only one cluster of highly correlated variables (ManagementNM/Manure), and thus the other variables can all be used.

Logically, we cannot separate the NM level from the Management variable. So it would be a possibility to dispense with the variable Manure (organic fertiliser). This does not mean that fertilisation has no influence, but only that we cannot separate fertilisation and nature conservation management with this data set. In the subsequent interpretation, however, we must consider that a significant effect of nature conservation management may be due to low levels of fertilisation.

As a comparison, Fig. 15.12 shows the corresponding cluster diagram for the Appalachian data set.

From this analysis we would select only one variable from the cluster pH/ExchAc (perhaps pH because it is more familiar), as well as one from the cluster P/Zn and from the cluster Na/CECbuffer/Ca/Mg also only one (perhaps Ca, because this is the most common of these cations in the soil).

---

[12] This approach is termed "agglomerative". Alternatively, with "divisive" clusters, you could start with all of the variables and then break them up into smaller clusters.

[13] The cluster analysis actually works with distances or dissimilarities. In the case of correlation, this is simply $1 - \rho^2$, but it could also be one of a dozen other distance measures: Gower, Jaccard, Euclid, Manhattan, Bray-Curtis, Mahalanobis, Hoeffding, etc. From the combination of over 50 distance measures and another dozen linking methods, there is an unmanageable number of variations. From the literature cited above, a combination of Spearman's ρ + complete linkage seems to be the best in my opinion, or as an alternative, Hoeffding-distance + Ward-linkage. Defaults and personal preferences vary enormously between different cluster analysis implementation methods.

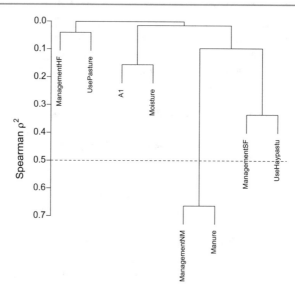

Fig. 15.11 Cluster diagram of an example data set from Jongman et al. [1987]. The variables are A1 (continuous: strength of the A1 horizon in cm), Management (categorical: BH: *biological farming*, HF: *hobby farming*, NM: *nature conservation management*, SF: *standard farming*), Use (*Hayfield, Haypasture, Pasture*) and Manure (categorical, but transformed here to a continuous predictor for illustration). Categorical and continuous variables can be mixed in the cluster analysis. The horizontal line at $\rho^2 = 0.5$ gives us the critical correlation value for this distance measure

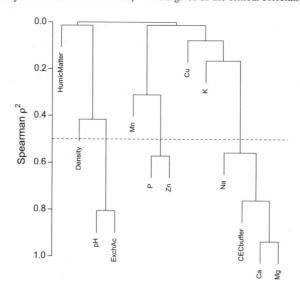

Fig. 15.12 Cluster diagram of soil variables from 20 forest plots in the Appalachian mountains. As compared to Fig. 15.9, this tree contains more information about the strength of the correlation and groups the variables more clearly. The one-dimensional format (all variables are next to each other) misleads us about possibly existing correlations between, for example, Ca and Cu. The two principal components can be recognised here as the first branching at the upper edge of the cluster diagram. The horizontal line at $\rho^2 = 0.5$ indicates the critical correlation value for this distance measure

15.4   Model Selection

As a rule, we can have only as many terms in our model as the number of data points will allow. The rule of thumb here is at least 5–10 data points per model term (Harrell 2001). This means that if we have 100 data points, we can easily include 10 predictors (so 4 main effects, 3 quadratic effects and three interactions).[14] We call such a model the "full model", because we include all of the predictors without actively selecting any.

[14]This rule is a bit different for binary data. Here it is not simply the number of data points that is important, but rather the number of zeros and ones, which even we have fewer off in our data set.

Model selection refers to any activity that reduces the complexity of the full model. Traditionally, such model selection proceeded in a stepwise fashion. A *forward stepwise selection* starts with an intercept-only model, then each predictor is added in turn, and the model offering most improvement (e.g. in terms of lowering AIC) is taken as the basis for the next step. In this next step, the procedure is repeated, leading to more and more predictors being included in the model, until no further improvement is possible.

Backwards stepwise model selection starts with the full model, and in turn trials the removal of each predictor in the model, leading to the deletion of the least important (and so forth). Forward model selection is *strongly discouraged*: if a predictor is relevant only as part of an interaction, or in a non-linear way, it typically will not be picked up, leading to suboptimal models. Backwards stepwise model selection is less sensitive, but, as we will see in the next chapter, may also yield suboptimal solutions.

Finally, we can construct all possible sub-models of the full model and trial all of them (called "best subset regression" (Draper and Smith 1998) or "model dredging"). The big advantage of really testing each combination of predictors comes at the cost of much higher computational burden (potentially thousands of models have to be computed).

One problem is that model selection *inevitably* leads to a distortion of parameter estimators! The forward stepwise model selection approach in particular can lead to serious mistakes (which is why we don't even introduce this method in this book). But even the better backwards stepwise selection can increase the type 1 error, because every step corresponds to a test, for which we would have to correct (see multiple tests, p. 157).

The full model is often the best (Harrell 2001): parameters can be estimated the most accurately (i.e. without bias), we do not lose any information by leaving out predictors, predictions have the least systematic error. In short, the more variables we fit, the better the model fit. But there is also a problem: what we are actually doing is just tracing our data set, whereas we might actually be interested in a transferable, general statement!

This problem is known as the variance-bias-trade-off (Hastie et al. 2009). *Variance* represents the share of declared variance, so how well our model describes the data; *bias* designates (in this case) deviation of model predictions for a new, independent data set, i.e. an error in generalisability. So the variance-bias-trade-off tells us that we can make good use of our data and minimise the variance, but then we are not able to make reasonable predictions. Or we accept a more moderate model fit, but on average are not out of line when making predictions to new data. The goal, of course, is to fit the data as well as possible without having a systematic error.

With model selection, this is exactly the problem we try to grapple with: from all the possible combinations of variables, we want to find the "best" combination.[15]

The most important approach to finding the "best" model (or the best models) is based on information-theoretical considerations (Burnham and Anderson 2002; Link and Barker 2006): Fit is determined by the log-likelihood value, model complexity by the number of fit parameters, and the quality criterion is then e.g. the AIC value of the model (for definition and refresher see p. 49).

We will approach the topic with the Titanic data set as an example. In the first step, we compare two models and learn the perspectives of each. In the subsequent second step, we follow the typical path from highly complex to minimally adequate models.

To explain the survival of Titanic passengers we have three variables available: `age`, `sex` and `passenger class` (1st, 2nd and 3rd).

So now we go through the following steps:

1. Formulate alternative models. We do not formulate only a single regression model, but in this case we will have two competing models. These two potential models are also called candidate models.
2. Fit the models and compare their AIC/BIC values. Fitting can be achieved using a GLM, whose log-likelihood can be used to calculate values for AIC/BIC. The model with the lowest AIC/BIC value has the better fit.
3. Compare the models using a *likelihood-ratio*-test. Instead of the AIC/BIC comparison, the change in likelihood-ratio can also be compared by means of a statistical test, the likelihood-ratio test (LRT). As the logarithm of a quotient of the likelihood is equal to the difference of the log-likelihoods, it is actually a log-likelihood-difference test. This difference

---

[15]Why is "best" in quotation marks here? Because there are different criteria for what "good" is. Accordingly, it could just so happen that we might choose one model based on criteria 1 and a completely different model based on criteria 2. Typical criteria include AIC, BIC, R^2 or log-likelihood (see Ward 2008 for a comparison of different criteria). The faith put into the single "best" model has been fundamentally questioned in ecology at least since the paper from Hilborn and Mangel [1997], but in general for much longer ever since the legendary publication from the geologist Chamberlin [1890].

Table 15.1 Comparison of both statistical models for explaining the survival probability of Titanic passengers. In addition to the degrees of freedom (df), the log-likelihood, and the AIC/BIC of the fit for the training data, Pearson's R^2, biserial R^2 and AUC of the test data are shown. The likelihood-ratio-test checks for significant differences of the fitted models and in this case finds such a difference

Model	df	log lik	AIC/BIC	R^2	Biserial R^2	AUC	$\Delta(-2\ell)$	$P(\chi^2)$
Model 1	3	−457.1	920/935	0.421	0.670	0.830		
Model 2	4	−448.7	905/924	0.418	0.664	0.836	16.92	3.91e-05

(more precisely the difference of -2ℓ) is χ^2-distributed, with as many degrees of freedom as the models differ in their number of parameters.

4. Compare predictions on the test data using R^2.[16]

Here are our two candidate models

Modell 1:   **survived** $\sim$ Binom($p = \text{logit}^{-1}(a + b \cdot \textbf{sex} + c \cdot \textbf{age})$)
Modell 2:   **survived** $\sim$ Binom($p = \text{logit}^{-1}(a + b \cdot \textbf{sex} + c \cdot \textbf{age} + d \cdot \textbf{sex} \cdot \textbf{age})$)

In this case, `sex` is a dummy variable with the value 0 for *female* and 1 for *male* (in alphanumerical order).

Table 15.1 gives us the result of this model comparison. As we can see, the interpretation depends on the choice of criterion. Model 2 (with the interaction between age and gender) fits the data better and has a better AUC value for the training data. Model 1, on the other hand, is a step ahead when considering both R^2 values. In model selection, the AIC value (or for some statisticians, the BIC value) is currently the most important criterion, and here, AIC and the likelihood-ratio test clearly speak in favour of the somewhat more complicated model 2.

Now to step 2, the typical simplification from complex to minimal adequate model. To this end, we formulate the model using all reasonable predictors:

Model 3:   **survived** $\sim$ Binom($p = \text{logit}^{-1}(a + b \cdot \textbf{sex} + c \cdot \textbf{age} + d \cdot \textbf{pClass} + e \cdot \textbf{sex} \cdot \textbf{age} + f \cdot \textbf{sex} \cdot \textbf{pClass} + g \cdot \textbf{age} \cdot$ **pClass** $+ h \cdot \textbf{sex} \cdot \textbf{age} \cdot \textbf{pClass}$))

This model has eight parameters (a to h), the three main effects, three two-way interactions and one three-way interaction. When we fit this model we get the following (analogous to the entries in Table 15.1):

Model	df	Log lik	AIC/BIC	R^2	Biserial R^2	AUC	$\Delta(-2\ell)$	$P(\chi^2)$
Model 3	8	−369.0	762/819	0.421	0.669	0.861	159.4	0.0000

This model is much better than Model 2!

Here is the corresponding ANOVA table:

```
                    Df Deviance Resid. Df Resid. Dev P(>|Chi|)
NULL                                844    1142.08
sex                  1  227.166     843     914.91 < 2.2e-16 ***
age                  1    0.587     842     914.32   0.44346
passengerClass       2  110.787     840     803.54 < 2.2e-16 ***
sex:age              1   19.048     839     784.49 1.275e-05 ***
sex:passengerClass   2   33.978     837     750.51 4.186e-08 ***
age:passengerClass   2    5.805     835     744.71   0.05489 .
sex:age:passengerClass 2  6.736     833     737.97   0.03447 *
---
Signif. codes:  0 `***' 0.001 `**' 0.01 `*' 0.05 `.' 0.1 ` ' 1
```

---

[16]Or in this case, with binary data, using the AUC values. AUC stands for *area under curve*, and *curve* means a so-called *receiver-operator characteristic*. What AUC is exactly and how to calculate it is described in Harrell [2001] or Hastie et al. [2009]. AUC values are between 0.5 (very poor) and 1 (perfect). We simply use it as a comparative number: the larger, the better the prediction for the test data set.

Can we simplify this model further? Well, apparently `age` is not significant, and the interaction between `age` and `passengerClass` is also not significant. So maybe we can take them out?

No, we cannot! Both non-significant terms are still part of higher interaction. *As long as a term is part of an interaction, it must remain in the model.* We must remove neither `age` nor the interaction, as both are contained in the three way interaction. This rule is called the "marginality theorem" (see Nelder 1977; Venables 2000).

In this case, the full model is actually the best and the simplest adequate model! In the next chapter we will deal with a model that we can also simplify even further.

A final remark: If we have so many predictors or so few data points that the model consumes more degrees of freedom than we have data points, then it cannot be fit. This is often due to the fact that we have not checked for collinearity. Sometimes we can merge correlated variables through a PCA, or in some cases only use the main effects plus first order interactions.

15.4.1 Two-Way ANOVA by Hand

In a variance analysis, the test for the significance of the variables involved is carried out somewhat differently than described for regression. It is an extension of the ideas, which were executed in Sect. 11.2 for only one factor.[17]

When we now calculate the two-way ANOVA (so named because it has two explanatory factors) by hand, we do so to see that the combination of predictors in a model leads to a higher test sensitivity than simply calculating two different models, one for each factor. We will use the cormorant data from Sect. 15.1.1 as an example.

The calculations of ANOVA require the calculation of different means and the deviations of the data points from these means.

Let's first calculate the grand mean value and its deviation squares for the null model (only one mean value for all data points) using a calculator (or R). In the example, the grand mean over all data points is $\bar{y} = 17.4$ (`mean(divetime)`), and the sum of the deviation squares $SS_{total} = 959.1$ (`sum((divetime-mean(divetime))^2)`).

Now for each subspecies we will calculate a mean value (a group mean) ($\bar{y}_C = 19.03$, $\bar{y}_S = 15.77$), and the $SS_{residuals}$ for the respective group mean:

```
> sum((divetime[subspecies == "C"] - mean(divetime[subspecies == "C"]))^2)

[1] 532.1055
```

and

```
> sum((divetime[subspecies == "S"] - mean(divetime[subspecies == "S"]))^2)

[1] 320.0655
```

We get $SS_C = 532.11$ and $SS_S = 320.07$, so that the $SS_{residual}$ when considering the factor subspecies = 532.11 + 320.07 = 852.18. The effect of the factor "subspecies" is therefore $SS_{subspecies} = 959.1 - 852.18 = 106.92$.

Let's go through the same procedure for the factor "season" (with four means and SS values). We then get $\bar{y}_{spring} = 11.86$, $\bar{y}_{summer} = 15.09$, $\bar{y}_{autumn} = 19.23$, $\bar{y}_{winter} = 23.42$, $SS_{spring} = 29.22$, $SS_{summer} = 33.87$, $SS_{autumn} = 45.36$ and $SS_{winter} = 94.48$. If we sum it all up, we get 202.93 unexplained SS for the residuals of the season effect. Accordingly, the factor "season" explains $SS_{season} = 959 - 203 = 756$ of 959 units.

Together, subspecies (106.9) and season (756) explain 862.9 units. With that, the remaining unexplained variance is $SS_{residual} = SS_{total} - SS_{subspecies} - SS_{season} = 959.1 - 862.9 = 96.2$.

So now we have calculated the majority of the ANOVA. We still need to calculate the degrees of freedom and then the mean deviation square, which will allow us to further calculate the F-value and the corresponding P-value. The F-statistic for `subspecies` alone can be calculated so:

$$F_{subspecies} = \frac{SS_{subspecies}/df_{subspecies}}{SS_{residuals}/df_{residuals}} = \frac{107/1}{(959-107)/(40-1-1)} = \frac{107}{22.42} = 4.77$$

[17]You could argue that the ANOVA is just another formulation of the linear model, and therefore does not deserve more attention. On the other hand, no other modern statistical method has made its way into biological statistics as quickly and fundamentally as the ANOVA.

the associated P-value is 0.035.

The same calculation for (four!) `seasons` looks like this:

$$F = \frac{SS_{season}/df_{season}}{SS_{residuals}/df_{residuals}} = \frac{756/3}{(959 - 756)/(40 - 3 - 1)} = \frac{252}{5.34} = 47.17$$

the associated P-value is smaller than 0.001:

```
> pf(47.17, 1, 38, lower.tail = F)
```

```
[1] 3.711755e-08
```

What we just calculated are the individual effects of the factors `subspecies` and `season`. However, our experiment was designed in such a way that both of these factors were manipulated and we therefore want to use a combined analysis. The result of this is that the values for the residuals become dramatically lower.

Together, the SS_{Factor} of our two factors `subspecies` and `season` add up to $107 + 756 = 863$. Thus, only $959 - 863 = 96$ is left for the residuals. As we now recalculate the F-statistics, based on this much smaller value for SS_{res}, our factors become more strongly significant. However, it must be taken into account that the degrees of freedom for the residuals now refer to the overall model, not just to one factor. Instead of $40 - 1 - 1 = 38$ for the residuals of `subspecies` we now have $40 - 1 - 3 - 1 = 35$ for both factors:

$$F = \frac{SS_{subspecies}/df_{subspecies}}{SS_{res}/df_{res}} = \frac{107/1}{96/(40 - 4 - 1)} = \frac{107}{2.74} = 39.0$$

The associated P-value is 0.001.

Ditto for `season`:

$$F = \frac{SS_{season}/df_{season}}{SS_{res}/df_{res}} = \frac{756/3}{96/(40 - 4 - 1)} = \frac{252}{2.74} = 92.0$$

The associated P-value is also (much) lower than 0.001.

So we see that the combination of both factors in an analysis dramatically increases both the explanatory power (R^2) **and** the inferential power (significances) of our statistical model.

For the additive ANOVA model (i.e. with subspecies and season, but without interactions) we can now construct an ANOVA table.[18]

Effect	df	SS	MS	F	P
Subspecies	1	107	107	39.0	<0.0001
Season	3	756	252	92.0	<0.0001
Residuals	35	96	2.7		
Total	49	959			

Finally, we can also calculate the interaction in the same way by forming groups in which both `subspecies` and `season` are specified ($2 \cdot 4 = 8$ groups). This gives us only a small (non-significant) reduction of SS_{res} compared to the additive model: 175 versus 185.

Figure 15.13 clearly shows how much better the combined mean values for season and subspecies approximate the data points than do the respective effects individually. The inclusion of the interaction makes in this case no difference (bottom right panel).

We can extend this calculation method to many more factors. However, once significant interactions are added, interpretability is hindered. This example illustrates that there is no magic behind the ANOVA, but that these calculations can also be carried out with a simple calculator.

---

[18]Compare this with the result from R, Model `fm2`, Sect. 16.1.1.

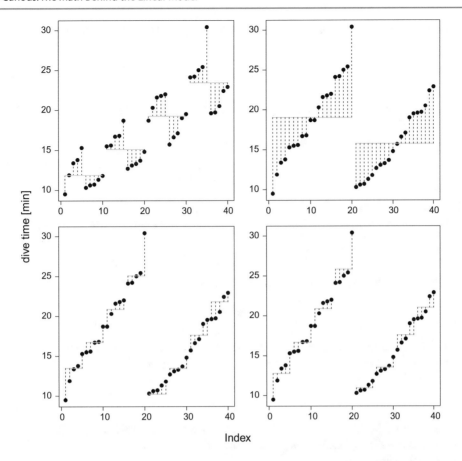

Fig. 15.13 Deviation of the data points from the mean for each season (upper left, data sorted by season then subspecies), from the mean for each subspecies (upper right, data sorted by subspecies, then season), from the mean of season and subspecies (corresponds to model fm, Sect. 16.1.1; lower left) and from the mean of the interaction of season and subspecies (fm2; lower right). The small difference between the lower two panels is the reason why the interaction is not significant. The grey lines correspond to the predictions of the corresponding models

15.5   For the Math Curious: The Math Behind the Linear Model

The linear model is calculated using algebraic methods, more specifically matrix operations. Imagine a linear model (regardless of whether the explanatory variables are categorical or continuous) consisting of three parts: (1) A vector $\mathbf{Y}$, which shows the n measured values of the dependent variables:

$$\mathbf{Y} = \begin{pmatrix} y_1 \\ y_2 \\ \vdots \\ y_n \end{pmatrix};$$

(2) multiple (p) vectors of explanatory variables $\mathbf{X}_1 \cdots \mathbf{X}_p$, each with n values, arranged as a matrix $\mathbf{X}$ (the matrix has $p + 1$ columns, the first ($x_{.0}$) is for the y-intercept):

$$\mathbf{X} = \begin{pmatrix} x_{10} & x_{11} & x_{12} & \cdots & x_{1p} \\ x_{20} & x_{21} & x_{22} & \cdots & x_{2p} \\ \vdots & & & \ddots & \vdots \\ x_{n0} & x_{n1} & x_{n2} & \cdots & x_{np} \end{pmatrix};$$

and (3) a vector $\boldsymbol{\beta}$ with the coefficients of the model, which connects $\mathbf{Y}$ and $\mathbf{X}$:

$$\boldsymbol{\beta} = \begin{pmatrix} \beta_0 & \beta_1 & \beta_2 & \ldots & \beta_p \end{pmatrix}^T.$$

With this information, the linear model looks so: $\mathbf{Y} = \mathbf{X}\boldsymbol{\beta} + \boldsymbol{\epsilon}$, where $\boldsymbol{\epsilon}$ is an error vector with the same structure as $\mathbf{Y}$.

From this equation, we can calculate the model coefficients (β_i) by solving the so-called normal equation[19]:

$$\mathbf{X}^T\mathbf{X}\mathbf{b} = \mathbf{X}^T\mathbf{Y},$$

where $\mathbf{b}$ is the ordinary least square-estimator for $\boldsymbol{\beta}$ and a vector of the partial regression coefficients. $\mathbf{X}^T$ is the transposed (diagonally flipped) matrix of $\mathbf{X}$.

It follows (by dividing both sides by $\mathbf{X}^T\mathbf{X}$), that

$$\mathbf{b} = (\mathbf{X}^T\mathbf{X})^{-1}(\mathbf{X}^T\mathbf{Y}),$$

where $\mathbf{X}^{-1}$ is the inverse matrix of $\mathbf{X}$.

The variance of the partial regression coefficients as well as the covariance between the explanatory variables can also be calculated as:

$$s^2 = MS_{\text{Residuals}}(\mathbf{X}^T\mathbf{X})^{-1}.$$

Another useful calculation uses the so-called *hat*-matrix $\mathbf{H}$. $\mathbf{H}$ is an $n \times n$-matrix, defined as

$$\mathbf{H} = \mathbf{X}(\mathbf{X}^T\mathbf{X})^{-1}\mathbf{X}^T.$$

The diagonal of $\mathbf{H}$ contains the influential value of the individual data points on the regression. Its main function is that we can use it along with the observed response $\mathbf{Y}$ to calculate the predicted values $\hat{\mathbf{Y}}$:

$$\hat{\mathbf{Y}} = \mathbf{H}\mathbf{Y}.$$

And finally, the effect squares of the ANOVA can also be calculated from these matrices. For this, we need a correction factor $K = \mathbf{Y}^T\mathbf{1}\mathbf{1}^T\mathbf{Y}/n$. Here, $\mathbf{1}$ is a vector of length n, where all entries are 1s.

The sum of squares can be calculated as:

$$SS_{\text{total}} = \mathbf{Y}^T\mathbf{Y} - \mathbf{K}.$$

The residual squares can be calculated as:

$$SS_{\text{resid}} = \mathbf{b}^T\mathbf{Y}^T\mathbf{Y} - \mathbf{K}.$$

The effect squares are the differences:

$$SS_{\text{Effect}} = \mathbf{Y}^T\mathbf{Y} - \mathbf{b}^T\mathbf{Y}^T\mathbf{Y}.$$

We will probably rarely have to calculate this ourselves in this manner. Nevertheless, to aid your understanding of the mathematics behind such calculations, an example from Crawley [2002] is shown here. It is a regression of wood volume against the circumference and height of the trees.

```
> trees <- read.csv("trees.metric.csv")
> attach(trees)
```

Now we create vector $\mathbf{X}$, by indicating an intercept value of 1 and combining this with the other explanatory variables in a matrix:

```
> X <- cbind(1, DBH, Height)
```

Then we calculate the transposed matrix of $\mathbf{X}$: xt.

```
> Xt <- t(X)
```

Now we have the matrix $\mathbf{X}^T\mathbf{X}$ (matrix multiplications must be put between percent symbols):

---

[19]Here it would be good to refresh your linear algebra skills, specifically matrix multiplication. A quick refresher can be found on https://en.wikipedia.org/wiki/Matrix_(mathematics).

```
> Xt
```

```
              DBH    Height
      31.0  1043.20   718.30
DBH  1043.2 37007.74 24413.47
Height 718.3 24413.47 16757.71
```

In the diagonals, we first find the sample size $n = 31$, then the SS for the circumference and then the SS for the height. The second and third value in column 1 are the sum of the values for circumference and height, respectively. The missing value (bottom middle or right middle) is the sum of the products of circumferences and heights. Let's do some calculations:

```
> sum(DBH)
```

```
[1] 1043.2
```

```
> sum(Height)
```

```
[1] 718.3
```

```
> sum(DBH^2)
```

```
[1] 37007.74
```

```
> sum(Height^2)
```

```
[1] 16757.71
```

```
> sum(Height * DBH)
```

```
[1] 24413.47
```

The next thing we are interested in are the coefficients **b**. The function `solve` inverts a matrix:

```
> b <- solve(Xt %*% X) %*% Xt %*% Volumen
> b
```

```
              [,1]
        -1.63766904
DBH      0.05255248
Height   0.03123067
```

We compare these with the coefficients from the linear model:

```
> lm(Volume ~ DBH + Height, data=trees)
```

```
Call:
lm(formula = Volume ~ DBH + Height, data = trees)
```

```
Coefficients:
(Intercept)        DBH       Height
   -1.63767    0.05255      0.03123
```

For the sake of completeness, we can now calculate the $SS_{\text{Residuals}}$ of the linear model[20]:

---

[20]The correctness of this calculation can be checked using `summary(aov(Volume - DBH + Height))`.

```
> t(Volume) %*% Volume - t(b) %*% Xt %*% Volume
```

```
            [,1]
[1,] 0.3425996
```

```
> detach(trees)
```

References

1. Anderson, T. (1958). *An Introduction to Multivariate Analysis*. New York: Wiley.
2. Bondell, H. D., & Reich, B. J. (2007). Simultaneous regression shrinkage, variable selection, and supervised clustering of predictors with OSCAR. *Biometrics*, *64*, 115–121.
3. Burnham, K. P. & Anderson, D. R. (2002). *Model Selection and Multi-Model Inference: a Practical Information-Theoretical Approach*. Berlin: Springer, 2nd edition.
4. Chamberlin, T. C. (1890). The method of multiple working hypotheses. *Science*, *15*, 92–96.
5. Chatfield, C., & Collins, A. J. (1980). *Introduction to Multivariate Analysis*. New York: Springer.
6. Crawley, M. J. (2002). *Statistical Computing. An Introduction to Data Analysis using S-Plus*. Chichester: John Wiley & Sons Ltd.
7. Dormann, C. F., Schymanski, S. J., Cabral, J., Chuine, I., Graham, C., Hartig, F., et al. (2012). Correlation and process in species distribution models: bridging a dichotomy. *Journal of Biogeography*, *39*, 2119–2131.
8. Draper, N. R. & Smith, H. (1998). *Applied Regression Analysis*. New York: Wiley, 3rd edition.
9. Evritt, B. S., Landau, S., & Morven, L. (2001). *Cluster Analysis*. London: Hodder Arnold, 4th edition.
10. Fahrmeir, L., & Künstler, R., Pigeot, I., & Tutz, G., (2009). *Statistik*. Berlin: Springer.
11. Fraley, C., & Raftery, A. E. (1998). How many clusters? Which clustering method? Answers via model-based cluster analysis. *The Computer Journal*, *41*, 578–588.
12. Harrell, F. E. (2001). *Regression Modeling Strategies-with Applications to Linear Models, Logistic Regression, and Survival Analysis*. New York: Springer.
13. Hastie, T., Tibshirani, R. J., & Friedman, J. H. (2009). *The Elements of Statistical Learning: Data Mining, Inference, and Prediction*. Berlin: Springer, 2nd edition.
14. Hilborn, R., & Mangel, M. (1997). *The Ecological Detective: Confronting Models with Data*. Princeton, NJ: Princeton University Press.
15. Izenman, A. J. (2008). *Modern Multivariate Statistical Techniques: Regression, Classification, and Manifold Learning*. New York: Springer.
16. Joliffe, I. T. (2002). *Principal Component Analysis*. Berlin: Springer.
17. Jongman, R., ter Braak, C., & van Tongeren, O. (1987). *Data Analysis in Community and Landscape Ecology*. Wageningen, NL: Pudoc.
18. Kaufman, L. & Rousseeuw, P. J. (2005). *Finding Groups in Data: An Introduction to Cluster Analysis*. Oxford: WileyBlackwell, 2nd edition.
19. Link, W. A., & Barker, R. J. (2006). Model weights and the foundations of multimodel inference. *Ecology*, *87*, 2626–2635.
20. Marsh, H. W., Wen, Z., Nagengast, B., & Hau, K.-T. (2012). Structural equation models of latent interaction. In R. H. Hoyle (Ed.), *Handbook of Structural Equation Modeling* (pp. 436–458). New York: Guilford Press.
21. Nelder, J. A. (1977). A reformulation of linear models. *Journal of the Royal Statistical Society*, *140*, 48–77.
22. Pearson, K. (1901). On lines and planes of closest fit to systems of points in space. *Philosophical Magazine*, *2*, 559–572.
23. Tabachnick, B., & Fidell, L. (1989). *Using Multivariate Statistics*. New York: Harper & Row.
24. Tufte, E. R. (1983). *The Visual Display of Quantitative Information*. Cheshire, CN: Graphics Press.
25. Venables, W. N. (2000). Exegeses on linear models. *S-PLUS User's Conference, Washington D.C.*, 1998.
26. Ward, E. J. (2008). A review and comparison of four commonly used Bayesian and maximum likelihood model selection tools. *Ecological Modelling*, *211*, 1–10.

Multiple Regression in R

16

I wish to perform brain surgery this afternoon at 4pm and don't know where to start. My background is the history of great statistician sports legends but I am willing to learn. I know there are courses and numerous books on brain surgery but I don't have the time for those. Please direct me to the appropriate HowTos, and be on standby for solving any problem I may encounter while in the operating room. Some of you might ask for specifics of the case, but that would require my following the posting guide and spending even more time than I am already taking to write this note.

—I. Ben Fooled (aka Frank Harrell)

At the end of this chapter . . .
. . . you will be able to fit and visualise regression models with two predictors.
. . . you will have two ideas for how to deal with correlated predictors: Principal component analysis and cluster analysis.
. . . you will be able to calculate a 2-way ANOVA by hand, if necessary.
. . . you will be able to execute a step-wise simplification of a model, both by hand and in an automated manner.
. . . this book will draw to a close.

In this chapter we will deal with three topics. First, we will begin where the last chapter left off and learn how to implement multiple regression in R. Related to this, we will learn how to visualise and interpret statistical interactions. The second topic we will deal with is how to handle correlated predictors, which we will approach using principal component analysis (PCA). In addition, we will also consider an alternative to this approach, namely cluster analysis. Finally, the third topic will deal with model simplification through the selection of explanatory variables.

16.1 Visualising and Fitting Interactions

16.1.1 Two Categorical Predictors: Regression and ANOVA

We will look at the simplest case of a model with two predictors using the cormorant diving data as an example. First, we will execute a multiple regression and try to understand the results. Then we look at the same data set from the perspective of an ANOVA approach.

Let us first have another look at the cormorant data (Fig. 16.1):[1]

---

[1] The `xyplot` can be adjusted to suit most situations, but the syntax takes some getting used to. In this case, the effect before the vertical line ("|") determines the *x*-axis, for each level of the factor after the line, there is an individual panel. Arguments are applied unless they are specifically addressed: With `cex = 1.5` we can increase the size of the points, with `scales = list(cex = 1.5)` we can change the size of the axis labels (F, S, H, W). Accordingly, arguments passed behind `xlab` affect the *x*-axis labels and `par.strip.text` on the text in the upper strip.

© Springer Nature Switzerland AG 2020
C. Dormann, *Environmental Data Analysis*,
https://doi.org/10.1007/978-3-030-55020-2_16

227

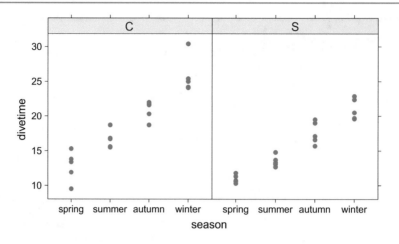

Fig. 16.1 Diving time (in minutes) of two European cormorant species *Phalacrocorax carbo carbo* (c) and *sinensis* (s) in four seasons with 5 measurements each

```
> cormorant <- read.table("../../data/cormorant.txt")
> cormorant$season <- factor(cormorant$season, levels=c("spring", "summer",
+    "autumn", "winter"))
> library(lattice)
> xyplot(divetime ~ season | subspecies, data=cormorant, cex=1.5, pch=16,
+    scales=list(cex=1.5, alternating=F), xlab=list(cex=2),
+    ylab=list(cex=2), par.strip.text=list(cex=2))
```

We see a clear seasonal effect and the larger subspecies (c) seems to dive a little longer than the smaller one (s).

Now we will first execute a multiple regression *without interaction*, which will calculate the effects for season and subspecies. Specifically, we fit the following model:

$$\text{divetime} \sim N(\mu = a + b\ \textbf{SeasonSummer} + c\ \textbf{SeasonAutumn} + d\ \textbf{SeasonWinter} + e\ \textbf{SubspeciesS}, \sigma = f)$$

Here, a represents the estimator for spring and subspecies C, all other parameters are the deviations from this. All of these variables are dummies, which are derived from the factors of `season` or `subspecies`. In R, it looks like this:

```
> fm <- glm(divetime ~ season + subspecies, family=gaussian, data=cormorant)
> summary(fm)
```

```
Call:
glm(formula = divetime ~ season + subspecies, family = gaussian,
data = cormorant)

Deviance Residuals:
Min       1Q  Median      3Q      Max
-3.995   -0.965   0.025   0.970   5.345

Coefficients:
Estimate Std. Error t value Pr(>|t|)
(Intercept)    13.4950     0.5855   23.047   < 2e-16  ***
seasonsummer    3.2300     0.7407    4.361  0.000109  ***
seasonautumn    7.3700     0.7407    9.951  9.65e-12  ***
seasonwinter   11.5600     0.7407   15.608   < 2e-16  ***
subspeciesS    -3.2700     0.5237   -6.244  3.69e-07  ***
```

```
---
Signif. codes:  0 `***' 0.001 `**' 0.01 `*' 0.05 `.' 0.1 ` ' 1

(Dispersion parameter for gaussian family taken to be 2.742886)

Null deviance:     959.100  on 39  degrees of freedom
Residual deviance:  96.001  on 35  degrees of freedom
AIC: 160.53

Number of Fisher Scoring iterations: 2
```

Let's attempt to interpret this output: First we have the intercept, which represents the first of all levels (for season `spring` and for subspecies `c`) with a value of 13.5. For seasons there are then three "slopes": respectively from `spring` to `summer`, from `spring` to `autumn` and from `spring` to `winter` with values of 3.23, 7.37 and 11.56. The subspecies `s` is then shown relatively to subspecies `c` in spring and lies 3.27 units lower (see Fig. 16.1, lower left).

The regression coefficients we just calculated do not take possible interactions between subspecies and season into account. They only show the results of the simultaneously performed bivariate regressions. Since subspecies and season can interact with each other, i.e. the subspecies perhaps have the same dive times in summer but different in winter, we have to determine three further coefficients: `seasonsummer:subspeciesS`, `seasonautumn:subspeciesS` and `seasonwinter:subspeciesS`. These three coefficients will show us the deviation for the three seasons from the value calculated from `seasonspring` and `subspeciesC`:

$$divetime \sim N(\mu = a + b\ \mathbf{SeasonSummer} + c\ \mathbf{SeasonAutumn} + d\ \mathbf{SeasonWinter}$$
$$+ e\ \mathbf{SubspeciesS} + f\ \mathbf{SeasonSummer} \cdot \mathbf{SubspeciesS}$$
$$+ g\ \mathbf{SeasonAutumn} \cdot \mathbf{SubspeciesS} + h\ \mathbf{SeasonWinter} \cdot \mathbf{SubspeciesS}, \sigma = i)$$

```
> summary(fm2 <- glm(divetime ~ season * subspecies, family=gaussian,
+   data=cormorant))

Call:
glm(formula = divetime ~ season * subspecies, family = gaussian,
data = cormorant)

Deviance Residuals:
Min      1Q   Median      3Q      Max
-3.280  -0.905  -0.290   0.945   4.580

Coefficients:
Estimate Std. Error t value Pr(>|t|)
(Intercept)              12.7800    0.7288  17.535  < 2e-16 ***
seasonsummer              3.8800    1.0307   3.764 0.000676 ***
seasonautumn              8.1000    1.0307   7.859 5.76e-09 ***
seasonwinter             13.0400    1.0307  12.651 5.38e-14 ***
subspeciesS              -1.8400    1.0307  -1.785 0.083720 .
seasonsummer:subspeciesS -1.3000    1.4577  -0.892 0.379139
seasonautumn:subspeciesS -1.4600    1.4577  -1.002 0.324051
seasonwinter:subspeciesS -2.9600    1.4577  -2.031 0.050671 .
---
Signif. codes:  0 `***' 0.001 `**' 0.01 `*' 0.05 `.' 0.1 ` ' 1

(Dispersion parameter for gaussian family taken to be 2.656)

Null deviance: 959.100  on 39  degrees of freedom
Residual deviance:  84.992  on 32  degrees of freedom
```

```
AIC: 161.66
```

```
Number of Fisher Scoring iterations: 2
```

None of the interaction coefficients are significantly different from 0, which suggests that there is no interaction between season and subspecies. In order to suss out this information, we can produce an ANOVA table. It tests the significance of the effects across all levels.

```
> anova(fm2, test="F")
```

```
Analysis of Deviance Table
```

```
Model: gaussian, link: identity
```

```
Response: divetime
```

```
Terms added sequentially (first to last)
```

	Df	Deviance	Resid. Df	Resid. Dev	F	Pr(>F)	
NULL			39	959.10			
season	3	756.17	36	202.93	94.9009	5.185e-16	***
subspecies	1	106.93	35	96.00	40.2594	4.013e-07	***
season:subspecies	3	11.01	32	84.99	1.3817	0.2661	

```
---
Signif. codes:  0 `***' 0.001 `**' 0.01 `*' 0.05 `.' 0.1 ` ' 1
```

Indeed, the interaction is not significant according to the ANOVA-table. In this case, we would simplify the model by eliminating the interaction and return to our earlier model fm. Here is the ANOVA table for the earlier model[2]:

```
> anova(fm, test="F")
```

```
...
```

	Df	Deviance	Resid. Df	Resid. Dev	F	Pr(>F)	
NULL			39	959.10			
season	3	756.17	36	202.93	91.895	< 2.2e-16	***
subspecies	1	106.93	35	96.00	38.984	3.691e-07	***

```
---
Signif. codes:  0 `***' 0.001 `**' 0.01 `*' 0.05 `.' 0.1 ` ' 1
```

Once we have calculated all these coefficients, we can use them to reconstruct the predicted value of an observation. This is practically a three-dimensional reconstruction of the data. Dimensions 1 and 2 are subspecies and season, while dive time is the third dimension. If we want to know the predicted mean value for subspecies S in autumn, we add the following coefficients of the above model: (Intercept)+SubspeciesS+SeasonAutumn+SubspeciesS:SeasonAutumn = $12.78 - 1.84 + 8.10 - 1.46 = 17.58$. In R this can be done directly using the predict function:

```
> predict(fm, newdata=data.frame(subspecies="S", season="autumn"))
```

```
       1
17.595
```

---

[2]We could have used lm instead of glm in this case. With an lm-object we can construct the ANOVA table using either anova or summary(aov(.)). The path we select here is more consistent with the mindset of this book: everything is a GLM.

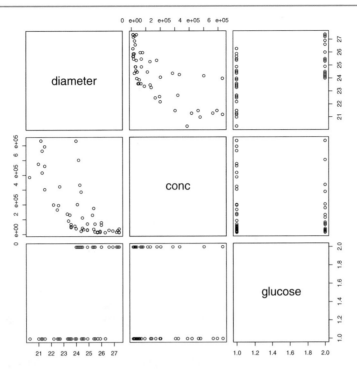

Fig. 16.2 Scatter plot of the variables cell diameter, cell concentration and glucose presence. We see a clear non-linear relationship between cell diameter and cell density. Note that glucose levels are represented numerically, with 1 for the alphabetically first level ("No")

So in summary, we can interpret the analysis as follows: The two subspecies differ significantly in their dive times, with *sinensis* having a dive time around 3.3 min shorter than *carbo* (13.5 ± 0.59 min). In addition, there is a clear effect of the season on dive time: In spring, the diving time is the shortest (13.5 min) and becomes longer through summer (16.7 min), fall (20.8 min) and winter (25.1 min). In total, we can explain $R^2 = 89\%$ of the variance, whereby the effect of season (partial $R^2 = 79\%$) is larger than the species effect (partial $R^2 = 11\%$).[3]

16.1.2   One Continuous and One Categorical Predictor

In the next-most complicated step, we will look at a multiple regression with one continuous predictor and one categorical predictor. For normally distributed data, the term ANCOVA, or analysis of co-variance is used (the continuous variable is the covariate to the categorical predictor). The reason behind this designation is that ANOVA was used in the 1980s primarily for the analysis of manipulative experiments. The manipulated factors were almost exclusively categorical (with and without fertilisers, herbivores or competition, for example). Variables that described the experimental units in more detail without being part of the manipulation (such as age or pH) were then included in the analysis as possible explanatory variables, but ones that were not really the primary interest. In the meantime, experiments have become more complex (10 levels of fertilisation, investigation of aging effects) and the co-variables have also become the subject of research. So it's really only for nostalgic reasons we are still attached to the term ANCOVA.

As an example, let's assume that we have two explanatory variables, one categorical and one continuous. As we did in the previous section, we can perform calculations first for one variable and then for the other. The only difference is in how we calculate the predicted values. With the continuous variable, this is done in the same way as regression (using the regression line equation), and for the categorical variable (as shown above) by using the mean of the groups. A specific mathematical equation will not be shown here, since it does not really differ from the equation in Sect. 15.4.1 on page 219 interested in the mathematical derivation of each individual SS-value, see Underwood [1997].

---

[3]Both are calculated from the *deviances* = SS: Season: $756.17/(756.17 + 106.93 + 96)$ and subspecies: $106.93/(756.17 + 106.93 + 96)$.

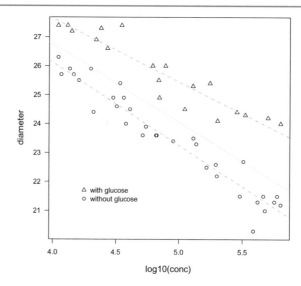

Fig. 16.3 Cell diameter of *Tetrahymena* depending on the log of the cell concentration for cultures with and without added glucose. The dotted line shows the regression for all data points, whereas the dashed lines shows the individual regressions for each glucose treatment

In the previous examples, the interactions were not significant. In order to understand interactions between variables, we should have a look at an appropriate example.[4] In an experiment, the diameter of *Tetrahymena* cells (a freshwater ciliate) were measured in cultures of different cell concentrations and with/without added glucose. First, let's load the data and plot all variables against each other.

```
> ancova <- read.table("ancova.data.txt", header = T)
> attach(ancova)
> plot(ancova)
```

We can see (Fig. 16.2) that the relationship between `diameter` and `conc` is non-linear, and therefore we will use the log of the concentration in the analysis. We can now look at the data for both of the explanatory variables together in a single plot (Fig. 16.3).

```
> plot(diameter ~ log10(conc), pch = as.numeric(glucose), cex.lab=1.25, las=1)
> legend(4.2, 22, legend = c("with glucose", "without glucose"), pch = 2:1,  bty = "n")
> abline(lm(diameter ~ log10(conc)), lty=3, lwd=2, col="grey")
```

Now we use the command `lm` to fit a regression for each glucose concentration treatment and add these lines to the plot.

```
> tethym.gluc <- ancova[glucose == "Yes", ]
> tethym.nogluc <- ancova[glucose == "No", ]
> lm.nogluc <- lm(diameter ~ log10(conc), data = tethym.nogluc)
> lm.gluc <- lm(diameter ~ log10(conc), data = tethym.gluc)
> abline(lm.nogluc, lty=2, lwd=2, col="grey")
> abline(lm.gluc, lty=2, lwd=2, col="grey")
```

Are these regressions different from each other? To answer this question, we look at the interaction between `log10(conc)` and `glucose`. This implies the following model:

$$\textbf{diameter} \sim N(\mu = a + b\log_{10}(\textbf{conc}) + c\textbf{glucoseYes} + d\log_{10}(\textbf{conc}) \cdot \textbf{glucoseYes}, \sigma = e) \tag{16.1}$$

---

[4]This example has been modified from Dalgaard [2002].

```
> fm3 <- aov(diameter ~ log10(conc) * glucose)
> summary(fm3)
```

```
                   Df Sum Sq Mean Sq F value  Pr(>F)
log10(conc)         1 115.22  115.22  530.80  <2e-16 ***
glucose             1  53.09   53.09  244.56  <2e-16 ***
log10(conc):glucose 1   1.55    1.55    7.13  0.0104 *
Residuals          47  10.20    0.22
---
Signif. codes:  0 '***' 0.001 '**' 0.01 '*' 0.05 '.' 0.1 ' ' 1
```

In fact, there is a significant interaction between cell concentration and glucose treatment. Let's take a look at the linear model to obtain the coefficients:

```
> summary(lm(fm3))
```

```
Call:
lm(formula = fm3)
```

```
Residuals:
    Min      1Q  Median      3Q     Max
-1.2794 -0.1912  0.0118  0.2656  0.9552
```

```
Coefficients:
                         Estimate Std. Error t value Pr(>|t|)
(Intercept)               37.5594     0.7138  52.618   <2e-16 ***
log10(conc)               -2.8610     0.1444 -19.820   <2e-16 ***
glucoseYes                -1.1224     1.2187  -0.921   0.3617
log10(conc):glucoseYes     0.6621     0.2479   2.670   0.0104 *
---
Signif. codes:  0 '***' 0.001 '**' 0.01 '*'0.05 '.' 0.1 ' ' 1
```

```
Residual standard error: 0.4659 on 47 degrees of freedom
Multiple R-Squared: 0.9433,      Adjusted R-squared: 0.9397
Signif. codes:  0 '***' 0.001 '**' 0.01 '*' 0.05 '.' 0.1 ' ' 1
```

What does a significant interaction between a categorical and a continuous predictor actually mean? It simply means that the slope of the covariate (i.e. the continuous variable) for the respective levels of the categorical variable is significantly different.

Let's look at the `lm` table above by means of plotting the each effect separately: First, we have a significant `(Intercept)`, which is caused by the fact that the values for cell diameter are not centred on 0 (Fig. 16.4a). This corresponds to a regression in which only one intercept is fit[5]:

```
> par(mfrow = c(2, 2), las=1, mar=c(2,3,1,1), oma=c(3,3,0,0))
> interceptOnly <- lm(diameter ~ 1)
> plot(diameter ~ log10(conc), pch = as.numeric(glucose), ylab = "")
> abline(h = coef(interceptOnly))
> text(5.7, 27, "a", cex = 2)
```

Next, we see that the effect of the cell concentration is significant (`log10(conc)`, Fig. 16.4b). So the cell diameter is dependent on the cell concentration. This corresponds to a model with the intercept and cell concentration.

---

[5]The `par(mfrow(...))` command allows us to show four plots in a single graphic; with the `mtext` command we create the labels of the x and y axes.

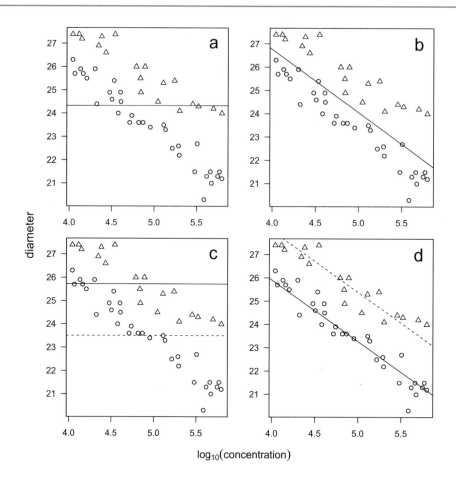

Fig. 16.4 Cell diameter of *Tetrahymena* plotted against the log of the cell concentration for cultures with and without added glucose. The lines show the regression for the respective model: **a** intercept only, **b** only $\log_{10}$(cell concentration), **c** only glucose effect, **d** $\log_{10}$(cell concentration) and glucose effect. The interaction of cell concentration and glucose is shown in Fig. 16.5. The dashed lines show models without the glucose, whereby the thin solid lines show models with added glucose

```
> concEffect <- lm(diameter ~ log10(conc))
> plot(diameter ~ log10(conc), pch = as.numeric(glucose), ylab = "")
> abline(concEffect)
> text(5.7, 27, "b", cex = 2)
```

The effect of glucose is also significant (glucoseYes, Fig. 16.4c). This means that cells in cultures with added glucose become significantly larger. Our corresponding model now contains the intercept and the glucose effect:

```
> glucEffect <- lm(diameter ~ glucose)
> plot(diameter ~ log10(conc), pch = as.numeric(glucose), ylab = "")
> abline(h = coef(glucEffect)[1] + coef(glucEffect)[2])
> abline(h = coef(glucEffect)[1], lty = 2)
> text(5.7, 27, "c", cex = 2)
```

In the next step, we put in both the cell concentration term and the glucose effect in an additive manner:

```
> additive <- lm(diameter ~ log10(conc) + glucose)
> plot(diameter ~ log10(conc), pch = as.numeric(glucose), ylab = "")
> abline(a = additive$coef[1], b = additive$coef[2], lty = 1)
> abline(a = (additive$coef[1] + additive$coef[3]), b = additive$coef[2], lty = 2)
```

```
> text(5.7, 27, "d", cex = 2)
>
> mtext("diameter", side = 2, line = 0, outer = T, las=3, cex=1.5)
> mtext(expression(log[10](concentration)), side = 1, line = 1, outer = T, cex=1.5))
```

And finally, the interaction is significant, which means that the effect of the cell concentration is different in cultures with or without glucose (Fig. 16.3). In this way, the addition of glucose does have an effect, but this effect is only detectable if we take the cell concentration into account.

In other words, the test of the glucose treatment represents a test on different intercepts of the cell concentration regression line. If it is significant, the regression lines are parallel but not identical (Fig. 16.4d). The interaction effect, on the other hand, is a test for the difference in slope of the regression line. If it is significant, both have the same intercept but different slopes. If both the categorical variable and the interaction are significant, we will have both different intercepts and different slopes (Fig. 16.3).

In the end, however, we want to be able to read the linear equations for these regressions from the table. How does that work? First, let's remember that the linear equation consists of a y-intercept and the slope. Additionally, we must remember that `glucoseYes` and `log10(conc):glucoseYes` in the upper `summary` represent the *difference* from the level `glucoseNo` ($\hat{a}$) and the effect of `log10(conc)` ($\hat{b}$).[6] Then the linear equation for the `glucoseNo` regression is:

$$\hat{y}_{\text{without glucose}} = \hat{a} + \hat{b} \cdot \log_{10}(\textbf{conc}) = 37.6 - 2.9 \cdot \log_{10}(\textbf{conc})$$

For the `glucoseYes` regression, the intercept and the slope differ, respectively, from the estimated value of the previous regression by `glucoseYes` ($\hat{c}$) and `log10(conc):glucoseNo` ($\hat{d}$):

$$\hat{y}_{\text{with glucose}} = (\hat{a} + \hat{c}) + (\hat{b} + \hat{d}) \log_{10}(\textbf{conc})$$
$$= 37.6 - 1.1 + (-2.9 + 0.7) \log_{10}(\textbf{conc})$$

Using this same logic, we have just drawn the linear equation into the figure.

While Fig. 16.3 is useful for helping us to understand what the different models mean and what an interaction actually is, it contradicts good practice by extending the regression line beyond the value range of the data points. It would also make sense to display the confidence intervals for this regression. Since the mathematics behind the calculation of the confidence intervals would be too much of a digression, we will restrict ourselves to the necessary R code for inserting a line only for the value range of data points and including the confidence intervals.[7]

We start with the simple plot of the data points, then use the model with the interaction (`fm3`) to plot the regression lines and their confidence intervals. Then we calculate the predicted value and confidence interval for new data points in the range of values. With the function `matlines` we can then put several lines into the figure at once. We do this procedure once for one line, then again for the other. The (purely aesthetic) effects of the options `las` and `tcl` can be found under `?par`.

```
> par(mar=c(5,5,1,1))
> plot(diameter ~ log10(conc), pch = as.numeric(glucose), ylab="diameter",
+   las=1, tcl=0.5)
> newconc <- seq(min(conc), max(conc), len=50)
> # Yes line:
> newdat.Yes <- data.frame("conc"=newconc, glucose=c("Yes"))
> pred.Yes <- predict(fm3, newdata=newdat.Yes, interval="confidence")
> matlines(log10(newdat.Yes$conc), pred.Yes, lty=c(1,2,2), lwd=c(2,1,1),
+   col="grey30")
```

---

[6]Have a look back at Eq. 16.1 to remind yourself of what a to d are. Also note that in Eq. 16.1 the parameters are just variables, while here they are actually estimated using the data; as a consequence, they gain a hat. Thus, $\hat{a}$ is the estimate of parameter a in Eq. 16.1. We get their values from the regression summary output.

[7]There are confidence intervals for different things (e.g. for parameter estimates, for predictions, for new data points)! Here we will limit ourselves to the confidence range with which the data allow the estimation of the regression line (option `interval='confidence'`; see Fig. 16.5). This indicates, speaking loosely, where the regression line would be under repeated sampling. In addition, there is a (larger) prediction interval, which shows us the range in which a *new value* would fall, with a 95% probability (option `interval='prediction'`).

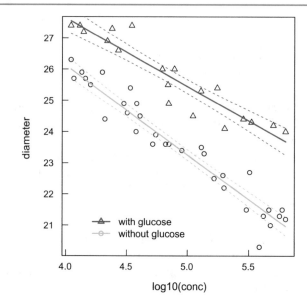

Fig. 16.5 Cell diameter of *Tetrahymena* depending on the log of the cell concentration for cultures with (circles) or without (triangles) glucose. The lines show the regression for the respective model, each with a 95% confidence interval

```
> # No line:
> newdat.No <- data.frame("conc"=newconc, glucose=c("No"))
> pred.No <- predict(fm3, newdata=newdat.No, interval="confidence")
> matlines(log10(newdat.No$conc), pred.No, lty=c(1,2,2), lwd=c(2,1,1),
+   col="grey70")
> legend(4.2, 21.5, col=c("grey30", "grey70"), legend=c("with glucose",
+   "without glucose"), lty=1, lwd=2, bty="n", pch=2:1)
```

16.1.3   Multiple Regression with Two Continuous Variables

Let's tackle this next section in the same way: by using an example. Paruelo and Lauenroth [1996] did vegetation surveys at 73 locations in North America. Here we are looking at the square root of the proportion of C3-species recorded in the survey.[8] As predictors, we have mean annual temperature (MAT), mean annual precipitation (MAP), the proportion of summer precipitation (JJAMAP) and of winter precipitation (DJFMAP), as well as the longitude and latitude of the survey location (LONG, LAT).

As the very first step, we must have a look at whether our environmental predictor variables are correlated with each other (Fig. 16.6):

```
> paruelo <- read.delim("paruelo.txt")
> paruelo$sqC3 <- sqrt(paruelo$C3)
> attach(paruelo)
> library(psych)
> pairs.panels(paruelo[,2:5])
```

---

[8]This example is taken from the excellent book by Quinn and Keough [2002], and as in this book we will simply and without further explanation transform the proportion of C3-species. Quinn and Keough [2002] use a logarithmic transformation, but in my opinion, the square root is more appropriate here (i.e. the data are a bit more "normal" than they are with the logarithm).

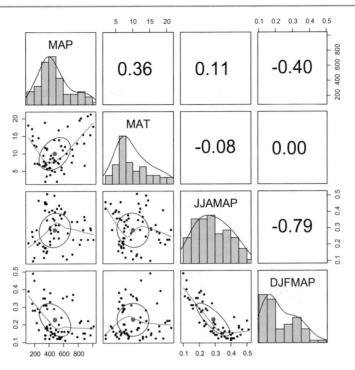

Fig. 16.6 Pair-wise scatter plot (lower left plots) and Pearson correlation coefficients (upper right) together with the histograms of the predictors (diagonal). The strong negative correlation between JJAMAP and DJFMAP can be clearly seen

Summer and winter precipitation are strongly negatively correlated with each other. Therefore, we will only include one of these predictors in our model. If we use the logic that in winter, there is less growth than there is in summer, and that precipitation in winter is probably less important, we choose to include the summer precipitation (JJAMAP).

Now we put all predictors into our full model of maximum complexity: All three predictors, all three 2-way interactions, and the single 3-way interaction.[9]

$$\textbf{sqC3} \sim N(\mu = a + b\,\textbf{MAP} + c\,\textbf{MAT} + d\,\textbf{JJAMAP} + e\,\textbf{MAP}\cdot\textbf{MAT} + f\,\textbf{MAP}\cdot\textbf{JJAMAP}$$
$$+ g\,\textbf{MAT}\cdot\textbf{JJAMAP} + h\,\textbf{MAP}\cdot\textbf{MAT}\cdot\textbf{JJAMAP}, \sigma = i)$$

The R syntax saves us the trouble of having to enter all of these interactions by hand:

```
> fm <- lm(sqC3 ~ MAP*MAT*JJAMAP)
> anova(fm)

Analysis of Variance Table

Response: sqC3
```

	Df	Sum Sq	Mean Sq	F value	Pr(>F)	
MAP	1	0.0000	0.00002	0.0003	0.98594	
MAT	1	1.9615	1.96153	35.7762	1.043e-07	***
JJAMAP	1	0.1003	0.10034	1.8301	0.18080	
MAP:MAT	1	0.0297	0.02966	0.5409	0.46470	
MAP:JJAMAP	1	0.0158	0.01584	0.2888	0.59281	
MAT:JJAMAP	1	0.2818	0.28176	5.1389	0.02673	*
MAP:MAT:JJAMAP	1	0.0002	0.00017	0.0032	0.95536	

[9]Quinn and Keough [2002] also use latitude and longitude as predictors. I find this to be logically problematic, due to the fact that longitude and latitude can't directly affect a plant. Furthermore, they are strongly correlated with climatic predictors, which could lead to collinearity issues (see Sect. 15.3 on page 211). As a little exercise, include them in the pairs.panels plot (Fig. 16.6) to confirm this.

```
Residuals       65 3.5638 0.05483
---
Signif. codes:  0 `***' 0.001 `**' 0.01 `*' 0.05 `.' 0.1 ` ' 1
```

Apparently, only one interaction is significant, so we make a new model that only includes this effect in addition to the main predictors.[10]

```
> fmfinal <- update(fm, .~ MAP*JJAMAP + MAT)
> anova(fmfinal)

Analysis of Variance Table

Response: sqC3
           Df Sum Sq Mean Sq F value    Pr(>F)
MAP         1 0.0000 0.00002  0.0003   0.98567
MAT         1 1.9615 1.96153 37.1705 5.756e-08 ***
JJAMAP      1 0.1003 0.10034  1.9014   0.17244
MAT:JJAMAP  1 0.3028 0.30279  5.7379   0.01936 *
Residuals  68 3.5884 0.05277
---
Signif. codes:  0 `***' 0.001 `**' 0.01 `*' 0.05 `.' 0.1 ` ' 1
```

We cannot simplify the model any further: the non-significant term JJAMAP is part of a significant interaction. How good is our model now and what do the effects look like?

```
> summary(fmfinal)

Call:
lm(formula = sqC3 ~ MAP + MAT + JJAMAP + MAT:JJAMAP)

Residuals:
     Min      1Q   Median      3Q     Max
-0.53269 -0.16411 -0.00075 0.15794 0.42541

Coefficients:
             Estimate Std. Error t value Pr(>|t|)
(Intercept)  0.3444225  0.2081817   1.654   0.1026
MAP          0.0002565  0.0001380   1.859   0.0674 .
MAT          0.0070555  0.0201479   0.350   0.7273
JJAMAP       1.1290855  0.6740891   1.675   0.0985 .
MAT:JJAMAP  -0.1506714  0.0629007  -2.395   0.0194 *
---
Signif. codes:  0 `***' 0.001 `**' 0.01 `*' 0.05 `.' 0.1 ` ' 1

Residual standard error: 0.2297 on 68 degrees of freedom
Multiple R-squared: 0.3972, Adjusted R-squared: 0.3618
F-statistic:  11.2 on 4 and 68 DF,  p-value: 4.864e-07
```

So, we can explain 36% of the variance, which is not too shabby for ecological data. We can only directly interpret MAP (since MAT and JJMAMAP are significant in an interaction, their interpretation is not quite so simple): The more it rains in a year, the higher the proportion of plants that use C3-photosynthesis.

---

[10]This is known as model selection, which will be correctly and comprehensively detailed in the next section. The casual approach here only serves to have a model with an interaction for visualisation purposes!

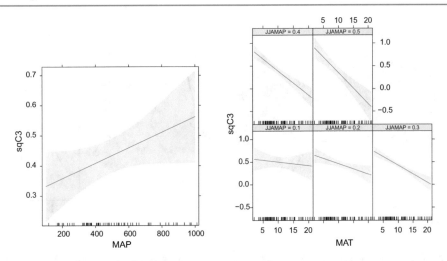

Fig. 16.7 The influence of mean annual precipitation (MAP, left) and the interaction of mean annual temperature (MAT) and the proportion of summer precipitation (JJAMAP) on the proportion of C3 plant species (square root transformed: sqC3) as a panel plot (right). In this type of plot, the total range of JJAMAP is broken up into 5 parts, and for each of these values (given as panel header), the effect of MAT is plotted. In other words, each panel is a slice through the 3-D-plot at a given value of JJAMAP

In order to interpret the interaction, we need to visualise it first. This is no simple task if we have two continuous variables. Figure 16.7 shows a 2-D variant of such a visualisation. Using so-called effect plots, variables are shown in the simplest way possible. In the case here, we can show the effect of MAP without even considering MAT or JJAMAP, since these effects do not interact with MAP. This is shown in Fig. 16.7 (left). The R code for this approach looks as follows:[11]

```
> library(effects)
> MAP.fmfinal <- effect("MAP", fmfinal)
> plot(MAP.fm4, main="")
> int.fmfinal <- effect("MAT:JJAMAP", fmfinal)
> plot(int.fmfinal, main="")
```

effect takes the mean of the model terms that are not displayed, which makes it so that the value range on the *y*-axis is not directly interpretable. However, we can now compare the values between the plots. In this case, this means that over the range of values for MAP, the root of C3-plant species increases by around 0.2 units (from 0.35 to 0.55). These 0.2 units can be compared with the figure on the right, in which the value range is quite different (from 0.5 to 1). The MAP effect in Fig. 16.7 (left) is comparable in its strength to the MAT effect in the last row of the panels in Fig. 16.7 (right).

The interaction of MAT and JJAMAP is shown in the panel plot on the right of Fig. 16.7. To avoid a 3-D figure, the effects function chooses for the variable multiple values (default is 5, and changing it is awkward) and shows the effect of the other variables for this value. While the slope of MAT continually increases with increasing JJAMAP values, this is broken up into individual linear pieces in the panels. The selected values are depicted as panel header. The value starts from the minimum in the very bottom left panel and goes to the top right. This type of figure takes some getting used to! The advantage of such a figure is that we can also depict more than three dimensions by showing multiple smaller panels.

An alternative to this approach can be seen in Fig. 16.8, which is based on the following (seemingly complex) R code:

```
> # One vector for each variable range:
> temperature <- seq(min(MAT), max(MAT), length=50)
> summerrain <- seq(min(JJAMAP), max(JJAMAP), length=50)
> # one prediction for all combinations using outer:
> z <- outer(X=temperature, Y=summerrain, FUN=function(X,Y) predict(fmfinal,
+   newdata=data.frame("MAP"=mean(MAP), "MAT"=X, "JJAMAP"=Y)))
```

---

[11]This is even easier if you use the command plot(allEffects(fm4)). We would then be asked to choose which term we would like to have displayed. effects (and allEffects) is actually smart enough to know not to show main effects if they are also part of an interaction. If we force it to do so, it will give us a warning message.

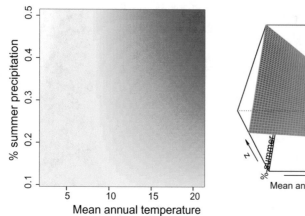
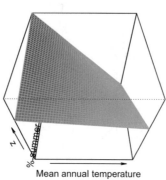
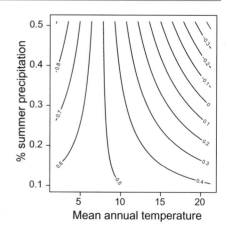

Fig. 16.8 The interaction of mean annual temperature (MAT) and the proportion of summer precipitation (JJAMAP) and its effect on the proportion of C3 plant species (square root transformed: sqC3). In the left plot, warmer colours represent *lower* (!!) values ("heat map"), in the middle plot, a true 3-D plot with perspective, and the right plot is a contour line plot. All three show that the effect of mean annual temperature is highest at high summer precipitation. As opposed to Fig. 16.7, we see a continuous change of the slope

```
> # Now to make the figures:
> par(mfrow=c(1,3))
> image(temperature,summerrain,z, cex.lab=1.5, ylab="prop. summer rainfall",
+   xlab="Mean annual temperature", col=heat.colors(50))
> persp(temperature,summerrain,z, theta=0, phi=50, cex.lab=1.5, ylab="prop.
+   summer rainfall", xlab="Mean annual temperature", shade=0.9, border="grey")
> contour(temperature, summerrain, z, cex.lab=1.5, ylab="prop. summer
+   rainfall", xlab="Mean annual temperature", las=1)
```

More details can be found in the help for the respective function, and personal preferences here tend to vary widely. A combination of image and contour is usually fairly easy to interpret, whereas the persp plot requires manual manipulation of the viewing angle, the grid colors and other parameters.[12]

All of these figures hide some very important information: where do the actual data points lie on this surface? Do we really have all combinations of annual temperature and proportion of summer precipitation? Or is the entire upper right hand corner missing, because perhaps there are no hot areas with high summer rainfall in North America? If we want to avoid drawing conclusions about conditions that do not actually exist, then we should include the points in the figures and identify the ranges where they are defined[13]:

```
> par(mar=c(5,5,1,1))
> image(temperature, summerrain,z, cex.lab=1.5, ylab="prop. of summer rainfall",
+   xlab="Mean annual temperature [Ã‚ÂºC]", col=heat.colors(50), las=1)
> contour(temperature, summerrain, z, add=T)
> points(MAT, JJAMAP, pch="+", cex=1.5)
> ch <- chull(cbind(MAT, JJAMAP))
> polygon(cbind(MAT, JJAMAP)[ch,])
```

In Fig. 16.9, we see that the environment range is fairly well covered and only the extreme corners in the upper right and lower left are not found in our data set. We should avoid making model predictions for these areas, since they would have no basis in real data.[14]

---

[12]Furthermore it is extremely difficult to depict the data points in a 3-D space, even with the **rgl** package (e.g. plot3d function), which offers a nice interactive variation.

[13]Here, we are *not* talking about geographic ranges, but rather regions in parameter or environmental space.

[14]This problem becomes more and more dramatic with each dimension that we add. In this example, we have $> 90\%$ coverage for each individual dimension (from the smallest to largest value), but in 2-D, we have less than 80% (the area within the polygon compared to the total area depicted). With 5 interacting predictors, even if we had similarly good coverage, we would only have $0.9^5 = 0.59$ or 60% of the space covered by our data! Nearly half of our predictions would be for areas in which we have no data in our data set! This problem has been called the "curse of dimensionality" (Bellman 1957).

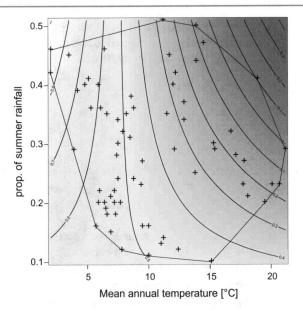

Fig. 16.9 Combined `image-contour`-plot for the interaction of mean annual temperature (MAT) and the proportion of summer precipitation (JJAMAP) and its effect on the proportion of C3 plant species (square root transformed: `sqC3`). The measured value combinations are depicted with a "+" and surrounded by a minimum convex polygon (convex hull)

16.2 Collinearity

Before we issue any treatment, we must make a diagnosis. Before we learn how to deal with collinearity by means of principal component analysis or cluster analysis, we first need to determine whether our predictors are so correlated with each other that it will be problematic.

As a rule of thumb, a correlation of 0.7 is our limit. For the soil data from the Appalachian mountains (from Sect. 15.3.1 on page 211) we can calculate this as follows (as we have done multiple times before):

```
> apa <- read.csv("soil.csv")
> round(cor(apa[,-c(1,2,16)]), 2)
```

	CECbuffer	Ca	Mg	K	Na	P	Cu	Zn	Mn	...
CECbuffer	1.00	0.94	0.93	0.61	0.78	0.64	0.52	0.41	0.22	
Ca	0.94	1.00	0.98	0.62	0.64	0.47	0.61	0.29	0.16	
Mg	0.93	0.98	1.00	0.71	0.67	0.51	0.63	0.37	0.23	
K	0.61	0.62	0.71	1.00	0.36	0.22	0.29	0.15	0.13	
Na	0.78	0.64	0.67	0.36	1.00	0.80	0.43	0.56	0.24	
P	0.64	0.47	0.51	0.22	0.80	1.00	0.15	0.62	0.34	
Cu	0.52	0.61	0.63	0.29	0.43	0.15	1.00	0.29	0.16	
Zn	0.41	0.29	0.37	0.15	0.56	0.62	0.29	1.00	0.49	
Mn	0.22	0.16	0.23	0.13	0.24	0.34	0.16	0.49	1.00	
HumicMatter	0.15	0.06	-0.03	-0.31	0.14	0.08	0.21	-0.18	0.01	
Density	0.12	0.35	0.32	0.22	-0.15	-0.25	0.21	-0.10	0.00	
pH	0.10	0.41	0.40	0.24	-0.23	-0.26	0.36	-0.13	-0.01	
ExchAc	0.03	-0.31	-0.29	-0.19	0.32	0.41	-0.35	0.26	0.11	

We remove columns 1 (BaseSat), 2 (SumCation) and 16 (Diversity) from the correlation matrix. BaseSat and SumCation are the result of a linear combination (the sum in the case of SumCation) of the cation concentration, so that there is no new information coming from these columns. Diversity is our response variable, and is therefore not of interest to us in this context. To save some space, we round the correlation coefficients to two decimal places.

Multiple variables are highly correlated here (such as Ca and Mg). If we were to now calculate a simple model with all of these predictors, then the parameter estimators would be very instable and overestimated ("inflated") in their value. This variance inflation can be calculated using variance inflation factors (VIF):

```
> fm <- glm(Diversity ~ ., data=apa[,-c(1,2)])
> library(car)
> vif(fm)
```

```
  CECbuffer          Ca           Mg            K          Na            P          Cu
65655.6319  49093.1419     2096.2032      78.3152      7.0172       5.4517     10.0659
         Zn          Mn   HumicMatter      Density          pH        ExchAc
     8.8790      3.1326       31.7383      23.3705     22.2211    7609.6002
```

The notation `Diversity ~ .` is a shorthand way to use all variables in a dataset. Then we need to also add the following specification: `data=apa[,-c(1,2)]`, in order to exclude the first two columns.

Indeed, some of these VIF values are well above the acceptable threshold of 10 (Fox 2002). The VIF does not identify which pairs are problematic, but rather identifies the parameters where there was inflation. We see now that problems arose with CECbuffer and ExchAc, for example, but not how. This information is contained in the correlation matrix: CECbuffer correlates very ($|r| > 0.7$) strongly with Ca, Mg and Na and somewhat strongly ($|r| > 0.5$) with P and K.

With the diagnosis of "collinear" and the knowledge that multiple variable pairs are highly correlated, we can now start with our first antidote: Principal component analysis.

16.2.1 Principal Component Analysis in R

Principal component analysis (PCA) can be implemented in R in many ways. Here, we will use the function `prcomp`, which is the most often used variant.[15]

With the PCA there is an important detail to consider: should the *correlation* matrix of the variables be used, or rather their *covariance* matrix? In the covariance matrix, the absolute values of the variables matter, i.e. a variable with values between 2 and 5 dominates over a variable with values between 0.2 and 0.5. In the correlation matrix, on the other hand, all values are normalised beforehand (centred to a mean of 0 and scaled to a standard deviation of 1). I do not know of a single case in ecology where the covariance matrix was useful. Nevertheless, the default for `prcomp` is `scale.=FALSE`, despite an explicit recommendation in the help to set this value to `TRUE`.

So let's have a look at how the soil analysis data is processed in a PCA by R.

```
> apa <- read.csv("soil.csv")
> pca <- prcomp(apa[,-c(1,2,16)], scale=T)
> str(pca)
```

```
List of 5
 $ sdev    : Named num [1:13] 2.269 1.838 1.204 1.027 0.805 ...
  ..- attr(*, "names")= chr [1:13] "1" "2" "3" "4" ...
 $ rotation: num [1:13, 1:13] 0.414 0.413 0.429 0.291 0.344 ...
  ..- attr(*, "dimnames")=List of 2
  .. ..$ : chr [1:13] "CECbuffer" "Ca" "Mg" "K" ...
  .. ..$ : chr [1:13] "PC1" "PC2" "PC3" "PC4" ...
 $ center  : Named num [1:13] 0.7621 0.1934 0.0469 0.0257 0.0129 ...
  ..- attr(*, "names")= chr [1:13] "CECbuffer" "Ca" "Mg" "K" ...
 $ scale   : Named num [1:13] 0.20528 0.18461 0.02716 0.00649 0.00308 ...
```

---

[15]For alternative implementations, see also `princomp` (uses eigenvalue decomposition with the function `eigen` instead of the supposedly more reliable singular value decomposition using `svd` in `prcomp`), `FactoMineR::PCA` (no typos!) or `vegan::rda`, while `labdsv::pca` or `rrcov::PcaClassic` along with many other packages internally use `prcomp`.

```
..- attr(*, "names")= chr [1:13] "CECbuffer" "Ca" "Mg" "K" ...
$ x       : num [1:20, 1:13] -1.439 -2.433 2.165 0.685 0.924 ...
..- attr(*, "dimnames")=List of 2
.. ..$ : NULL
.. ..$ : chr [1:13] "PC1" "PC2" "PC3" "PC4" ...
- attr(*, "class")= chr "prcomp"
```

With the argument `scale=T` we ensure that the PCA is performed on the correlation matrix and thus remains uninfluenced by the different value ranges of the variables.[16] The function `str` shows us the structure of the object. We do not need to know this, *per se*, but it will help us to understand the next steps in this case.

With `pca`, we are dealing with a list of the `prcomp` class with five entries: first the standard deviations of the principal components (`$sdev`), then the rotation matrix, which shows the weights of the individual variables on the principal components (`$rotation`), then the means (`$center`), and standard deviations (`$scale`) of the original data (these are standardised by the argument `scale=T`), and finally the new values in the new orthogonal PCA (`$x`).

We now use the `summary` function to call up some important information about our object "pca".

```
> summary(pca)

Importance of components:
                          PC1    PC2    PC3    PC4     PC5     PC6    PC7     PC8
Standard deviation     2.2687 1.8377 1.2036 1.02749 0.80509 0.70591 0.6410 0.46839
Proportion of Variance 0.3959 0.2598 0.1114 0.08121 0.04986 0.03833 0.0316 0.01688
Cumulative Proportion  0.3959 0.6557 0.7671 0.84835 0.89821 0.93654 0.9681 0.98502
                          PC9   PC10   PC11    PC12     PC13
Standard deviation     0.33506 0.21766 0.1802 0.05158 0.002838
Proportion of Variance 0.00864 0.00364 0.0025 0.00020 0.000000
Cumulative Proportion  0.99365 0.99730 0.9998 1.00000 1.000000
```

The summary of the PCA tells us how much variance is realised from each axis (or principal component: PC). More precisely, the first line shows the root of the variance = the standard deviation, the second line shows the proportion of the total variance and the third line shows the cumulative proportion of variance (additive for each principal component). In this case, the first axis explains 40% and we need six axes in order to capture over 90% of the variance.

This type of information is typically shown in a screeplot (Fig. 16.10).

```
> names(pca$sdev) <- as.character(1:13) # Trick to display PC labels
> par(mar=c(5,5,1,1))
> screeplot(pca, las=1, main="", cex.lab=1.5, xlab="Principal components")
```

The row commented with "Trick" is necessary if we want to provide names for the principal components in the figure.[17] Now we can have a look at the rotation matrix:

```
> round(pca$rotation,2)

            PC1   PC2   PC3  PC4  PC5   PC6   PC7   PC8   PC9  PC10  PC11  PC12  PC13
CECbuffer  0.41 -0.08 -0.13 0.16 0.06 -0.16  0.15 -0.31 -0.04  0.01 -0.21 -0.24  0.73
Ca         0.41  0.10 -0.17 0.08 0.01 -0.20  0.06 -0.25  0.03 -0.22 -0.13 -0.46 -0.63
Mg         0.43  0.09 -0.08 0.08 0.06 -0.05 -0.04 -0.09  0.01 -0.13 -0.28  0.82 -0.12
K          0.29  0.11  0.11 0.43 0.56  0.43 -0.16  0.08 -0.08  0.12  0.40 -0.06 -0.02
```

---

[16]According to the syntax in the help, `prcomp` requires the argument "scale." (with a full stop after `scale`!). It works just as well if you omit the full stop. The reason is that `scale` is a function, but in the call of `prcomp` it is used as an argument. The full stop can be omitted because arguments can be abbreviated.

[17]Since we cannot provide names in the form of an argument, we can have a look at the R code of this function (via `$get Anywhere("screeplot.default")$`) and find out that the names of the `$sdev`-component of the `prcomp` object is used for labelling. We then provide this with the desired names and get proper labelling.

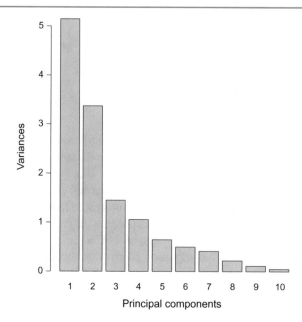

Fig. 16.10 *Screeplot* of the explained variance per principal component, using the soil analysis example. By default, only the first 10 principal components are shown here

```
Na          0.34 -0.26 -0.01  0.10 -0.25 -0.02  0.13  0.45  0.69  0.17  0.13 -0.03  0.00
P           0.28 -0.32  0.16  0.04 -0.19 -0.40 -0.31  0.39 -0.56 -0.03  0.18 -0.02  0.00
Cu          0.29  0.10 -0.30 -0.37 -0.24  0.59  0.22  0.27 -0.34  0.07 -0.14 -0.09  0.01
Zn          0.23 -0.22  0.42 -0.31 -0.31  0.29 -0.28 -0.53  0.11 -0.05  0.27  0.02 -0.01
Mn          0.15 -0.12  0.34 -0.61  0.60 -0.15  0.22  0.16  0.07 -0.04 -0.11 -0.04  0.00
HumicMatter -0.01 -0.24 -0.68 -0.28  0.18 -0.16  0.00 -0.19  0.01  0.07  0.53  0.14 -0.01
Density     0.11  0.43  0.25  0.02 -0.21 -0.24  0.60 -0.08 -0.14  0.16  0.45  0.12 -0.01
pH          0.12  0.47 -0.06 -0.23  0.00 -0.20 -0.46 -0.01  0.09  0.66 -0.11 -0.07  0.01
ExchAc     -0.07 -0.51  0.07  0.18  0.05  0.06  0.28 -0.22 -0.19  0.65 -0.23  0.06 -0.25
```

This rotation matrix converts the original values to the new PCA values.[18] The values show how important an original variable is for the new position. For example, if Mg has a value of 0.43 on PC1 (we say "has a loading of 0.43 on PC1"), then Mg is much more important for this PC1 than Zn with a loading of only 0.23. The absolute value is important, while the direction of the axis contains no information and can change from one function to another (even between different versions of R!).

The last important information we need are the new values. These so-called scores are saved in the list x:

```
> round(pca$x,2)

        PC1    PC2    PC3   PC4    PC5    PC6    PC7    PC8    PC9   PC10   PC11   PC12  PC13
[1,]  -1.44   2.06   1.53  0.83  -0.41  -0.21   0.41  -0.06  -0.18   0.20  -0.01   0.03  0.00
[2,]  -2.43   2.62   1.56  1.02  -1.13  -0.86   0.73   0.46  -0.08  -0.13  -0.04   0.00  0.00
[3,]   2.16   1.71   1.00 -0.42   0.15   0.18  -1.68  -0.10  -0.31  -0.17   0.08   0.04  0.00
[4,]   0.69   1.60   1.14 -0.64  -0.83   0.48  -0.99  -0.21   0.92   0.10  -0.05  -0.04  0.00
[5,]   0.92  -3.31   1.59  0.48   0.46  -0.01  -0.05  -0.09  -0.16   0.02   0.46  -0.02  0.00
[6,]   0.42  -2.90   1.19  1.54   1.24   0.66  -0.02   0.65   0.09  -0.17  -0.33  -0.03  0.00
[7,]   3.22   2.65  -1.03  1.11   0.80   0.50   0.39  -0.26   0.17  -0.25   0.18   0.07  0.00
[8,]   2.89   1.53  -1.62  0.61   0.35   0.67   0.79   0.38   0.21   0.10   0.01  -0.04  0.00
```

---

[18]This is a simple matrix multiplication of the (standardised) data matrix with the rotation matrix: `as.matrix(scale(apa[,-c(1,2,16)]))` %*% `pca$rotation`.

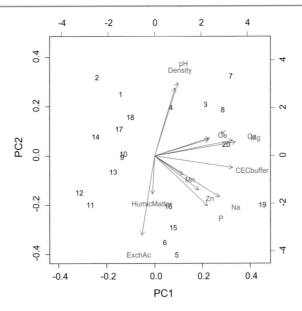

Fig. 16.11 Biplot of the PCA for the Appalachian soil analysis data (repeat of Fig 15.9). Note that there are two different scalings. The upper and right axes correspond to the values in `pca$x`, while the lower and left axes correspond to the loadings of the individual variables (arrows) or `pca$rotation`

```
 [9,] -1.38 -0.05  0.49 -1.27  0.94 -0.59  0.28  0.24  0.54  0.05 -0.04 -0.04  0.00
[10,] -1.32  0.06  1.02 -1.67  1.42 -0.77  0.40  0.01 -0.01 -0.10  0.07  0.07  0.00
[11,] -2.72 -1.64 -1.75  0.16 -0.01 -0.53 -0.01 -0.34  0.44  0.00  0.06  0.02  0.00
[12,] -3.18 -1.23 -1.92  0.53 -0.37 -0.39 -0.57 -0.42 -0.02 -0.40 -0.06  0.00  0.00
[13,] -1.75 -0.54  0.03  1.05  0.53  0.49 -0.40 -0.70 -0.21  0.57 -0.22  0.06  0.00
[14,] -2.50  0.63 -0.30  1.26 -0.39  0.32 -0.32  0.08 -0.18 -0.13  0.14 -0.07 -0.01
[15,]  0.78 -2.40 -0.52 -0.31 -1.46  0.50  0.49  0.41  0.21  0.17  0.14  0.09  0.00
[16,]  0.57 -1.69  0.91 -1.62 -1.14  1.27  0.71 -0.47 -0.28 -0.25 -0.13 -0.02  0.00
[17,] -1.51  0.89 -1.20 -1.05  0.24  0.29 -0.09  0.33 -0.38  0.30  0.23 -0.09  0.00
[18,] -1.03  1.28 -1.28 -1.50  0.29  0.46 -0.24  0.64 -0.39 -0.05 -0.22  0.03  0.00
[19,]  4.58 -1.65 -0.70  0.14 -0.78 -1.63 -0.58  0.54 -0.18  0.12 -0.13  0.00  0.00
[20,]  3.03  0.37 -0.16 -0.28  0.11 -0.85  0.74 -1.10 -0.19  0.01 -0.15 -0.07  0.00
```

Here we find the new coordinates for the original 20 data points. Point 1 receives the new coordinates $(-1.44, 2.06)$ if we consider the two first principal components.

We plot this new space as a biplot, which plots the first two axes by default, because the largest variance comes from there (Fig. 16.11).

```
> biplot(pca)
```

16.2.2 Cluster-Analysis in R

PCA is very limited in its applicability due to its assumptions (multivariate, normally distributed data). Although there are more recent developments that remove both the assumption of normal distribution and the limitation to continuous variables,[19] here we want to look at a structurally different approach: the cluster analysis.

---

[19] Such as multidimensional scaling (MDS); see function `metaMDS` in the **vegan** package as a starting point or Zuur et al. [2007].

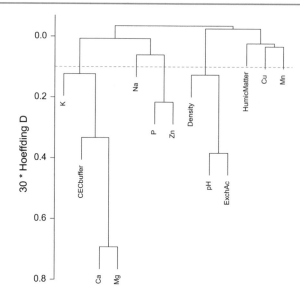

Fig. 16.12 Cluster diagram of the soil variables for 20 forest plots in the Appalachian mountains using the Hoeffding distance. The typical threshold value here is 0.1 (dashed line). For a more detailed description, see Fig. 15.12 on page 216)

Of the many functions that implement a cluster analysis in R,[20] the function `varclus` (package **Hmisc**) is one of the simplest.[21] The cluster analysis for the Appalachian soil data set can be executed as follows, with the result shown in Fig. 15.12 on page 216.

```
> plot(varclus(~., data=apa[,-c(1,2,16)]), las=1, cex.lab=1.5)
```

The function `varclus` produces an object that we immediately plot. As with the PCA, we leave columns 1, 2, and 16 out. The syntax "~." takes some getting used to. The full stop (".") stands for "use all columns of the data", with the tilde (~), we create a "formula". Only when this formula notation is used will this function also accept categorical variables which are internally converted into dummy variables. If we only have continuous variables, as in the Appalachian soil data set, we can also pass the matrix directly. We now use Hoeffdings distance to consider the influence on the result (Fig. 16.12):

```
> par(mar=c(1,5,1,1), las=1, cex.lab=1.5)
> plot(varclus(as.matrix(apa[,-c(1,2,16)]), similarity="hoeffding"))
```

In the Spearman-based analysis (Fig. 15.12 on page 216), we identified the clusters pH/ExchAc, P/Zn and Na/CECbuffer/Ca/Mg. Here, we have K instead of Na in the third cluster and the appearance of the cluster diagram is completely different. this is typical of the variability between different cluster analyses.

16.3   Model Selection

We have already done some simple model selection in Sect. 16.1.3, by removing non-significant effects from the model. That was a rather informal approach. Now, we will go through the steps of model selection in a more statistically informed manner, using the same data that we just used for the PCA.

---

[20]See https://cran.r-project.org/web/views/Cluster.html for a complete picture of available functions.

[21]It uses the function `hclust`, which implements different linkage algorithms and requires a distance matrix as an entry.

Generally, it is ideal if we could compare all possible predictor combinations with each other.[22] Then we could see which model is the best based on the AIC, for example. Although, this could be many different models: with 4 main effects, 4 quadratic effects and 6 two-way interactions (so 14 model terms), there are $2^{14} = 16364$ different possible models! With really large models we can quickly end up with hundreds of thousands of options.

An alternative is backward stepwise model selection, in which we start with the full model and simplify this step by step. To this end, we remove each model term from the model, as far as this is permissible,[23] and compare the new model, e.g. with regard to the AIC. The term causes the AIC to decrease the most when it is removed is then deleted from the model. With the new (one term smaller) model, we repeat this process from the beginning.

It is important to note that we are comparing the old model with each of the new models. This can be different than considering the explained deviance (or its significance) for each term of the fitted model and then deciding which term should be eliminated! That is because the list in the `anova` of a GLM is sequential, i.e. the terms are brought into the model one after the other, not as just described for stepwise selection, where terms are dropped from the model one at a time. If all predictors are perfectly orthogonal (uncorrelated), then these methods do not differ. Otherwise, the order of the terms in the model has a meaning and the result of a manual simplification is a different (and wrong) one.

16.3.1 Model Selection by Hand

Since we only have 20 data points, we can not fit all possible combinations of the 13 main model components. We actually only take the first four, which cover over 85% of the variance. The individual steps are: Import data, perform PCA, extract and append principal components to the data set, perform multiple regression, create ANOVA table:

```
> apa <- read.csv("soil.csv")
> pca <- prcomp(apa[,-c(1,2,16)], scale=T)
> apa2 <- cbind(apa, pca$x)
> attach(apa2)
> fm <- glm(Diversity ~ PC1*PC2*PC3*PC4, family=gaussian)
> anova(fm, test="F")
```

```
...
```

	Df	Deviance	Resid. Df	Resid. Dev	F	Pr(>F)	
NULL			19	0.0205629			
PC1	1	0.0015481	18	0.0190148	2.5346	0.18659	
PC2	1	0.0011942	17	0.0178206	1.9552	0.23458	
PC3	1	0.0074263	16	0.0103943	12.1586	0.02519	*
PC4	1	0.0026012	15	0.0077931	4.2588	0.10800	
PC1:PC2	1	0.0016358	14	0.0061573	2.6781	0.17708	
PC1:PC3	1	0.0002563	13	0.0059010	0.4197	0.55239	
PC2:PC3	1	0.0001433	12	0.0057577	0.2346	0.65344	
PC1:PC4	1	0.0001282	11	0.0056295	0.2099	0.67064	
PC2:PC4	1	0.0011175	10	0.0045120	1.8296	0.24759	
PC3:PC4	1	0.0002301	9	0.0042819	0.3768	0.57251	
PC1:PC2:PC3	1	0.0014475	8	0.0028344	2.3698	0.19854	
PC1:PC2:PC4	1	0.0000309	7	0.0028035	0.0506	0.83303	
PC1:PC3:PC4	1	0.0000040	6	0.0027995	0.0066	0.93930	
PC2:PC3:PC4	1	0.0000002	5	0.0027993	0.0003	0.98773	
PC1:PC2:PC3:PC4	1	0.0003561	4	0.0024432	0.5831	0.48766	

---

[22]This is also known as "best subset regression". Although in doing so we calculate a large number of models, we arguably do not carry out any actual statistical tests. That is why we do not have to correct for the number of models. Nevertheless, many applied statisticians are concerned about this approach. They consider this to be "fishing for significance", "data dredging" or "data snooping", all variations of "p-hacking" (Munafé et al. 2017).

[23]We are not allowed to remove any main effects that are still parts of interactions that remain in the model.

```
---
Signif. codes:  0 `***' 0.001 `**' 0.01 `*' 0.05 `.' 0.1 ` ' 1
```

In this model, the first four principal components are retained in all interactions: as pairwise interactions, 3-way and one 4-way interaction. Now let's simplify this model manually. First we need to look at the 4-way interaction, because only if we can eliminate this term can we continue with our simplification. Remember: we can not remove a term that is part of an interaction. However, since all terms in the triple interactions are part of the quadruple interaction, everything depends on it.

In this case, the 4-way interaction is far from significant. We "update" the model by removing this interaction. As a criteria we use the AIC:

```
> AIC(fm)

[1] -89.44643

> AIC(fm2 <- update(fm, .~. -PC1:PC2:PC3:PC4))

[1] -88.72486
```

According to our AIC criteria, we can not simplify the model. When we remove the 4-way interaction, the AIC goes up – the model becomes worse. This brings us to the end of model simplification using AIC for this data.

This is highly unsatisfying! A huge model full of non-significant terms and we can't simplify it at all? Well, instead of using the AIC as a criteria, we could also use the BIC. The BIC takes the number of available data points into account, and therefore usually leads to smaller simpler models (e.g. Ward 2008).[24]

```
> BIC(fm)

[1] -72.51898

> BIC(fm2 <- update(fm, .~. -PC1:PC2:PC3:PC4))

[1] -72.79314
```

Aha! With the same approach, the BIC gets smaller (becomes more negative), so we can now remove the 4-way interaction and proceed. For a clear and understandable overview, the R entries and outputs are shown for each step. Next, we can look at all four of the three-way interactions:

```
> BIC(fm3a <- update(fm2, .~. -PC1:PC2:PC3))
> BIC(fm3b <- update(fm2, .~. -PC1:PC2:PC4))
> BIC(fm3c <- update(fm2, .~. -PC1:PC3:PC4))
> BIC(fm3d <- update(fm2, .~. -PC2:PC3:PC4))

[1] -70.56442
[1] -75.58801
[1] -75.75989
[1] -75.78771
```

The BIC decreases the most with the model variant fm3d, so we remove PC2:PC3:PC4. And further:

```
> BIC(fm4a <- update(fm3d, .~. -PC1:PC2:PC3))
> BIC(fm4b <- update(fm3d, .~. -PC1:PC2:PC4))
> BIC(fm4c <- update(fm3d, .~. -PC1:PC3:PC4))
```

---

[24] Alternatively, we could also use the AICc, the sample size corrected version of the AIC, available in packages such as **MuMIn** and **AICcmodavg**.

```
[1] -71.27534
[1] -78.55695
[1] -78.7548
```

We remove PC1:PC3:PC4. Since the 2-way interaction PC3:PC4 is not part of an interaction that is still in the model, we can then take a critical look at this term as well:

```
> BIC(fm5a <- update(fm4c, .~. -PC1:PC2:PC3))
> BIC(fm5b <- update(fm4c, .~. -PC1:PC2:PC4))
> BIC(fm5c <- update(fm4c, .~. -PC3:PC4))
```

```
[1] -74.24162
[1] -81.53122
[1] -77.54527
```

We remove PC1:PC2:PC4.

```
> BIC(fm6a <- update(fm5b, .~. -PC1:PC2:PC3))
> BIC(fm6b <- update(fm5b, .~. -PC1:PC4))
> BIC(fm6c <- update(fm5b, .~. -PC2:PC4))
> BIC(fm6d <- update(fm5b, .~. -PC3:PC4))
```

```
[1] -76.27569
[1] -83.29865
[1] -84.52452
[1] -80.4912
```

We remove PC2:PC4.

```
> BIC(fm7a <- update(fm6c, .~. -PC1:PC2:PC3))
> BIC(fm7b <- update(fm6c, .~. -PC1:PC4))
> BIC(fm7c <- update(fm6c, .~. -PC3:PC4))
```

```
[1] -76.20757
[1] -85.9539
[1] -82.80941
```

We remove PC1:PC4:

```
> BIC(fm8a <- update(fm7b, .~. -PC1:PC2:PC3))
> BIC(fm8b <- update(fm7b, .~. -PC3:PC4))
```

```
[1] -77.59436
[1] -85.67267
```

Both candidates have a higher BIC-value than model fm7b (-85.9). Accordingly, we have now reached the end of the line. Our final model according to BIC is:

```
> anova(fm7b, test="F")
```

```
...
        Df  Deviance Resid. Df Resid. Dev      F      Pr(>F)
NULL                      19  0.0205629
PC1      1 0.0015481      18  0.0190148  5.0498 0.0484117 *
```

```
PC2             1 0.0011942        17   0.0178206  3.8954 0.0766742 .
PC3             1 0.0074263        16   0.0103943 24.2242 0.0006032 ***
PC4             1 0.0026012        15   0.0077931  8.4850 0.0154876 *
PC1:PC2         1 0.0016358        14   0.0061573  5.3358 0.0435122 *
PC1:PC3         1 0.0002563        13   0.0059010  0.8362 0.3820082
PC2:PC3         1 0.0001433        12   0.0057577  0.4675 0.5096852
PC3:PC4         1 0.0003489        11   0.0054088  1.1380 0.3111505
PC1:PC2:PC3     1 0.0023431        10   0.0030657  7.6432 0.0199705 *
---
Signif. codes:  0 `***' 0.001 `**' 0.01 `*' 0.05 `.' 0.1 ` ' 1
```

No further simplification is possible: all terms are either significant or part of a significant interaction. Only PC3:PC4 messes with this picture. Depending on your preference, you could remove this interaction (and then subsequently cite Crawley 2007) or not. Since we decided at the start to use the BIC as our criteria and not "everything has to be significant", this should be the end, in my opinion.

We do not have to do the backward stepwise selection[25] by hand. There are functions in R that can help us. One more comment before we look at this approach with an example: our selected model contains a significant 3-way interaction, a significant 2-way interaction, and also three non-significant 2-way interactions. If we had only used the main effects, would the model have been much worse?

Let's go back to the BIC and specify a model only with main effects:

```
> fmhaupt <- glm(Diversity ~ PC1+PC2+PC3+PC4, familty=gaussian)
> BIC(fmhaupt)

[1] -82.27303
```

The result is clear: fm7b is better (has a BIC value that is 3.6 units lower). Remember that the AIC and BIC are calculated on the log-likelihood scale. 3.6 units corresponds to a $e^{3.6} = 37$-fold likelihood increase. Another rule of thumb: with a difference of at least 2 AIC or BIC units, we consider the model difference to be substantial.

16.3.2   Automated Model Selection

Automated model selection should not be viewed as a panacea (see Austin and Tu 2004). Using the step-by-step application could potentially create a selection effect that leads to inconsistent models. The good news is that with backward stepwise selection, this effect becomes smaller the more data points we have (Steyerberg et al. 1999).[26]

Let's have a look at how the stepwise simplification is implemented in R, and then apply it to our Appalachian PCA data. The relevant function here is called step[27] and is simply applied to the model object. The individual selection steps are then shown as output:

```
> fmstepAIC <- step(fm)
```

---

[25]There is also *forward* stepwise selection, in which the terms are added to the model one after the other. In such a case, the significance of a term is dependent on other terms already included in the model. Furthermore, variables that are only significant in an interaction (such as PC1 and PC2 in the example) don't have a chance. Forward stepwise selection has been criticised as a bad approach by many (Wilkinson and Dallal 1981) and should *not* be used (even when some scientists still unfortunately use it). Critiques have also been made on backward stepwise selection (Steyerberg et al. 1999, Whittingham et al. 2006), but comparisons between *best-subsets* and *backward selection* did not show any remarkable differences (see, e.g., Dormann et al. 2008).

[26]To be more exact: how many data points we have *per variable*. This relationship is known as EPV (events per variable), and it should not be less than 10 in the final model (Harrell 2001). Steyerberg et al. [1999] found consistent models for an EPV of at least 40. More recently (van Smeden et al. 2016) argued that the small sample issue is mostly related to the separation problem for binary data (which is that the 0s and 1s can be perfectly predicted by the model, leading to run-away estimates of the model parameters), and hence tackling that may allow for lower EPVs. One way to handle the 0–1-separation is Firth' correction, as implemented as logistf in package **logistf**.

[27]In the **MASS** package, the function stepAIC is available, which can also be used on model types other than glm.

```
Start:  AIC=-89.45
Diversity ~ PC1 * PC2 * PC3 * PC4

                    Df  Deviance      AIC
<none>                 0.0024431  -89.446
- PC1:PC2:PC3:PC4  1  0.0027993  -88.725

> anova(fmstepAIC, test="F")


...
                 Df  Deviance Resid. Df Resid. Dev       F  Pr(>F)
NULL                            19  0.0205629
PC1               1  0.0015481  18  0.0190148   2.5346 0.18659
PC2               1  0.0011942  17  0.0178206   1.9552 0.23458
PC3               1  0.0074263  16  0.0103943  12.1586 0.02519 *
PC4               1  0.0026012  15  0.0077931   4.2588 0.10800
PC1:PC2           1  0.0016358  14  0.0061573   2.6781 0.17708
PC1:PC3           1  0.0002563  13  0.0059010   0.4197 0.55239
PC2:PC3           1  0.0001433  12  0.0057577   0.2346 0.65344
PC1:PC4           1  0.0001282  11  0.0056295   0.2099 0.67064
PC2:PC4           1  0.0011175  10  0.0045120   1.8296 0.24759
PC3:PC4           1  0.0002301   9  0.0042819   0.3768 0.57251
PC1:PC2:PC3       1  0.0014475   8  0.0028344   2.3698 0.19854
PC1:PC2:PC4       1  0.0000309   7  0.0028035   0.0506 0.83303
PC1:PC3:PC4       1  0.0000040   6  0.0027995   0.0066 0.93930
PC2:PC3:PC4       1  0.0000002   5  0.0027993   0.0003 0.98773
PC1:PC2:PC3:PC4   1  0.0003561   4  0.0024432   0.5831 0.48766
---
Signif. codes:  0 `***' 0.001 `**' 0.01 `*' 0.05 `.' 0.1 ` ' 1
```

The `step` function does not simplify our model at all! This is consistent with the result that we got when we did model selection by hand using the AIC. As we can see in the output, `step` tests the elimination of the 4-way interaction first, finds that the AIC increases and then leaves that term in the model. Accordingly, no further simplification can take place. In short: the full model has the lowest AIC and can therefore not be simplified.

In the simplification we did by hand, we decided to use the BIC as a criterion. Another thing we could have used is the sample size corrected version of the AIC, namely the AICc. Let's start by using the BIC. According to the definition (Equation 3.14 on page 59), the penalty term for the BIC is $k = \ln(n) \cdot$number of data points.[28] In the `step` function, we can define the penalty term as an argument, `k`:

```
> fmstepBIC <- step(fm, k=log(20))

Start:  AIC=-73.51
Diversity ~ PC1 * PC2 * PC3 * PC4

                    Df  Deviance      AIC
- PC1:PC2:PC3:PC4  1  0.0027993  -73.789
<none>                 0.0024431  -73.515

Step:  AIC=-73.79
Diversity ~ PC1 + PC2 + PC3 + PC4 + PC1:PC2 + PC1:PC3 + PC2:PC3 +
```

---

[28]More precisely: the number of effective data points. For normal and Poisson distributed data, this is n, for binary data it is the minimum (number of 0s, number of 1s), or in R: `min(table(y))`. This means, if for the 1300 data points from the Titanic survival data, there were only 100 survivors, the number of effective data points would only be 100.

```
        PC1:PC4 + PC2:PC4 + PC3:PC4 + PC1:PC2:PC3 + PC1:PC2:PC4 +
        PC1:PC3:PC4 + PC2:PC3:PC4


                    Df  Deviance     AIC
- PC2:PC3:PC4   1 0.0027995 -76.783
- PC1:PC3:PC4   1 0.0028034 -76.756
- PC1:PC2:PC4   1 0.0028276 -76.584
<none>            0.0027993 -73.789
- PC1:PC2:PC3   1 0.0036349 -71.560

Step:  AIC=-76.78
Diversity ~ PC1 + PC2 + PC3 + PC4 + PC1:PC2 + PC1:PC3 + PC2:PC3 +
        PC1:PC4 + PC2:PC4 + PC3:PC4 + PC1:PC2:PC3 + PC1:PC2:PC4 +
        PC1:PC3:PC4


                    Df  Deviance     AIC
- PC1:PC3:PC4   1 0.0028035 -79.751
- PC1:PC2:PC4   1 0.0028313 -79.553
<none>            0.0027995 -76.783
- PC1:PC2:PC3   1 0.0040748 -72.271

Step:  AIC=-79.75
Diversity ~ PC1 + PC2 + PC3 + PC4 + PC1:PC2 + PC1:PC3 + PC2:PC3 +
        PC1:PC4 + PC2:PC4 + PC3:PC4 + PC1:PC2:PC3 + PC1:PC2:PC4


                    Df  Deviance     AIC
- PC1:PC2:PC4   1 0.0028344 -82.527
<none>            0.0028035 -79.751
- PC3:PC4       1 0.0034595 -78.541
- PC1:PC2:PC3   1 0.0040808 -75.237

Step:  AIC=-82.53
Diversity ~ PC1 + PC2 + PC3 + PC4 + PC1:PC2 + PC1:PC3 + PC2:PC3 +
        PC1:PC4 + PC2:PC4 + PC3:PC4 + PC1:PC2:PC3


                    Df  Deviance     AIC
- PC2:PC4       1 0.0028347 -85.520
- PC1:PC4       1 0.0030139 -84.294
<none>            0.0028344 -82.527
- PC3:PC4       1 0.0034681 -81.487
- PC1:PC2:PC3   1 0.0042819 -77.271

Step:  AIC=-85.52
Diversity ~ PC1 + PC2 + PC3 + PC4 + PC1:PC2 + PC1:PC3 + PC2:PC3 +
        PC1:PC4 + PC3:PC4 + PC1:PC2:PC3


                    Df  Deviance     AIC
- PC1:PC4       1 0.0030657 -86.950
<none>            0.0028347 -85.520
- PC3:PC4       1 0.0035876 -83.805
- PC1:PC2:PC3   1 0.0049907 -77.203

Step:  AIC=-86.95
```

```
Diversity ~ PC1 + PC2 + PC3 + PC4 + PC1:PC2 + PC1:PC3 + PC2:PC3 +
    PC3:PC4 + PC1:PC2:PC3


                 Df  Deviance      AIC
<none>               0.0030657  -86.950
- PC3:PC4        1   0.0036115  -86.668
- PC1:PC2:PC3    1   0.0054088  -78.590
```

So now we do get a model simplification! In each step, all terms that could potentially be eliminated are separately removed from the model and compared to the current model. In the output, the current model shows up as <none> and is always in the list according to AIC value (from largest to smallest; in our case actually BIC since we defined it using k) with the terms that are up for possible elimination. Terms above <none> lead to a better model once they are removed (lower AIC or BIC), whereas terms below would lead to worse models. The uppermost term is eliminated and then the next step is taken. Only when there are no more terms above <none> is the simplification complete.

The final model is identical with the model that we simplified and selected by hand, fm7b:

```
> anova(fmstepBIC, test="F")
```

```
...
              Df  Deviance Resid. Df Resid. Dev       F     Pr(>F)
NULL                            19   0.0205629
PC1           1  0.0015481       18   0.0190148   5.0498  0.0484117  *
PC2           1  0.0011942       17   0.0178206   3.8954  0.0766742  .
PC3           1  0.0074263       16   0.0103943  24.2242  0.0006032  ***
PC4           1  0.0026012       15   0.0077931   8.4850  0.0154876  *
PC1:PC2       1  0.0016358       14   0.0061573   5.3358  0.0435122  *
PC1:PC3       1  0.0002563       13   0.0059010   0.8362  0.3820082
PC2:PC3       1  0.0001433       12   0.0057577   0.4675  0.5096852
PC3:PC4       1  0.0003489       11   0.0054088   1.1380  0.3111505
PC1:PC2:PC3   1  0.0023431       10   0.0030657   7.6432  0.0199705  *
---
Signif. codes:  0 `***' 0.001 `**' 0.01 `*' 0.05 `.' 0.1 ` ' 1
```

Now to the other AIC-alternative, namely the use of the small sample correction for the AIC in the step function. This is not available as a standard function in R, but Christoph Scherber rewrote the stepAIC function from the **MASS** package, so that it uses the AICc instead.[29] We can load this R script with the source function and use it to preform model simplification. For space reasons, we will suppress the output of the selection steps (trace=F) and have a look straight to the final model:

```
> source("stepAICc.r") ## Christoph Scherber's R-script
> fmstepAICc <- stepAICc(fm, trace=F)
> anova(fmstepAICc, test="F")
```

```
...
              Df  Deviance Resid. Df Resid. Dev       F     Pr(>F)
NULL                            19   0.0205629
PC1           1  0.0015481       18   0.0190148   4.4048  0.0576715  .
PC2           1  0.0011942       17   0.0178206   3.3979  0.0901051  .
PC3           1  0.0074263       16   0.0103943  21.1302  0.0006144  ***
PC1:PC2       1  0.0018695       15   0.0085248   5.3194  0.0397359  *
PC1:PC3       1  0.0000017       14   0.0085231   0.0048  0.9459541
PC2:PC3       1  0.0000647       13   0.0084584   0.1841  0.6754855
PC1:PC2:PC3   1  0.0042410       12   0.0042174  12.0669  0.0045991  **
```

---

[29]http://wwwuser.gwdg.de/~cscherb1/statistics.html under "R scripts".

```
---
Signif. codes:  0 `***' 0.001 `**' 0.01 `*' 0.05 `.' 0.1 ` ' 1
```

stepAICc eliminates the unnecessary interaction PC3:PC4, but also gets rid of the main effect PC4. If PC4 was part of the design of the experiment, then we would leave it in the model. To do this, we could also use update(fmstepAICc, . . + PC4) to enter it back in, but since this is not the case, we will not do this here.

Automated model simplification is thus to a certain extent also dependent on the selection of the criterion.

We can display the AIC value for each of these three models (stepfmAIC, stepfmBIC and stepfmAICc) created by stepwise simplification:

```
> AIC(fmstep, fmstep2, fmstep3)


            df         AIC
fmstepAIC   17  -89.44643
fmstepBIC   11  -96.90695
fmstepAICc   9  -94.52760
```

The best model (with the lowest AIC) is – the BIC selected model! That seems surprising. Even though we also tried selecting based on AIC, the AIC is better if we select a model using AICc or BIC as a selection criterion. How can that be?

Well, we're doing a stepwise simplification. It can happen be that a model would become better, if not *one* but *two* predictors would be removed. If we remove a single one, the AIC gets worse, but a joint removal would improve it. For this, the step function would have to look two steps ahead, so to say. BIC (and AICc) penalise more severely, and so a predictor is removed more easily than with the AIC. This can lead to BIC-selected models having better AIC values, simply because the way to get there for the AIC was "impaired".

How can we be sure that the BIC-based, stepwise model selection does not suffer from the same problem? We can't be sure. For this we would have to evaluate all possible models!

16.3.3   Best-Subset Regression

erWith best-subset regression, all combinations of terms are tried one-by-one and then sorted according to AIC or a similar criterion. With our small model without any quadratic terms this can be done relatively quickly: there are only 167 possibilities.

In order for this to work properly, we have to change an option in R that defines what should happen if the model runs into a data record with an NA in it.[30] As a criterion we choose the AIC and let R lead us to the best models (all models within two AIC units of the best model).

```
> library(MuMIn)
> options("na.action"= "na.fail")
> dm <- dredge(fm, rank="AIC")  # Needs less than 2 seconds!
> signif(subset(dm, delta < 2), 3)  # To slim down the output a bit
```

Intercept	PC1	PC2	PC3	PC4	PC1:PC2	PC1:PC3	PC1:PC4	PC2:PC3	PC2:PC4	PC3:PC4	PC1:PC2:PC3	df	logLik	AIC	delta	weight
0.23	-0.0082	-0.00045	0.013	-0.0092	0.00210	-0.0067	NA	0.0029	NA	0.0073	0.0036	11	59	-97	0.00	0.36
0.23	-0.0080	-0.00130	0.012	-0.0130	0.00310	-0.0081	-0.0037	0.0021	NA	0.0093	0.0035	12	60	-96	0.43	0.29
0.23	-0.0080	0.00190	0.012	-0.0065	0.00014	-0.0051	NA	0.0016	NA	NA	0.0034	10	58	-96	1.30	0.19
0.23	-0.0082	-0.00097	0.013	-0.0089	0.00240	-0.0064	NA	0.0027	0.0017	0.0087	0.0039	12	60	-95	1.70	0.16

(I left out all irrelevant 3- and 4-way interactions so that the results would fit on the page.)

The results are in and the best model is – the model that we have chosen before: fm7b = fmstepBIC. However, other models that we have not encountered in our model simplifications are not far behind. First the model without PC3:PC4 (we've seen this one), then one in which PC4 was also removed, etc.

---

[30]This was done very cleverly by the programmer behind the **MuMIn** package (Karmil Bartoń). Let's assume that we have an NA as a data record for one of our predictors. Then, all models with this predictor would have one data point less than those without this predictor. The likelihood is then no longer comparable, and neither are AIC or BIC. By specifying the option na.fail, we never run into such a situation. We have to remove this data point in advance.

In the last two columns, the differences in the AIC values are shown (`delta`), as well as a model weight, which sums up to 1 when we take all 167 models into account. Imagine that for each person alive today, we would calculate how much of the world's total wealth they have. Then this last column would indicate how much wealth goes to this particular model. Thus, the first five models have 75% of the explanatory power of all models.[31]

This practice, known as "dredging", is criticised by some statisticians, because in the end, we act as if we had only constructed this one final model, although in reality we tested many different variants. Therefore, the confidence intervals of the prediction as well as the significance values in this model are wrong (e.g. Regal and Hook 1991). However, this does not apply to the prediction of expected values ("point predictions"), which are computed correctly.

In summary, we can conclude that stepwise model selection can be misleading and might not always find the model with the lowest AIC value. The stricter BIC, or a complete model search can be useful in such situations.

16.4   Exercises

1. In the `barley` data set (**lattice** package) the barley yield for two years (1931 and 1932) and 10 varieties is shown. Fit a GLM with an interaction between variety and year. Simplify the model if possible. Calculate the estimated yield for the variety "Trebi" in 1932. Does this match with the mean of the measured values? Why or why not?
2. In the `logistic.txt` data set, the number of dead daphnia (`dead`) and the total number of daphnia (`n`) placed in different substances and their concentrations are provided. Calculate a binomial model with `product` and `logdose` as interacting predictors. Then try to visualise this interaction, perhaps including the *supersmoother*-line through the data.

References

1. Austin, P. C., & Tu, J. V. (2004). Automated variable selection methods for logistic regression produced unstable models for predicting acute myocardial infarction mortality. *Journal of Clinical Epidemiology, 57*, 1138–1146.
2. Bellman, R. E. (1957). *Dynamic Programming*. Princeton, NJ: Princeton University Press.
3. Burnham, K. P. & Anderson, D. R. (2002). *Model Selection and Multi-Model Inference: a Practical Information-Theoretical Approach*. Berlin: Springer, 2nd edition.
4. Crawley, M. J. (2007). *The R Book*. Chichester, UK: John Wiley & Sons.
5. Dalgaard, P. (2002). *Introductory Statistics with R*. Berlin: Spinger.
6. Dormann, C. F., Purschke, O., García Marquéz, J. R., Lautenbach, S., & Schröder, B. (2008). Components of uncertainty in species distribution analysis: a case study of the great grey shrike. *Ecology, 89*(12), 3371–86.
7. Fox, J. (2002). *An R and S-Plus Companion to Applied Regression*. Thousand Oaks: Sage.
8. Harrell, F. E. (2001). *Regression Modeling Strategies - with Applications to Linear Models, Logistic Regression, and Survival Analysis*. New York: Springer.
9. Munafé, M. R., Nosek, B. A., Bishop, D. V. M., Button, K. S., Chambers, C. D., Percie du Sert, N., et al. (2017). A manifesto for reproducible science. *Nature Human Behaviour, 1*(1), 0021.
10. Paruelo, J. M., & Lauenroth, W. K. (1996). Relative abundance of plant functional types in grassland and shrubland of north america. *Ecological Applications, 6*, 1212–1224.
11. Quinn, G. P., & Keough, M. J. (2002). *Experimental Design and Data Analysis for Biologists*. Cambridge, UK: Cambridge University Press.
12. Regal, R. R. & Hook, E. B. (1991). The effects of model selection on confidence intervals for the size o f a closed population. *Statistics in Medicine, 10*(September 1990), 717–721.
13. Steyerberg, E. W., Eijkemans, M. J. C., & Habbema, J. D. F. (1999). Stepwise selection in small data sets: a simulation study of bias in logistic regression analysis. *Journal of Clinical Epidemiology, 52*, 935–942.
14. Underwood, A. J. (1997). *Experiments in Ecology: Their Logical Design and Interpretation using Analysis of Variance*. Cambridge, UK: Cambridge University Press.
15. van Smeden, M., de Groot, J. A. H., Moons, K. G. M., Collins, G. S., Altman, D. G., Eijkemans, M. J. C., et al. (2016). No rationale for 1 variable per 10 events criterion for binary logistic regression analysis. *BMC Medical Research Methodology, 16*(1), 163.
16. Ward, E. J. (2008). A review and comparison of four commonly used Bayesian and maximum likelihood model selection tools. *Ecological Modelling, 211*, 1–10.
17. Whittingham, M. J., Stephens, P. A., Bradbury, R. B., & Freckleton, R. P. (2006). Why do we still use stepwise modelling in ecology and behaviour? *Journal of Animal Ecology, 75*(5), 1182–1189.
18. Wilkinson, L., & Dallal, G. (1981). Tests of significance in forward selection regression with an F-to enter stopping rule. *Technometrics, 23*, 377–380.
19. Zuur, A. F., Ieno, E. N., & Smith, G. M. (2007). *Analysing Ecological Data*. Berlin: Springer.

---

[31] These weights are derived from Burnham and Anderson [2002] and are used if you want to make a prediction using multiple models. Sometimes these weights are interpreted as the probability that this model is the "true" model. For this to be the case, all models would need to be equally plausible. Since the world is a complex place, more complex models are more probable than simpler models, which is not taken into account here. However, these are Bayesian thoughts that are creeping in now, and are going beyond the scope of this book.

Outlook

17

The covers of this book are too far apart. —Ambrose G. Bierce

This book set the foundations for the most common and widespread type of statistics: parametric statistics. It builds on assumptions about the distribution of the response variable, allowing us to leverage the framework of maximum likelihood. Of course, statistics is much wider, and in the following I would like to offer a few pointers towards other important areas.

From maximum likelihood, it is only a small step to **Bayesian statistics** (see for example the last chapter of Hilborn and Mangel (1997)), which extends parametric statistics in two directions: firstly, we can embrace prior knowledge, such as previous estimates of our model parameters. Secondly, Bayesian statistics answers "the right question"! While the likelihood tells you how probable data are given a model $P(\text{Data}|\text{Model})$, Bayesian statistics provide a probability of the model, given the data: $P(\text{Model}|\text{Data})$. Since in science we are typically interested in whether our hypothesis (= model) is correct, this is the "right" answer. The way there is somewhat tedious, however, not much aided by the fact that we probably need to learn a dialect of R.[1] Excellent practical introductions, starting off where we leave here, are Kéry (2010) and Kruschke (2015).

Another important field is *machine learning*, today often summarised under the more mystical term "artificial intelligence". As the Oxford statistician Brian Ripley once quipped: "Machine learning is statistics minus any checking of models and assumptions"' (useR! Konferenz 2004 in Vienna.[2]) Machine learning buzzwords comprise artificial neural networks, boosting and bagging. They are easily employed from R,[3] and the user needs to counteract the tendency to overfit the model on the data, typically through cross-validation. A comprehensive introduction is provided by Hastie et al. (2009), while James et al. (2013) and Kuhn and Johnson (2013) cover the more applied aspects.

Bayesian statistics and machine learning have not been married, yet, but that is an active field of statistical research. When our data have a more complicated structure (for example hierarchical models with several levels of nesting, or an observer model, or temporal or spatial structure), we are in the field of Bayesian statistics. For analyses of large data sets ("big data", e.g. in remote sensing, genetic data or GPS-telemetry), where inferential statistics are less important than predictions, then we should turn to machine learning. In a few years, Bayesian machine learning will be more common, as it is slowly emerging in fields such as astrophysics or consumer choice analysis. Using these tools properly requires understanding the basics presented in the this book—and quite a bit more.

Non-parametric statistics, as the obvious counterpart to the parametric statistics presented here, avoids making assumptions about distributions. Instead, permutations and randomisation of data are used to test for patterns. These techniques, as introduced for example in Manly (1997), show up time and again whenever data are not well-behaved and resist being squeezed into common distributions. It is also employed in machine learning, making the forefront of statistical computing a cocktail of parametric, non-parametric, Bayesian and machine-learning ingredients; Lewis (2016) is not the worst starting point for R-users.

---

[1] The free software WinBUGS (http://www.mrc-bsu.cam.ac.uk/bugs/winbugs), its open source-cousin OpenBUGS (www.openbugs.info), the most system-independent implementation JAGS (http://mcmc-jags.sourceforge.net), the newer and somewhat faster approach STAN (http://mc-stan.org) or R's NIMBLE (http://r-nimble.org/). There are additional Bayesian packages for specific tasks (see CRAN task view "Bayesian"), but these are the most flexible Bayesian extensions accessible from R.

[2] Retrievable through `fortune("machine learning")` in the package **fortunes**.

[3] A good starting point are the packages **caret** and **mlr**; see also the CRAN task view "Machine Learning & Statistical Learning".

© Springer Nature Switzerland AG 2020
C. Dormann, *Environmental Data Analysis*,
https://doi.org/10.1007/978-3-030-55020-2_17

Beyond these fields it becomes somewhat idiosyncratic; there are many important pockets of statistics with specific applications, such as **time-series analysis** (Cowpertwait and Metcalfe 2009; Cryer and Chan 2010; Shumway and Stoffer 2010), **spatial statistics and geostatistics** (Cressie 1993; Haining 2009; Dale and Fortin 2014) as well as **multivariate statistics** (Borcard et al. 2011; Legendre and Legendre 2013). All these topics justify separate treatises. I hope this book has prepared the reader for their consumption.

References

1. Borcard, D., Gillet, F., & Legendre, P. (2011). *Numerical Ecology with R*. Berlin: Springer.
2. Cowpertwait, P. S., & Metcalfe, A. V. (2009). *Introductory Time Series with R*. Berlin: Springer.
3. Cressie, N. A. C. (1993). *Statistics for Spatial Data*. New York: Wiley.
4. Cryer, J. D., & Chan, K.-S. (2010). *Time Series Analysis: With Applications in R*. Berlin: Springer.
5. Dale, M. R. T., & Fortin, M. J. (2014). *Spatial Analysis - A Guide for Ecologists*. Cambridge, UK: Cambridge Univ. Press.
6. Haining, R. P. (2009). Spatial autocorrelation and the quantitative revolution. *Geographical Review, 41*, 364–374.
7. Hastie, T., Tibshirani, R. J., & Friedman, J. H. (2009). *The Elements of Statistical Learning: Data Mining, Inference, and Prediction*. Berlin: Springer, 2nd edition.
8. Hilborn, R., & Mangel, M. (1997). *The Ecological Detective: Confronting Models with Data*. Princeton, NJ: Princeton University Press.
9. James, G., Witten, D., Hastie, T., & Tibshirani, R. (2013). *An Introduction to Statistical Learning: with Applications in R*. New York: Springer.
10. Kéry, M. (2010). *Introduction to WinBUGS for Ecologists: Bayesian Approach to Regression, ANOVA, Mixed Models and Related Analyses*. Salt Lake City, USA: Academic Press.
11. Kruschke, J. K. (2015). *Doing Bayesian Data Analysis. A Tutorial with R, JAGS, and Stan*. Cambridge, MA: Academic Press, 2nd edition.
12. Kuhn, M., & Johnson, K. (2013). *Applied Predictive Modeling*. Berlin: Springer.
13. Legendre, P. & Legendre, L. (2013). *Numerical Ecology*. Amsterdam: Elsevier, 3rd edition.
14. Lewis, N. (2016). *Deep Learning Made Easy with R: A Gentle Introduction for Data Science*. CreateSpace Independent Publishing Platform.
15. Manly, B. F. (1997). *Randomization, Bootstrap and Monte Carlo Methods in Biology*. New York: Chapman & Hall/CRC, 2nd edition.
16. Shumway, R. H., & Stoffer, D. S. (2010). *Time Series Analysis and Its Applications: With R Examples*. Berlin: Springer.

Index

© Springer Nature Switzerland AG 2020
C. Dormann, *Environmental Data Analysis*,
https://doi.org/10.1007/978-3-030-55020-2

Printed in the United States
by Baker & Taylor Publisher Services